Textbooks in Electrical and Ele

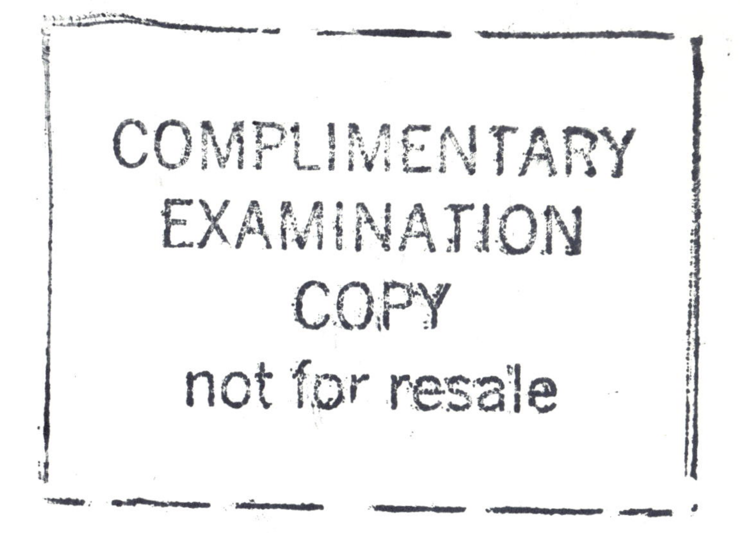

Series Editors

G. Lancaster E. W. Williams

---

1. Introduction to fields and circuits
GORDON LANCASTER

Introduction to Fields and Circuits

■

Gordon Lancaster

Electronic Engineering Group
Keele University

OXFORD NEW YORK TOKYO
OXFORD UNIVERSITY PRESS
1992

Oxford University Press, Walton Street, Oxford OX2 6DP

Oxford New York Toronto
Delhi Bombay Calcutta Madras Karachi
Petaling Jaya Singapore Hong Kong Tokyo
Nairobi Dar es Salaam Cape Town
Melbourne Auckland

and associated companies in
Berlin Ibadan

Oxford is a trade mark of Oxford University Press

Published in the United States
by Oxford University Press, New York

A catalogue record for this book is available from the British Library

Library of Congress Cataloging in Publication Data
Lancaster, Gordon
Introduction to fields and circuits/Gordon Lancaster
(Textbooks in electrical and electronic engineering: 1)
Includes index
1. Electric circuits. 2. Electronic circuits. 3. Electromagnetic fields. I. Title. II. Series
TK454.L37 1992 621.319'2—dc20 91-26459
ISBN 0-19-853932-0
ISBN 0-19-853931-2 (pbk.)

Typeset by Integral Typesetting, Gorleston, Norfolk
Printed in Hong Kong

Preface

Although much of circuit theory can be taught without explicit reference to the electromagnetic fields which exist in and around real circuits, a point is always reached at which ignorance of certain basic properties of electromagnetic fields hinders further progress. Additionally, knowledge of electromagnetism is necessary for the understanding of the operation of many other aspects of electronic systems in general, such as sensors, distributed elements and systems, telecommunications, radar, solid state devices/integrated circuits, optoelectronic systems, and the increasingly important topic of electromagnetic compatibility.

There is a quite intense debate currently concerning the degree of attention which should be devoted to electromagnetism in undergraduate courses in Electronic Engineering bearing in mind the pressure on students' time due to the increasing number of new topics which are being included in syllabuses. An argument which bears strongly on this debate is that students generally regard electromagnetism as a very challenging topic and this, coupled with the wide range of background experience of first year undergraduate students, in turn creates a significant challenge for the teacher. In Chapters 1 and 5 the topics in electromagnetism have been selected on a 'need to know' basis.

This text is aimed at first- and second-year courses in British universities and polytechnics and, as such, aims to provide a firm foundation in subsequent courses. The amount of formal theory has been kept relatively small but, inevitably, there are some passages of mathematical analysis. My suggestion is that such passages can be bypassed at a first reading, if necessary, and the reader's attention focused on the practical implications of the results of the analyses; in some instances the reader is explicitly pointed ahead to the conclusions of the analyses in question. Those equations that define important quantities and those that conclude important analyses are highlighted as also are the especially important passages of text.

In the practical world a vital skill is to be able to make sensible, justifiable approximations to mathematical expressions. For instance, this may be the only available way of making a problem tractable, or it may be possible to reduce a complicated mathematical result to a much simpler form by making approximations which can be justified in a particular technical context. Hence throughout the text emphasis is placed on making mathematical approximations.

In-text Examples are used to illustrate the formal results and practical situations, and Problems with solutions are also provided.

Keele G.L.
October 1991

Contents

1 Fundamental electromagnetism

2 Direct current circuits

3 Alternating currents

Fundamental electromagnetism

1

1.1 Introduction

This chapter is concerned with describing those physical phenomena which form the basis for understanding electrical circuits. Although circuits can be analysed and designed on an operational basis from an understanding of circuit theory, the development of solid-state electronic devices, microminiaturized integrated circuits, very high-frequency circuits, and the demand for strict standards of electromagnetic compatibility mean that at least a basic understanding of electromagnetic fields is essential to an electronic or electrical engineer.

It is anticipated that the readership will vary widely in respect of prior knowledge and practical experience, and so to some readers most of the topics may be quite familiar, whilst other readers may encounter many new concepts.

Electric charge and electric and magnetic fields are not tangible like solid objects and liquids, for example, but it must always be remembered that the force, energy, and power associated with electromagnetic phenomena are just as real as in the more tangible mechanical machines. Hence models have been invented to represent electromagnetic phenomena and to help to predict effects that are of practical importance in science and engineering. The nature of these models and the associated mathematical theory make electromagnetism a challenging subject for study, but its importance is so great that the effort devoted to understanding the basic concepts is well spent.

The logical sequence of development in the following description of the fundamentals of electromagnetism necessitates some mathematical passages that may prove rather indigestible at a first reading. Such passages may be bypassed provided that the important definitions, assumptions, and results are noted, provisionally, and that the practical implications of the conclusions are recognized (note the Summary given at the end of this chapter).

1.2 Force, field, and electric current

Electric current is the flow of an entity that we call electric charge. It is believed that the smallest electric charge it is possible to isolate has a magnitude equal to that of the charge associated with an electron, namely 1.602×10^{-19} coulomb (C). A current of one ampere (1 A) corresponds to a flow of charge at the rate of one coulomb per second, and so even a current as small as one nanoampere (10^{-9} A) corresponds to a flow (or 'flux') of about 6×10^{9} (six billion) electrons per second. Bearing in mind that electrons are sub-atomic in size, it can be appreciated why their particulate nature is not immediately apparent in electrical circuits and why for most practical purposes current can be treated as a smooth flow of charge.

Consider a current flowing in a conducting solid such as a metal wire. What physical features and factors determine the magnitude of the current? Clearly there must be available some charged entities that are free to move through the metal and, in addition, there must be an agency producing a force which causes the charges to move.

Within a metal, the constituent atoms interact with each other, and in the process some electrons (one, two, or three) become detached from each of the 'parent' atoms and are able to move through the metal if subjected to a force; these are the so-called **conduction electrons**. Hence, in a sample of a metal the number of conduction electrons is of the same order of magnitude as the number of atoms (10^{29} m^{-3}). In semiconductor materials the concentration of conduction electrons is smaller (10^{18}–10^{26} m^{-3} roughly). An electron can experience a force due to both an applied electric field and a magnetic field, but in this section attention will be focused on the effect of a constant electric field.

An **electric field** is said to exist in a region of space if a stationary test charge (i.e. a relatively small charged object) experiences a force; the magnitude of the electric field is defined as the magnitude of the force divided by the value of the charge. If the charge, force, and electric field are denoted by Q, F, and E respectively, then

$$E \equiv \frac{F}{Q} \tag{1.1}$$

The SI unit of Q is the coulomb (C) and of F is the newton (N), so the unit of E is the newton per coulomb, which is the same as the volt per metre (V m^{-1}).

If a moving test charge experiences a force in a region of space, then a **magnetic field** is said to exist. The expression for the force in terms of the strength of the magnetic field and the velocity of the test charge is given in Section 1.16 (eqn (1.48)). Of course, an electric field and a magnetic field can coexist in a region of space.

The simplest practical way to apply a force to the conduction electrons in a metal wire, for instance, is to connect the wire between the terminals of a battery. If the potential difference between the terminals of the battery is V volts and the length of the wire is l metres, then the electric field set up in the wire is V/l volts per metre. Hence, using eqn (1.1), the force on a conduction electron is eV/l newtons, where the magnitude of the charge on an electron is denoted by e.

Since acceleration is equal to force divided by mass, and the force and mass are constant in the situation being considered, it might be expected that the conduction electrons would accelerate continuously to higher and higher speeds, which in turn implies an ever-increasing current! Experience has shown that, in practice, the current settles at a constant value, which implies that the electrons are flowing at a constant speed and hence have constant kinetic energy (note that the constant value is attained in a time $\sim 10^{-13}$ seconds, which is virtually instantaneously for ordinary practical purposes). What happens is that the moving electrons interact with the atoms of the metal and hence lose energy to the metal as a whole, causing its temperature to increase; this effect is exploited in purpose-designed electric heating elements. The constant current situation corresponds to a state of (dynamic) equilibrium, in which the rate of gain of energy by the electrons from the agency providing the electric field is equal to the rate of loss of energy to the atoms of the metal wire (and ultimately to the surroundings).

Experience has shown that the magnitude of the current I is proportional to the magnitude of the applied electric field and to the area of cross-section of the conductor. For a wire of length l and uniform area of cross-section A in which the applied voltage is V (so that $E = V/l$) therefore,

$$I \propto \frac{V}{l} \cdot A$$

or

$$V = \rho \frac{Il}{A} \tag{1.2}$$

The constant of proportionality ρ (rho) in eqn (1.2) is called the **electrical resistivity** of the metal of the wire. Now

$$V = RI \tag{1.3}$$

where

$$R = \frac{\rho l}{A} \tag{1.4}$$

is the **resistance** of the wire. If V is measured in volts and I in amperes, then the unit of resistance is the ohm (Ω). Equation (1.3), in the form $R = V/I$, is the definition of electrical resistance.

It is often convenient to use the reciprocal of resistance, called the **conductance** (symbol G). From eqn (1.3),

$$I = GV \tag{1.5}$$

The reciprocal ohm is called the siemen (S).

The average speed of the electrons constituting the current in a wire can be easily estimated. If the speed is denoted by u then the electrons in a length of wire numerically equal to u will cross any particular section of the wire in one second (see Fig. 1.1); the total charge carried by these

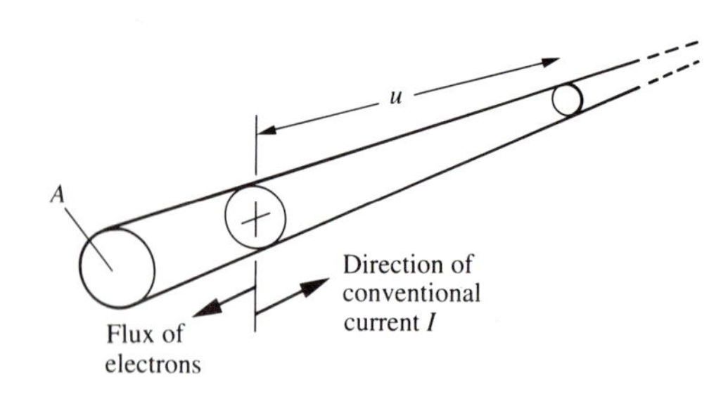

Fig. 1.1 The current in a wire in terms of the flow ('flux') of electrons.

electrons is the value of the current. For an area of cross-section A and concentration of electrons n, the current I is given by

$$I = \text{number of electrons per second} \times \text{charge on one electron}$$

or

$$I = nuAe.$$

Hence

$$u = \frac{I}{neA} \tag{1.6}$$

For $A = 1\,\text{mm}^2$ ($10^{-6}\,\text{m}^2$), $n = 10^{29}\,\text{m}^{-3}$, and $I = 1\,A$, this gives $u \approx 6 \times 10^{-5}\,\text{m s}^{-1}$. Note the small value of the speed at which the conduction electrons drift through the metal. This drift speed is superimposed on the random motion of the conduction electrons arising from their collisions with the thermally vibrating metal atoms; the root-mean-square value of the speed of this random motion is temperature-dependent and is $\sim 10^{-1}\,\text{m s}^{-1}$ at 300 K, i.e. much greater than the drift speed.

Note, however, that the *change* in the electric field in the wire (from zero to the value V/l) propagates along the wire at the speed of light in the material of the wire. Thus the electrons at the far end of a wire of length one metre will experience the change in value of the electric field after a time lapse of $\sim 10^{-8}$ s (assuming that the relevant value of the speed of light is $\sim 10^8\,\text{m s}^{-1}$).

For many conducting materials (metals and alloys) the resistance has been found to be independent of the value of the current for most practical purposes; this is *Ohm's law* (see Fig. 1.2(a)). A material whose resistance is a function of the value of the current is said to be **non-ohmic** (and non-linear) (see Fig. 1.2(b) and (c)). In this circumstance, a so-called **dynamic resistance** is defined through

$$R_{\text{dynamic}} \equiv \frac{\delta V}{\delta I} \tag{1.7}$$

where δV and δI are correlated changes in the current and the voltage (see Fig. 1.2(c)). Non-linear materials form the basis of many important electronic devices.

The term **resistor** is reserved for a material object which has resistance as its most important physical property in the context of an electrical circuit.

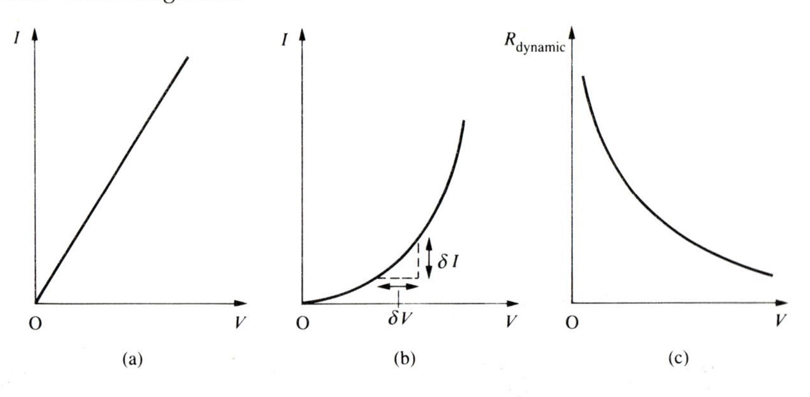

Fig. 1.2 The relation between voltage and current for (a) an ohmic (linear) material and (b) a non-ohmic (non-linear) material; (c) illustrates the dynamic resistance for a non-ohmic material with a characteristic as shown in (b).

Electric charges are classified as positive or negative, and, through an accident of history, the charge on an electron is negative. Hence, in a conductor the direction of a current, conventionally considered to be a flow of positive charges, is in the opposite sense to the direction of flow of the electrons (see Fig. 1.1; also Figs 1.3(b) and 1.4(a)).

Although the situation depicted in Fig. 1.1 is very simple, it does serve to draw attention to some very important concepts. In the calculation above to estimate the speed of the electrons in the wire, it was assumed that all the electrons entering the length of wire at section X_1 also left at section X_2, i.e. no electrons escaped through the surface of the wire. The electrons remain in the wire because it is energetically favourable for them to do so. Considerable energy has to be supplied to electrons to cause them to be emitted from a wire, for example by heating the wire to white heat as in the filament of the electron gun in a cathode ray tube. However, the circuits under consideration in this text will not be operating at white heat and so electron emission will be assumed to be negligible!

Hence electric charge and current will be assumed to be conserved in the circuits to be considered.

The agency providing the electric field does work on the electrons in moving them through the metal, and in the situation of dynamic equilibrium where the drift speed of the electrons is constant, the energy corresponding to this work is dissipated in the metal (predominantly as heat).

The work W done on an electron by a force F in moving the electron through a small distance δx, say, is given by the product of force and distance:

$$W = F\,\delta x$$

Since power p is equal to the rate at which work is done, then

$$p = \frac{F\,\delta x}{\delta t}$$

where δt is the time taken for the electron to travel the distance δx. In the mathematical limit of infinitesimally small increments δx, δt it follows that

$$\lim_{\delta t \to 0} \frac{\delta x}{\delta t} \to \frac{\mathrm{d}x}{\mathrm{d}t} = u \quad \text{(the speed of the electron).}$$

Hence

$$p = Fu.$$

If the wire is uniform, with area of cross-section A, and hence volume Al, the total number of electrons is equal to nAl and, since $F = eE$ where $E = V/l$, and using eqn (1.6) for u, the total power $P\ (= nAl \cdot p)$ is given by

$$P = nAl \cdot \left(e\frac{V}{l}\right) \cdot \frac{I}{neA}$$

or

$$P = VI \tag{1.8}$$

Using eqn (1.3) it follows that

$$P = I^2R$$

and

$$P = \frac{V^2}{R} \tag{1.9}$$

The SI unit of energy is the joule (J) and of power is the joule per second or watt (W).

Example 1.1 In a lightning strike a current flows to earth for a duration of 70 μs. If the average current in the strike is 8 kA, calculate the total electrical charge flowing to earth.

Current (I) is equal to rate of flow of charge (Q):

$$I = \frac{dQ}{dt} \qquad \text{or} \qquad dQ = I\,dt$$

Thus

$$Q = \int I\,dt,$$

where the limits of the integration are the time interval during which the current flows. So, in this case,

$$Q = I_{\text{average}} \int dt, \qquad \text{or} \qquad Q = I_{\text{average}} \times \Delta t$$

where Δt is the time interval. Therefore,

$$Q = 8 \times 10^3 \times 70 \times 10^{-6} \text{ coulomb (C)}$$

i.e.

$$Q = 0.56 \text{ C}.$$

□

Example 1.2 A light bulb is rated at 12 V, 24 W. What is the resistance of the bulb when it is in operation and how much charge passes through the bulb per second?

From eqns (1.9) $R = V^2/P = 144/24 = 6\,\Omega$.

From eqn (1.8) $I = P/V = 24/12 = 2$ A (i.e. 2 C s^{-1}).

Note that the filament in a tungsten filament light bulb is a non-linear resistive element. The resistance of the filament when cold is much less than when it is hot. □

Example 1.3 A 0.75 kW heater is needed to operate from the 240 V mains. The available nichrome wire has a diameter of 0.5 mm and a resistivity of $1.3 \times 10^{-6}\,\Omega$ m. Calculate the length of wire required for the heater element.

From eqn (1.4) $R = 4\rho l/\pi d^2$, where d is the diameter of the wire, and also, from eqns (1.9), $R = V^2/P$. Hence

$$l = \frac{\pi d^2 V^2}{4\rho P}^{*} \qquad \text{or} \qquad l = \frac{\pi \times (5 \times 10^{-4})^2 \times (240)^2}{4 \times 1.3 \times 10^{-6} \times 750}$$

(remember to put the quantities into consistent units). Therefore the required length $l = 11.6$ m. □

1.3 Electric field and electromotive force

As a result of experiments in the eighteenth century it was concluded that electric charge could be divided into two classes (so-called positive and negative charges) and that the force F between two 'point' charges Q_1, Q_2 in vacuum is proportional to the product of the magnitudes of the two charges and inversely proportional to the square of the distance of separation, i.e.

$$F \propto \frac{Q_1 Q_2}{x^2} \quad \text{(Coulomb's law)}$$

where x is the distance apart of the two charges. From experiments it was deduced that if Q_1, Q_2 are both positive charges or both negative charges, then the force is repulsive, whereas if they are of opposite signs, then the force is attractive. The constant of proportionality is expressed in the form $(4\pi\varepsilon_0)^{-1}$, where ε_0 is the so-called 'permittivity of free space', which has the value 8.854×10^{-12} farads per metre, i.e.

$$F = \frac{1}{4\pi\varepsilon_0} \cdot \frac{Q_1 Q_2}{x^2} \tag{1.10}$$

From dimensional considerations the constant of proportionality has the dimensions of (capacitance per metre)$^{-1}$ and the numerical value is a consequence of choosing SI units for distance, force and electric charge; the numerical factor 4π is incorporated for convenience. The concept of capacitance will be discussed later (see Section 1.5) and the wider significance of the permittivity of free space will also be explained.

* In all but the very simplest problems it is a good policy to assign symbols to all of the quantities involved and then to solve the problem algebraically **before** substituting numerical values. It is much easier to check an algebraic analysis than an arithmetic calculation.

The experimental observation that an electric charge Q_1 experiences a force F due to another charge Q_2 situated some distance away is represented by introducing the concept of an **electric field** E in the space in which the charges reside. The field E at the site of Q_1 due to Q_2 is such that the force on Q_1 is given by

$$F = Q_1 E \tag{1.11}$$

Hence by using eqn (1.10) the electric field due to Q_2 is

$$E = \frac{1}{4\pi\varepsilon_0} \cdot \frac{Q_2}{x^2} \tag{1.12}$$

It should always be remembered that this equation relates to a 'point' charge Q_2 or, in practice, a charged body whose lateral dimensions are much smaller than the distance of separation x. It is common practice to use loosely a term like 'a charge Q', where the implication is that the charge resides on an entity that is relatively small in the situation under consideration. The physical entity in question may be a sub-atomic particle such as an electron or a proton, or a dust particle or a droplet of liquid, or an even larger object such as an electrode in a piece of equipment or machinery. If the dimensions of the charged body are not 'small', then the field at an external point has to be calculated by summing (integrating) the contributions of the elements of charge that are distributed on the surface and through the volume of the body. In a few special cases the symmetry of the charged body greatly simplifies the calculation (e.g. for a spherical body), but in general, simplifying assumptions and approximations have to be made; for instance, 'edge effects' are often neglected in calculating the electric field between two parallel charged metal plates (see Section 1.5).

It is vital to grasp the concept of the existence of an electric field in the space around an electric charge or a system of charges. The force on a 'small' test charge Q situated at a point in the region of space where the field E exists is given by eqn (1.11).

It was indicated in Section 1.2 that electrons lose energy in flowing through a metal. Since energy is conserved overall, where did their initial energy come from? For the moment consider two charges of magnitude Q separated initially by a distance x, and assume that the charges are of opposite sign (such as an electron and a positively ionized atom) so that the force of interaction F is attractive. If an agency increases the separation between the charges by an amount δx (where $\delta x \ll x$), then the agency must do work $F\delta x$ against the attraction, and the potential energy W of the system of charges increases by an amount δW equal to $F\delta x$. If the agency were to be switched off, then the two charges would move back towards each other under the influence of the attractive force and the potential energy would be converted into the kinetic energy associated with the motion of the charges.

Solids, liquids, and gases are constituted of atoms and molecules that generally contain equal numbers of positive charges (protons) and negative charges (electrons) arranged in configurations having the lowest potential energy. This means that matter is generally electrically neutral when viewed or sampled on a macroscopic scale, that is, on a scale much larger than atomic/molecular dimensions.

In a source of **electromotive force** (e.m.f.) an agency does the work required to separate positive and negative charges and so produces a system in a state of high potential energy. If subsequently the system is free to do so, it will relax, with the charges returning to their configuration of lowest potential energy. The provision of an external conducting path, that is, a continuous circuit, is a familiar route by which the system can relax through the flow of mobile charges (usually electrons); see Fig. 1.3(a) and (b). Familiar examples of sources of e.m.f. are batteries (dry cells and accumulators), in which the positive and negative charges are separated by an electrochemical reaction. In a van de Graaff generator, positive and negative charges are separated by being transported on a moving belt which is driven mechanically (see Fig. 1.7(a)) and in a thermocouple the charges are separated by a thermoelectric effect. If the current in the circuit of Fig. 1.3(b) is described in terms of positive charges, then the potential energy of a positive charge $+e$ in the circuit is as shown in Fig. 1.4.

So these qualitative discussions have introduced the idea that a source of e.m.f. and the flow of current in a circuit are linked with **changes** in the potential energy of the mobile charges.

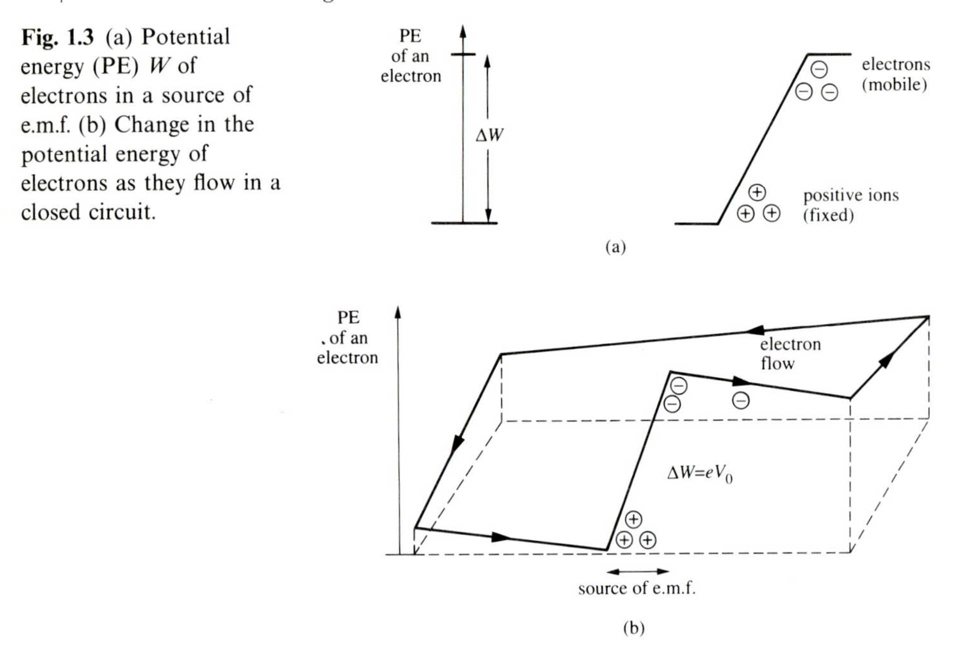

Fig. 1.3 (a) Potential energy (PE) W of electrons in a source of e.m.f. (b) Change in the potential energy of electrons as they flow in a closed circuit.

The concepts of potential energy, electrical potential, and 'voltage' will now be examined more closely.

If a charge Q is moved an incremental distance dx *against* the force QE due to a field E, then the work dW done on the charge is $-QE\,\mathrm{d}x$; the minus sign arises because the displacement dx is in the opposite direction to the field E (see Fig. 1.5):

$$\mathrm{d}W = -QE\,\mathrm{d}x$$

The variation of the potential energy of electrons in an electric field is exploited in **charge-coupled devices** (CCD), which are an important category of modern solid-state electronic devices used in integrated circuits. Consider a positively charged metal plate; if there are electrons in the space 'below' the plate, say (see Fig. 1.6(a)), then there will be a force of attraction between the (free) electrons and the (fixed) plate. Conversely, work would have to be done on an electron in order to move it away from the plate against the attractive force. Hence it can be seen that the region close to the plate is a region of low potential energy (or a **potential well**) for the electrons, since, if they are free to do so, they will move towards the plate. In a charge-coupled device an array of metal electrodes is separated from a layer of silicon by a thin layer ($\sim 1\,\mu$m thick) of highly insulating silicon dioxide, SiO_2 (see Fig. 1.6(a)). The silicon will normally contain very few conduction electrons (i.e. 'free' electrons),

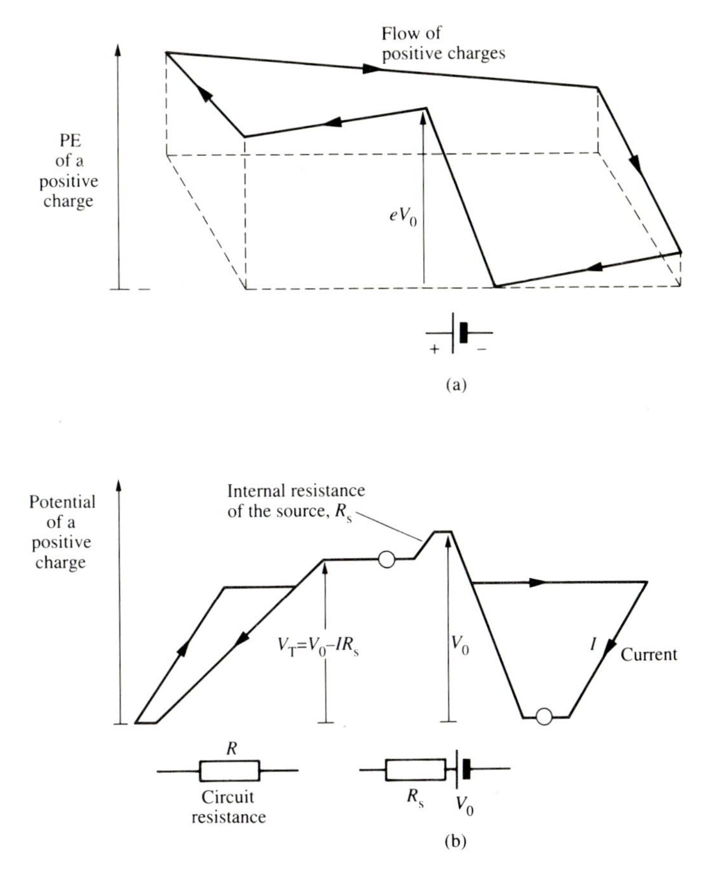

Fig. 1.4 (a) Potential energy of a positive charge $+e$ in a circuit which contains a source of e.m.f. V_0; note the sense of the flow of conventional current, i.e. the direction of flow of the positive charges $+e$. (b) Variation in potential energy of a charge $+e$ flowing in a circuit containing a resistance R and assuming that the connecting wires have negligible resistance so that the potential gradient along them is zero. The effect of the internal resistance R_s of the source means that the terminal voltage is now $V_T = V_0 - IR_s$.

but imagine that a packet of electrons can be injected into the silicon at one end of the device, and further assume that one of the metal electrodes is connected to a +10 V supply. The packet of electrons will migrate to the potential well under this electrode and become trapped there. Since the depth of a potential well increases with the magnitude of the positive voltage applied to an electrode, a packet of electrons can be caused to move along the line if each electrode of the array is connected successively to 10 V, 5 V, 0 V in a synchronized sequence (see Fig. 1.6(b)). Devices

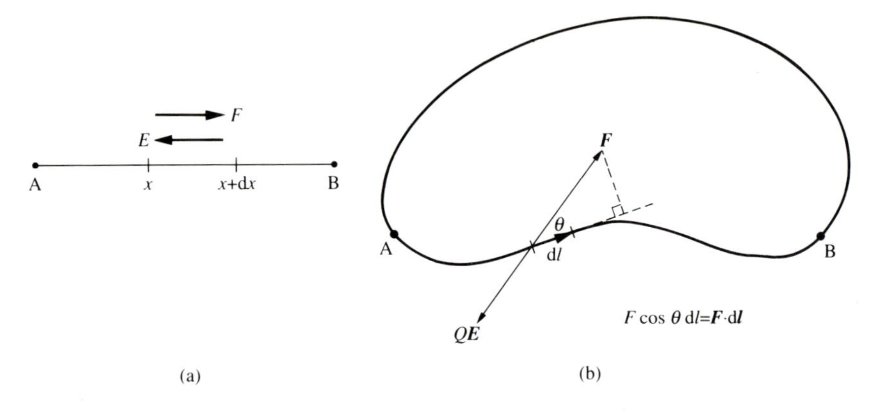

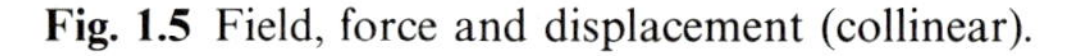

Fig. 1.5 Field, force and displacement (collinear).

based on this principle are used as delay lines, shift registers, and dynamic memories.

The total work done ΔW in moving a charge Q from A to B is given by summing (integrating) the elements of work $\mathrm{d}W$ as the point of application of the force moves from A to B:

$$\Delta W = \int_{\mathrm{A}}^{\mathrm{B}} -QE\,\mathrm{d}x \qquad \text{or} \qquad \Delta W = -Q\int_{\mathrm{A}}^{\mathrm{B}} E\,\mathrm{d}x$$

The electrical **potential difference** ΔV between A and B is defined by

$$\Delta V \equiv \Delta W/Q \tag{1.13}$$

$$\Delta V = -\int_{\mathrm{A}}^{\mathrm{B}} E\,\mathrm{d}x \tag{1.14}$$

The unit of ΔV is the joule per coulomb, which is called the volt (V).

In the context of electrical circuits it is usual to refer all potentials to a common reference potential ('common' or 'earth' or 'ground'), so from now on the term *potential* (symbol V) will be used where, strictly speaking, a difference of potential relative to a defined level of potential is implied.

The converse of eqn (1.14) is

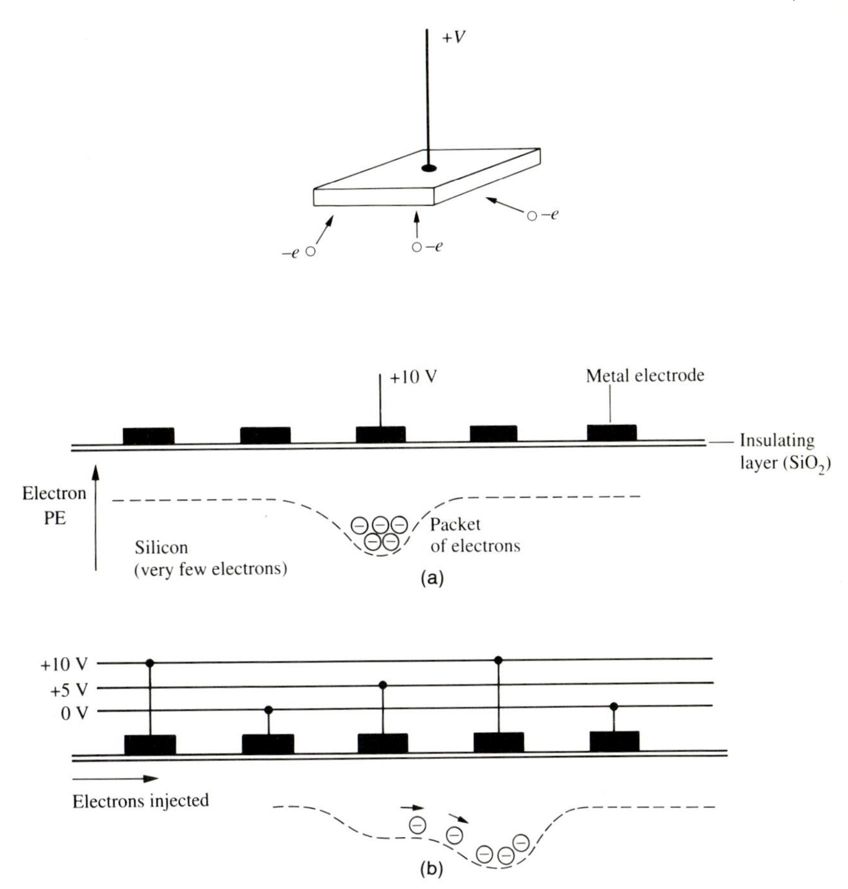

Fig. 1.6 Charge-coupled devices: (a) potential well for electrons 'under' a positively charged metal plate; (b) electrons spill from a shallow well into a deeper well and so can be shifted along the line by a sequence of synchronized voltage pulses applied to the array of electrodes.

$$E = -\frac{dV}{dx} \tag{1.15}$$

i.e. electric field is equal to (minus) the potential gradient. A useful analogy can be drawn with contour lines on a topographical map. The lines join points of equal altitude and therefore of equal gravitational potential energy; the closer together are the contour lines, the greater is the gradient and the greater is the gravitational force acting on an object lying on the slope.

In the introductory discussion of the flow of current in a wire in Section 1.2 it was implicitly assumed that the wire was uniform along its length in respect of its cross-section and its physical properties, so that the potential gradient (and hence the electric field) had a constant value, which was obtained simply by dividing the potential difference V between the ends of the wire by the length l of the wire. The discussions so far have involved situations where the displacement of charges was in the same or opposite direction to the electric field, i.e. collinear situations. In general the direction of an element of displacement will not be collinear with E (see Fig. 1.5(b)) and the element of work $\mathrm{d}W$ is given by the product of $\mathrm{d}l$ and the *component of F in the direction of* $\mathrm{d}l$:

$$\mathrm{d}W = F\cos\theta\cdot\mathrm{d}l. \tag{1.16}$$

Since both F and $\mathrm{d}l$ have directional properties, they can be written as vector quantities; in vector notation, eqn (1.16) is written as a scalar product, or 'dot product':

$$\mathrm{d}W = \boldsymbol{F}\cdot\mathrm{d}\boldsymbol{l}. \tag{1.17}$$

Note that the result of the scalar product of two vectors is a *scalar* quantity (i.e. a non-directional quantity), in this case the energy (or work) $\mathrm{d}W$.

If the point of application of the force F moves from A to B, then the total work done ΔW is given by

$$\Delta W = \int_{\mathrm{A}}^{\mathrm{B}} \boldsymbol{F}\cdot\mathrm{d}\boldsymbol{l}$$

or

$$\Delta W = -Q\int_{\mathrm{A}}^{\mathrm{B}} \boldsymbol{E}\cdot\mathrm{d}\boldsymbol{l} \quad \text{and} \quad \Delta V = -\int_{\mathrm{A}}^{\mathrm{B}} \boldsymbol{E}\cdot\mathrm{d}\boldsymbol{l}. \tag{1.18}$$

the 'line integral' of ***E*** from A to B

If the path of integration of the electric field is a *closed path*, then the line integral is zero; this is a characteristic of a so-called **conservative field**.

Some important general aspects of sources of e.m.f. can be illustrated by referring to the schematic diagram of a van de Graaff generator in Fig. 1.7(a). A moving belt made from a good insulating material transports charges in the directions shown and so produces separated assemblies of positive and negative charges; these separated charges produce an electric

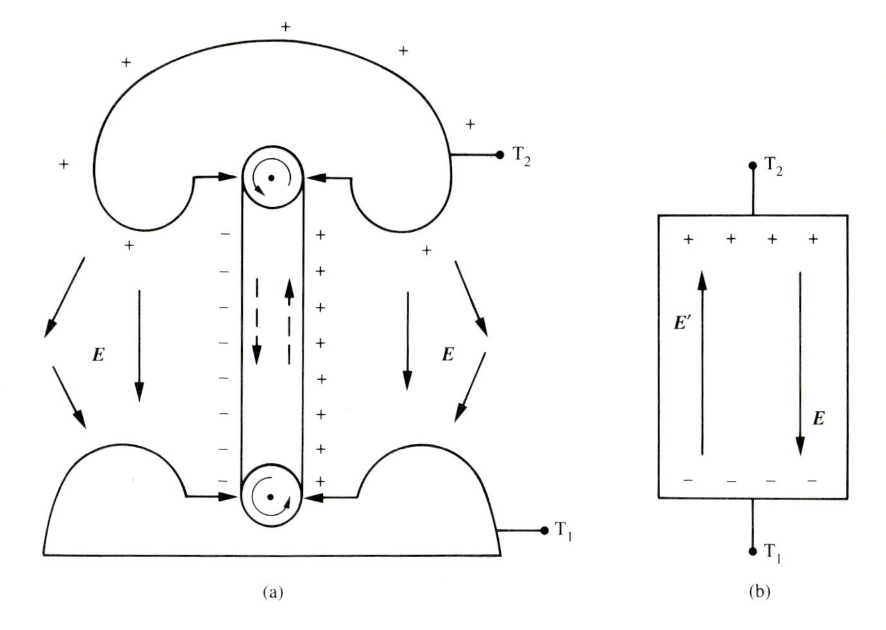

Fig. 1.7 (a) Van de Graaff generator; the separation of charges and the direction of the electric field $\boldsymbol{E}$ are shown schematically. (b) Schematic representation of a general source of e.m.f.

field $\boldsymbol{E}$. The external agency that moves the belt must exert a force $\boldsymbol{F}'$, say, on the charges and do work in order to move them against the force $\boldsymbol{F}$ due to the field $\boldsymbol{E}$.

Now consider a more general physical system B (see Fig. 1.7(b)) which could be a van de Graaff generator, a battery or a thermocouple, for instance, and assume that there is no external conducting path between T_2 and T_1, i.e. the source is in the 'open circuit' condition. From the definition of eqn (1.1) a **non-electrostatic field** $\boldsymbol{E}'$ can be defined through $\boldsymbol{E}' \equiv \boldsymbol{F}'/Q$. Imagine that such a field exists inside B; charges of opposite sign will continue to be separated by the agency until the resultant net force on a charge is zero, i.e. $\boldsymbol{F}' + \boldsymbol{F} = 0$, or $\boldsymbol{F}' = -\boldsymbol{F}$. In this limiting situation the work done ΔW by the agency in moving a charge Q from T_1 to T_2 is

$$\Delta W = \int_{T_1}^{T_2} \boldsymbol{F}' \,\mathrm{d}\boldsymbol{l} = -\int_{T_1}^{T_2} \boldsymbol{F}\,\mathrm{d}\boldsymbol{l} \qquad \text{or} \qquad \Delta W = -Q\int_{T_1}^{T_2} \boldsymbol{E}\,\mathrm{d}\boldsymbol{l}.$$

So

$$\Delta W/Q = -\int_{T_1}^{T_2} \boldsymbol{E}\,\mathrm{d}\boldsymbol{l}$$

which, from eqns (1.13) and (1.14), is the equal to the difference in electrical potential between the terminals T_2 and T_1, i.e. $\Delta W/Q = V_{T_2} - V_{T_1}$. This amount of work done per unit charge, when the source is on open circuit, is called the **electromotive force** E (e.m.f.) and is equal to the potential difference between the terminals:

$$E = -\int_{T_1}^{T_2} \boldsymbol{E}\,\mathrm{d}\boldsymbol{l}. \tag{1.19}$$

If the terminals T_1, T_2 are connected via an external circuit, then a current will flow, the charges at T_2, T_1 being replenished by $\boldsymbol{F}$. Since $\Delta W = QE$, it follows that the rate of working (power P) of the source is

$$P = \frac{\mathrm{d}(\Delta W)}{\mathrm{d}t} = E\frac{\mathrm{d}Q}{\mathrm{d}t}$$

Now $\mathrm{d}Q/\mathrm{d}t$ is equal to the current I, so

$$P = EI \tag{1.20}$$

The energy gained by the charges through the work done on them in the source is dissipated in the resistance of the circuit, as was described in Section 1.2 and illustrated in Fig. 1.4.

The e.m.f. of a van de Graaff generator may be millions of volts, but that of an electrochemical cell is only of the order of one volt, since the changes in energy of an atom or molecule involved in the chemical reaction that transports charges are of the order of one electron-volt.

The sources mentioned so far have been generators of a constant (or 'direct') e.m.f. In the majority of non-digital circuits there will be one or more sources of alternating e.m.f., i.e. sources whose time variation is sinusoidal, so that the current will flow out of and into a source in successive half-cycles of the sinusoidal waveform (see Fig. 1.8). The frequency of the alternating voltage/current may be that of the electrical mains (50 or 60 Hz) or, in the case of signal generators, may lie in a range extending from a fraction of a hertz to $\sim 10^{11}$ Hz. The fundamental aspects of alternating current circuits will be treated in detail in Chapter 3, so it suffices to say here that the concepts of the supply of energy and the dissipation of energy that are exemplified by Fig. 1.4 and eqns (1.8) and (1.9) apply equally to alternating current circuits.

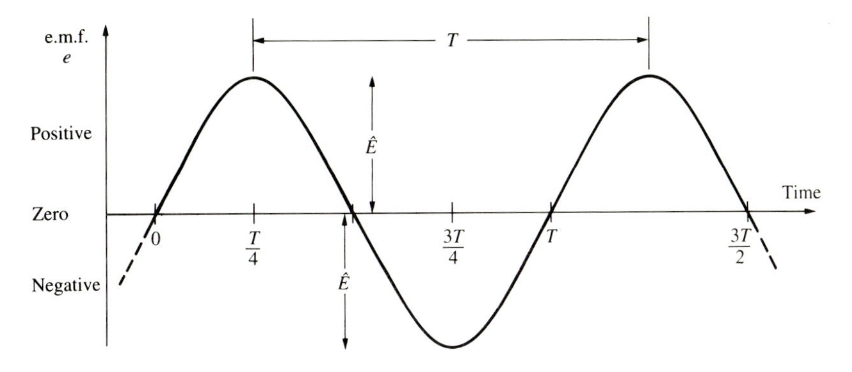

Fig. 1.8 The waveform of an alternating, sinusoidal electromotive force drawn as a function of time: $E = \hat{E}\sin(2\pi \cdot t/T)$.

1.4 Gauss's law

Although circuits can be analysed and designed on the basis of an operational understanding of circuit theory, the development of solid-state devices, integrated circuits, and very high-frequency circuits, and the imposition of strict standards of electromagnetic compatibility between electronic systems (see Chapter 9) mean that it is very useful to have a basic understanding of electromagnetic fields at least.

Gauss's law is a very useful guide to the relation between an electric field and the charges that are the source of the field.

Gauss's law

The integral of the flux of electric field E out of a closed surface S (either a real or a hypothetical surface) is proportional to the total electric charge Q contained in the volume that is enclosed by the surface.

If an infinitesimally small element of area of the closed surface S* is represented by a vector $\mathrm{d}\boldsymbol{S}$ which is perpendicular ('normal') to the element and is directed outwards (see Fig. 1.9), then the element of electric flux over the element is defined to be $E\cos\theta \cdot \mathrm{d}S$ (i.e. the normal component of E multiplied by $\mathrm{d}S$) or $\boldsymbol{E}\cdot\mathrm{d}\boldsymbol{S}$ in vector notation.

If SI units are used, the constant of proportionality in the mathematical expression of Gauss's law is $1/\varepsilon_0$ and in symbolic notation the law is

* If the vector nature of a vector quantity must be invoked, then in mathematical expressions a boldface character is used, e.g. electric field $\boldsymbol{E}$. If only the magnitude of the particular vector quantity is of relevance, the ordinary character is used to denote this (scalar) quantity, e.g. E.

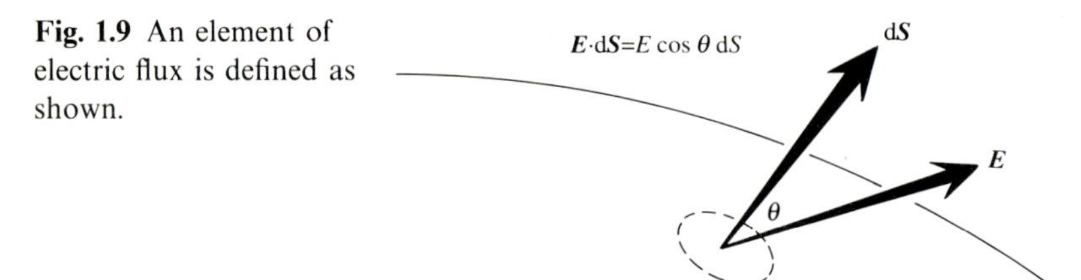

Fig. 1.9 An element of electric flux is defined as shown.

written as

$$\int_S \mathbf{E}\cdot d\mathbf{S} = \frac{Q}{\varepsilon_0} \quad \text{or} \quad \int E\cos\theta \, dS = \frac{Q}{\varepsilon_0}. \tag{1.21}$$

A simple illustrative example of Gauss's law is the calculation of the electric field at a distance x from a point charge Q. Consider a hypothetical spherical surface of radius x centred on the site of Q. By symmetry, the field $E(x)$ must have the same magnitude at all points on the surface and must be directed radially (i.e. everywhere normal to the surface so that $\theta = 0^\circ$ and $\cos\theta = 1$). Hence the constant value of $E(x)$ can be taken outside the integral sign in eqn (1.21) and the equation simply becomes the value of $E(x)$ of the electric field at distance x multiplied by the surface area of the sphere of radius x, namely $4\pi x^2$; that is,

$$E(x)\int_S dS = \frac{Q}{\varepsilon_0} \quad \text{or} \quad E(x)\cdot 4\pi x^2 = \frac{Q}{\varepsilon_0}.$$

Hence

$$E(x) = \frac{Q}{4\pi\varepsilon_0 x^2}$$

in agreement with eqn (1.12), which followed from Coulomb's law.

If the charge Q is distributed uniformly over a real spherical surface, then it can be shown that the electric field at points external to the sphere is the same as if the total charge Q were situated at the centre of the sphere.

Since the charge per unit area σ (sigma) is equal to $Q/4\pi x^2$, it follows that

$$E(x) = \frac{\sigma}{\varepsilon_0}. \tag{1.22}$$

This result holds for the electric field at a position close enough to any surface that has charge distributed over it; the normal component of the electric field in a particular locality has a magnitude equal to σ/ε_0. What qualifies as 'close enough' depends on the magnitude of the curvature of the surface in the locality in question.

For a spherical conductor of radius r carrying charge Q, the electric field E at the surface is equal to $Q/4\pi\varepsilon_0 r^2$ and the potential V at the surface (and hence of the conductor as a whole) is equal to $Q/4\pi\varepsilon_0 r$; so $E = V/r$. This relation says that, for a spherical conductor at a given potential V, the electric field just outside the surface of the conductor is inversely proportional to the radius; the smaller the radius, the higher the field strength. This argument can be extended qualitatively to conductors of a general shape; the electric field will be highest over the most highly curved parts of the surface of the conductor and particularly at corners and points. If the conductor is in air, then, if its potential is gradually increased, the field at its surface will eventually exceed the breakdown field ('breakdown strength' $\sim 30\ \mathrm{kV\ cm^{-1}}$) of air and a violent discharge (spark) will occur; if the surface has sharp corners, then the breakdown field strength will be attained first at these corners. In the case of points the discharge tends to be relatively gentle ('corona discharge') rather than a violent spark; this phenomenon is exploited in lightning conductors. On the other hand, the electrodes and other surfaces of high-voltage equipment should have a large radius of curvature and be free of points (including dust particles) to preclude corona discharge (see Chapter 9).

1.5 Capacitance

In general, a passive circuit element has resistance, capacitance, and inductance. If the dominant feature as far as its circuit properties are concerned is resistance, then the element is called a resistor; if capacitance, a capacitor; and if inductance, an inductor. Resistance has been discussed already in Section 1.2; inductance will be discussed later in Section 1.7.

It will become apparent later that capacitance (and inductance) are important in a circuit when the current and/or the voltage are changing with time, i.e. transient currents and voltages, repetitive voltages and currents (sinusoidal and non-sinusoidal). Important examples of transient situations are the changing currents and voltages when a source of e.m.f.

is switched into, or out of, a circuit, and the pulsed currents and voltages in digital circuits.

The capacitance of a body is defined in terms of the change in the electrical potential of the body when charge is transferred to or from it; an isolated spherical conductor serves as a useful starting point to the discussion.

Consider a spherical conductor* of radius R with a total net charge Q_1 distributed uniformly over its surface. In order to bring a charge Q_2, of the same sign as Q_1, from an infinite distance and add it to Q_1, work must be done against the repulsive force. Since the field E due to Q_1 is the same as if this charge were concentrated as a point charge at the centre of the spherical surface, the work $\mathrm{d}W$ done in moving Q_2 from a distance x to a distance $(x - \mathrm{d}x)$ is $F\,\mathrm{d}x$, where the repulsive force $F = Q_2 E$ and E is given by eqn (1.11); that is,

$$\mathrm{d}W = -Q_2 \cdot \frac{Q_1}{4\pi\varepsilon_0 x^2}\,\mathrm{d}x$$

(the minus sign appears because the displacement $\mathrm{d}x$ is in the opposite direction to the field E).

So the change (increase) in potential energy ΔW of the system of charges as Q_2 is brought from infinity to $x = R$ (i.e. at the end of the process Q_2 resides on the surface of the spherical conductor) is given by

$$\Delta W = -\frac{Q_2 Q_1}{4\pi\varepsilon_0}\int_{x=\infty}^{x=R}\frac{\mathrm{d}x}{x^2}$$

$$= -\frac{Q_2 Q_1}{4\pi\varepsilon_0}\left(-\frac{1}{x}\right)_{x=\infty}^{x=R}$$

or

$$\Delta W = \frac{Q_1 Q_2}{4\pi\varepsilon_0 R}.$$

If the charge Q_2 is now regarded as a test charge being used to probe the electric field around the charged spherical body, then it can be said that the potential V in the field at $x = R$ (i.e. the potential of the conductor) is given by

* Strictly it does not need to be a conductor, but in electrical circuits capacitance will arise in relation to parts of the circuit composed of conducting or semiconducting material.

$$V = \frac{Q_1}{4\pi\varepsilon_0 R} \tag{1.23}$$

(remember that $\Delta V \equiv \Delta W/Q$; see eqn (1.13)).

In using V rather than ΔV to denote the change in potential, it is implicit that a zero level for V has been assumed at infinity.

The quantity Q_1/V $(=4\pi\varepsilon_0 R)$ is a characteristic property of the spherical body, independent of its charge and potential, and is called the **capacitance** of the body; i.e. for a sphere,

$$C = 4\pi\varepsilon_0 R \tag{1.24}$$

The equivalent calculation for an isolated non-spherical body is not so easy; nevertheless the capacitance is still defined through

$$C \equiv \frac{Q}{V} \tag{1.25}$$

where Q is the charge on the conductor (in coulombs) and V is its electric potential in volts relative to an assumed zero of potential at infinity.

The SI unit of capacitance is the farad (F) and it follows that the unit of ε_0 is the farad per metre (F m^{-1}).

In practice, charged bodies can never be completely isolated but are always influenced by other bodies to some degree. For example, if one conducting body A has charge Q and would have potential V if completely isolated, then in a situation where it is influenced by other conductors having charges of the same sign as Q, the potential of A will be raised and its capacitance will be lowered (see Fig. 1.10). Hence to increase the capacitance of A the proximity of another conducting body (or bodies) having charges of the opposite sign to Q is required.

Suppose that conductor A, which is positively charged, is influenced by only one other conductor B, and that the total charge on B is equal and opposite to that on A; see Fig. 1.11. If a hypothetical closed surface is

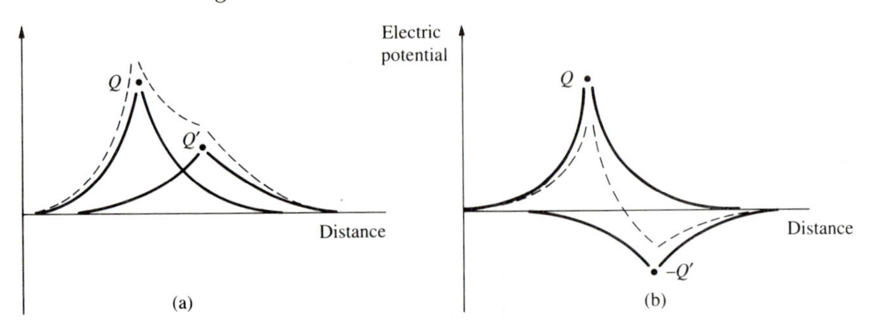

Fig. 1.10 A schematic representation of the raising or lowering of the electric potential of a charged conductor, depending on the sign of the charge on a nearby conductor: (a) same sign; (b) opposite signs. Dashed lines show the resultant potential.

drawn which includes both A and B, then, by Gauss's law, the net outward flux of the electric field E over the closed surface must be zero, since the net enclosed charge is zero. This means that all the electric field lines originating on the positive charges constituting the charge on A must terminate on the negative charges constituting the charge on B. The positive charges on A are the 'sources' of the electric field between A and B, and the negative charges on B are the 'sinks' of the field. Of course, field lines are an abstraction, but the idea of sources and sinks for field lines is a very useful aid to picturing the distribution of the electric field around and between charged bodies.

An arrangement of two oppositely charged conductors such as in Fig. 1.11 is called a capacitor, where the capacitance C is defined by

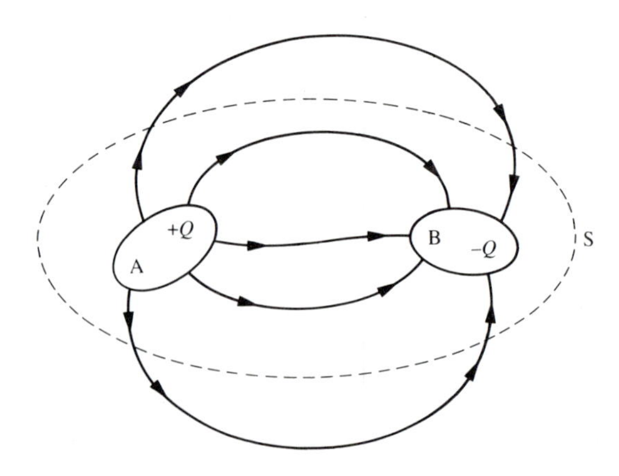

Fig. 1.11 The electric field lines between two charged bodies A and B (schematic). S is a hypothetical closed surface (often called a Gaussian surface). Notice that all field lines leaving the closed surface at one point enter it at another; this is consistent with Gauss's law, which requires $\int_S \boldsymbol{E} \cdot \mathrm{d}\boldsymbol{S} = 0$ in this situation.

$$C \equiv \frac{Q}{V_A - V_B}.$$

Strictly speaking this is the definition of the *mutual* capacitance of the two conductors; in contrast, the capacitance of an isolated body is its *self*-capacitance (refer to the discussions leading to eqns (1.24) and (1.25)). In circuits it is always mutual capacitance that is relevant and, for brevity, the adjective 'mutual' will be dropped.

An ideal geometry for a capacitor is one in which one conductor is completely surrounded by the other so that their charges are equal and opposite. Using eqn (1.23) for an idealized arrangement of two concentric spherical conducting shells of radii a and b ($a < b$),

$$V_A - V_B = \frac{Q}{4\pi\varepsilon_0 a} - \frac{Q}{4\pi\varepsilon_0 b}$$

and

$$C = \frac{4\pi\varepsilon_0 ab}{b - a}$$

Of course it is impossible in practice to have one conductor completely surrounding the other, because of the resulting inaccessibility of the inner conductor; the common practical forms are the parallel-plate capacitor and the concentric-cylinder capacitor.

Consider a parallel-plate capacitor as shown in Fig. 1.12. If the lateral extent of the plates is much greater than the separation t, then fringing effects such as depicted in Fig. 1.12(a) can be neglected and it can be assumed, without significant loss of accuracy, that the entire charge on each plate is distributed uniformly over that part of its surface which faces the other plate. Hence the electric field will be zero except in the space between the plates and, furthermore, will be uniform with value E, say.

Applying Gauss's law to the hypothetical closed surface shown in Fig. 1.12(b) and assuming that the area of the part of a plate over which the charge is distributed is A, then

$$E \cdot A = \frac{Q}{\varepsilon_0} \tag{1.26}$$

Now the uniformity of E implies that the charge is distributed uniformly with charge density $\sigma = Q/A$, and so $E = \sigma/\varepsilon_0$. The approximation that was made, that the lateral extent of the plates is much greater than their separation, means that this result gives the field outside a plane conductor that is infinite in extent. However, the usefulness of this expression for E is greatly enhanced by the fact that, to an approximation, it gives the normal electric field 'close to' a non-planar conductor; refer to eqn (1.22)

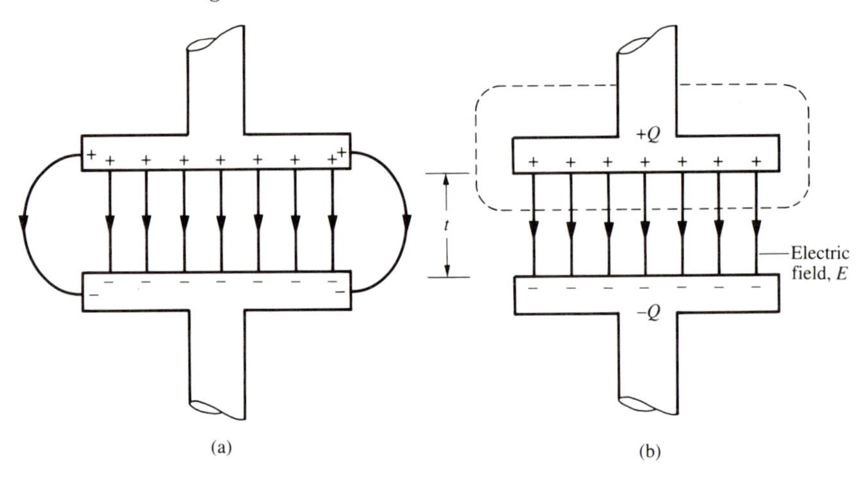

Fig. 1.12 (a) A charged parallel-plate capacitor with the fringing field shown schematically. (b) A hypothetical Gaussian surface drawn around one plate of the capacitor.

for the field outside a uniformly charged spherical surface. A good analogy is that of a person sitting in a small boat on a perfectly calm sea; the surface of the sea appears flat and (almost) infinite in extent.

Since the electric field between the capacitor plates is uniform, the potential difference V between the plates is simply given by

$$V = E \cdot t \qquad \text{(see eqn (1.15))},$$

and so, using eqn (1.26), it follows that

$$V = \frac{Qt}{\varepsilon_0 A}.$$

Thus the capacitance, which is equal to Q/V, is given by

$$C = \frac{\varepsilon_0 A}{t} \tag{1.27}$$

As has been stated, the SI unit of capacitance is the farad, so from eqn (1.27), if the required value of C is 1 F, and if $t = 1$ mm, then $A \approx 10^8$ m^2, which is not very practicable! This situation arises from the way in which the farad was defined historically. So in practice, values of capacitance much smaller than 1 F are the rule.

If the space between the capacitor plates is filled with an electrically insulating material (or **dielectric**), without altering the spacing, then it will be found that the capacitance is increased. Denoting the capacitance by

C_0 when there is air (or vacuum, strictly) between the plates and by C when the space is filled with the insulating material, then the **relative permittivity** ε_r of the material is defined through

$$\varepsilon_r \equiv \frac{C}{C_0} \tag{1.28}$$

or, using eqn (1.27),

$$C = \frac{\varepsilon_r \varepsilon_0 A}{t}. \tag{1.29}$$

Note that $\varepsilon_r \geqslant 1$.

For 'free space' (vacuum), $\varepsilon_r = 1$ and the origin of the name 'permittivity of free space' for the fundamental electric constant ε_0 can now be appreciated. The quantity $(\varepsilon_r \varepsilon_0)$ is called the **dielectric constant** of the medium.

Commonly used dielectric media are paper, mica and polyethylene, for which $\varepsilon_r \approx 3.7$, 6.0, and 2.3 respectively. The values of capacitance obtainable in capacitors (usually in concentric cylinder form) using such dielectrics range from about 10 picofarad (pF) to $\sim$10 microfarad (μF). The physical origins of the relative permittivity are discussed briefly in Section 5.1.

It is possible to obtain a large value of capacitance in a physically comparatively small capacitor by exploiting the electrolysis of certain salts to form an extremely thin insulating film ($\sim$0.02 μm) on aluminium foil electrodes. In this way, values of capacitance ranging up to $\sim$1000 μF can be obtained in capacitors that are cylindrical in external form ($\sim$2 cm long $\times$ 1 cm diameter). Two disadvantages of such **electrolytic capacitors** are their low working voltage (up to about 60 V commonly) and the need to ensure that the capacitor is always electrically polarized in a particular sense (usually indicated by + and − signs on the body of the capacitor, or by some other code). In non-electrolytic capacitors of the same order of external dimensions, the largest value of capacitance that is commonly available is $\sim$10 μF.

A system of practical importance in which a knowledge of the capacitance is of interest is the coaxial line or cable. Flexible coaxial cable is very widely used for making connections between electronic systems such as signal generators, amplifiers, and cathode ray oscilloscopes. Such a cable consists of a cylindrical outer conductor of radius b, say, and a coaxial inner conductor of radius a, say (see Fig. 1.13). Assume that there is charge $\pm Q$ per unit length on the inner and outer conductors respectively.

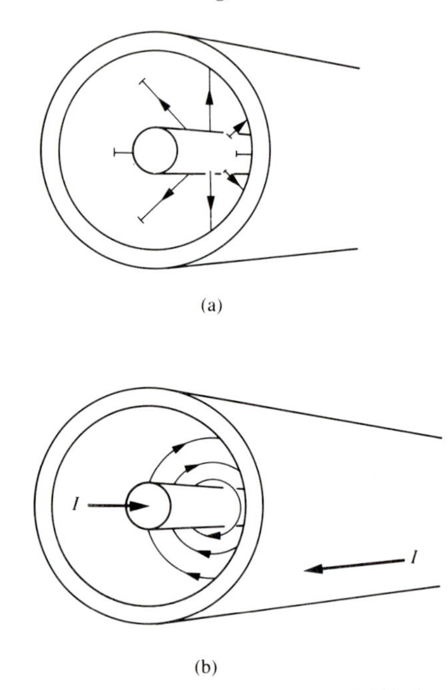

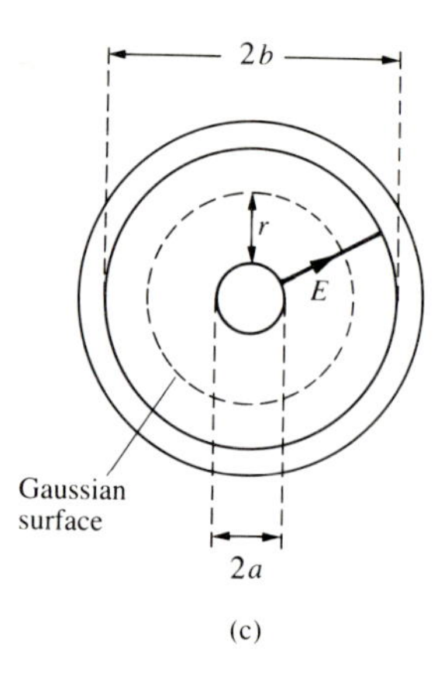

Fig. 1.13 (a) The electric field pattern and (b) the magnetic field pattern in the space between the conductors of a coaxial line. (c) The cylindrical Gaussian surface involved in the calculation of the radial electric field $E(r)$.

Consider a hypothetical closed Gaussian surface of radius r and unit length, which is concentric with the inner and outer conductors ($a < r < b$), and apply Gauss's law; see Fig. 1.13(c). Since $E(r)$ is radial, from symmetry it follows that $E(r)$ is everywhere normal to the curved portion of the Gaussian surface, and of constant magnitude. Hence $\int_S \boldsymbol{E}(\boldsymbol{r})\cdot \mathrm{d}\boldsymbol{S}$ in this case becomes

$$E(r)\cdot 2\pi r = \frac{Q}{\varepsilon_0}, \tag{1.30}$$

since the area of the curved surface of the cylinder of unit length is $2\pi r$. Note that there is no normal component of E over the flat end surfaces of the cylindrical Gaussian surface, which therefore make no contribution to $\int_S \boldsymbol{E}(\boldsymbol{r})\cdot \mathrm{d}\boldsymbol{S}$.

From eqn (1.18) the potential difference V between the outer and inner conductors is given by the 'line integral' of E from the outer to the inner conductor, which is simple to evaluate in this case, since the radial field is collinear with the radial path of integration:

$$V = -\int_{r=b}^{r=a} E\,\mathrm{d}r.$$

The minus sign arises because the direction of the path of integration is in the opposite direction to the electric field.

Using eqn (1.30),

$$\begin{aligned} V &= -\frac{Q}{2\pi\varepsilon_0}\int_{r=b}^{r=a}\frac{\mathrm{d}r}{r} \\ &= -\frac{Q}{2\pi\varepsilon_0}[\ln r]_b^a \\ &= -\frac{Q}{2\pi\varepsilon_0}[\ln a - \ln b] \end{aligned}$$

or

$$V = \frac{Q}{2\pi\varepsilon_0}\ln\left(\frac{b}{a}\right)$$

The capacitance per unit length (C) is equal to Q/V, so finally,

$$C = \frac{2\pi\varepsilon_0}{\ln(b/a)}\ \mathrm{F\,m^{-1}} \tag{1.31}$$

The insulation between the two conductors can take a variety of forms, depending on the frequency of the voltages and currents of interest, but in a commonly used type of cable the insulator is polyethylene, for which the value of the relative permittivity ε_r is 2.3; hence the expression for C (eqn (1.31)) must be multiplied by ε_r. Taking $2b \approx 3$ mm and $2a \approx 0.75$ mm, then $C \approx 100\ \mathrm{pF\,m^{-1}}$.

A consideration of the work done in establishing the charges $\pm Q$ on the plates of the parallel-plate capacitor (refer to Fig. 1.12) leads to a very important result concerning the energy associated with an electric field. Assume that charges $\pm q$ exist on the plates (where $q < Q$) and consider the work $\mathrm{d}W$ required to transfer an additional element of charge $\mathrm{d}q$ from the 'negative' plate to the 'positive' plate. If the capacitor is in a circuit, as shown in Fig. 1.14, then the additional charge $\mathrm{d}q$ would be transferred from one plate to the other via the external circuit, rather than directly across the gap between the plates, the required energy being supplied to

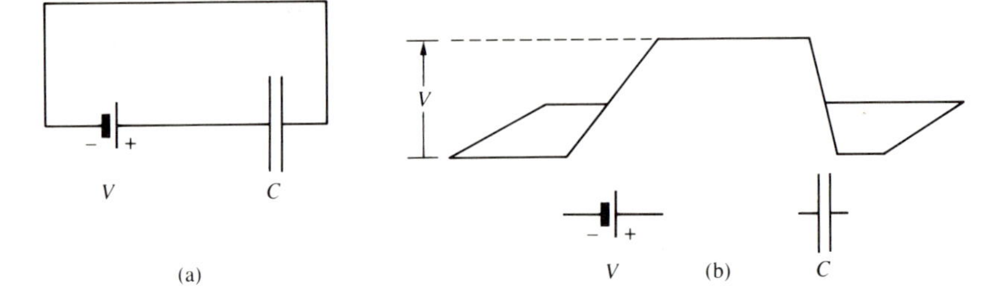

Fig. 1.14 (a) A capacitor charged to a potential difference V by a battery. (b) Energy diagram for the system.

the element of charge by the source of e.m.f. (Remember that the electrostatic field is 'conservative' and hence the work done in taking dq from one plate to the other is independent of the path taken.)

Using eqn (1.13), with V replacing ΔV and dq replacing Q,

$$\mathrm{d}W = V\,\mathrm{d}q$$

Now C $(=q/V)$ depends only on the geometry of the capacitor and is independent of the values of q, V. Hence

$$\mathrm{d}W = \frac{1}{C}q\,\mathrm{d}q$$

The total work ΔW required to establish the charge Q on the capacitor is then given by

$$\Delta W = \int_{q=0}^{q=Q} \mathrm{d}W = \frac{1}{C}\int_{q=0}^{q=Q} q\,\mathrm{d}q$$

or

$$\Delta W = \frac{Q^2}{2C}.$$

The amount of work done in separating the charges $\pm Q$ can be imagined as being stored as potential energy in the capacitor. Since $C = Q/V$, ΔW can be expressed in terms of V also, so it follows that

$$\Delta W = \frac{Q^2}{2C} \quad \text{or} \quad \Delta W = \frac{CV^2}{2} \quad \text{or} \quad \Delta W = \frac{QV}{2}. \tag{1.32}$$

In establishing the charges $\pm Q$ on the capacitor plates, an electric field E has been set up in the space between the plates, and it is very useful to imagine the potential energy as being stored in this electric field. Since the electric field is assumed to be uniform in the space between the plates (of separation t), eqn (1.15) can be used to give

$$V = Et.$$

Since the volume between the plates of the capacitor is equal to At, and using $\Delta W = CV^2/2$, it follows that **energy density in the electric field** is given by

$$\text{energy density} = \frac{\varepsilon_0 E^2}{2}\ \mathrm{J\,m^{-3}} \tag{1.33}$$

This very useful result applies also to time-dependent electric fields. The companion result for a magnetic field enables expressions to be obtained for the energy density in electromagnetic fields (see eqn (1.57)).

It is worth noticing that the electric field at the surface of a good conductor (strictly an ideal conductor of zero resistivity) is normal to the surface [see Figs. 1.12 and 1.13 for example]. For future reference it is also worth noticing that the magnetic field at the surface of a conductor is parallel to the surface (see Fig. 1.13(b)).

Example 1.4

A 12 volt battery supplies a current of 1 ampere continuously for 2 hours. How much energy does the battery deliver? To what voltage would a 1000 μF capacitor have to be charged in order to store an amount of energy equal to that supplied by the battery?

Since power P is equal to the rate at which energy W is supplied ($P = \mathrm{d}W/\mathrm{d}t$), $\mathrm{d}W = P\,\mathrm{d}t$ and $W = \int_t^{t+\Delta t} P\,\mathrm{d}t$, where Δt is the time interval over which energy is supplied. Now, from eqn (1.20), $P = EI$, and both E and I are constant in this case. So

$$W = P\int_t^{t+\Delta t} \mathrm{d}t = P\,\Delta t.$$

So

$$W = 12 \times 1 \times 2 \times 3600\ \mathrm{J} = 86.4\ \mathrm{kJ}$$

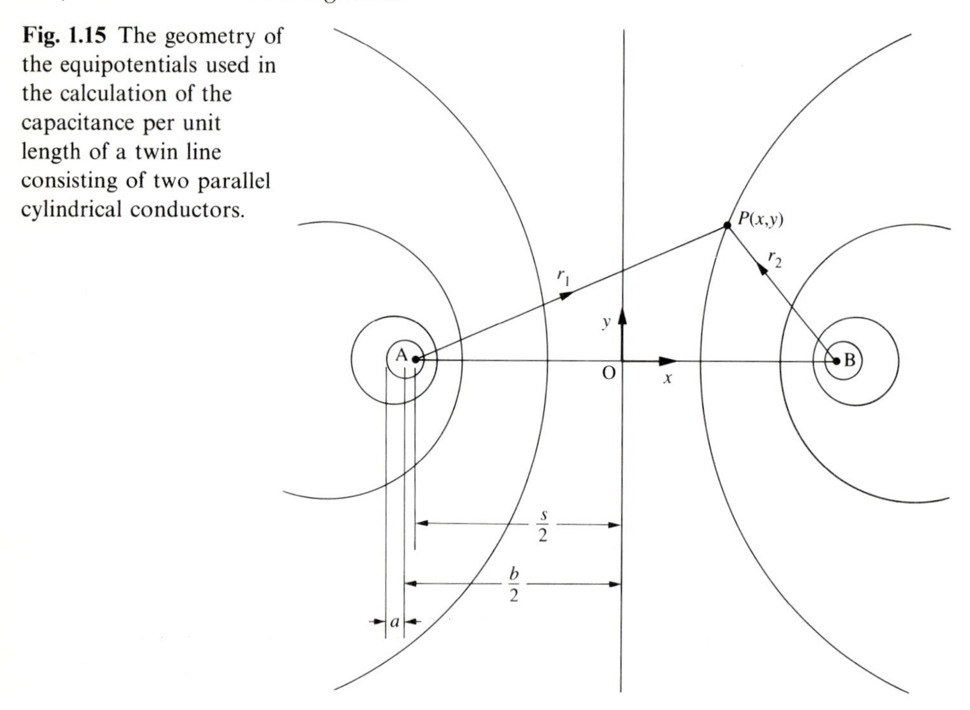

Fig. 1.15 The geometry of the equipotentials used in the calculation of the capacitance per unit length of a twin line consisting of two parallel cylindrical conductors.

From eqn (1.32), $V^2 = 2W/C$, i.e.

$$V^2 = \frac{2 \times 8.64 \times 10^4}{10^{-3}}, \qquad \text{or} \qquad V = 13.1\ \text{kV}. \qquad \square$$

Another physical system of great practical interest is the parallel twin line consisting of two parallel cylindrical conductors of radius a, say, spaced a distance b apart (see Fig. 1.15). The interest lies in the fact that pairs of parallel conductors ranging from purpose-designed transmission lines to conducting tracks on printed circuit boards, for instance, can be represented accurately (or at least to a useful degree of approximation) by the predictions of the theory of the ideal parallel twin line.

The derivation of the expression for the capacitance per unit length of the parallel lines involves mathematics more sophisticated than that used so far in this text. So you may find it appropriate to bypass the following analysis for the time being and just take note of the very useful result displayed in eqn (1.38).

The essence of the argument lies in the recognition that the surface of a conductor (strictly speaking an ideal conductor of infinite conductivity) is an **equipotential surface**, since the electric field, and hence the gradient of the electric potential, is zero in an ideal conductor. Now the solutions of electrostatic field calculations are given in the form of potential functions $V(x, y, z)$ and equipotential surfaces are specified by $V(x, y, z) =$ a constant. Consequently a conductor whose surface fits the form of a particular equipotential surface, and which is at the appropriate value of potential, can be placed in the field without disturbing in any way the form of the field over the remainder of the space. As will be seen shortly, the equipotential surfaces of two parallel lines of charges of opposite sign are cylindrical, so two parallel cylindrical conductors can be placed so that their cross-sections fit exactly to a pair of equipotential surfaces. It is then easy to determine the potential difference between the two conductors in terms of the charge per unit length and hence to determine the capacitance per unit length.

Imagine that two lines of charge, each having charge $\pm Q$ per unit length, are spaced a distance s apart as shown in Fig. 1.15, where line A is positively charged. First it is necessary to find the potential at a general point P; it will be assumed, for convenience, that the potential is zero on the plane equidistant from both lines of charge. In particular, the potential is zero at O ($x = 0$, $y = 0$).

By using the expression for the electric field, eqn (1.30), in eqn (1.14), the potential difference between A and O is

$$V_{\mathrm{A}} - V_{\mathrm{O}} = \frac{Q}{2\pi\varepsilon_0}\int_{r=0}^{r=s/2}\frac{\mathrm{d}r}{r} = \frac{Q}{2\pi\varepsilon_0}\,|\ln r|_{r=s/2}.$$

Similarly

$$V_{\mathrm{P}} - V_{\mathrm{A}} = -\frac{Q}{2\pi\varepsilon_0}\,|\ln r|_{r=r_1}.$$

Now $V_{\mathrm{P}} - V_{\mathrm{O}} = (V_{\mathrm{P}} - V_{\mathrm{A}}) + (V_{\mathrm{A}} - V_{\mathrm{O}})$ and on putting $V_{\mathrm{O}} = 0$, as was assumed earlier, the potential at P is

$$V_{\mathrm{P}} = \frac{Q}{2\pi\varepsilon_0}\ln(s/2r_1).$$

Similarly, the potential at P due to the negative charge on line B is

$$V_{\mathrm{P}} = -\frac{Q}{2\pi\varepsilon_0}\ln(s/2r_2),$$

and so the total potential at P is

$$V_P = \frac{Q}{2\pi\varepsilon_0} \ln\left(\frac{r_2}{r_1}\right). \tag{1.34}$$

An equipotential in the x–y plane is defined by

$$\frac{r_2}{r_1} = K \qquad \text{(a constant)}. \tag{1.35}$$

From the geometry of Fig. 1.15,

$$r_1^2 = \left\{\left(\frac{s}{2} + x\right)^2 + y^2\right\} \quad \text{and} \quad r_2^2 = \left\{\left(\frac{s}{2} - x\right)^2 + y^2\right\},$$

and so on substituting in eqn (1.33), the equipotential line for a given value of K is given by

$$x^2 - sx\frac{(1 + K^2)}{(1 - K^2)} + y^2 = -\frac{s^2}{4}.$$

By adding the term $\left\{\frac{s(1 + K^2)}{2(1 - K^2)}\right\}^2$ to both sides of the above equation it follows that

$$\left\{x - \frac{s(1 + K^2)}{2(1 - K^2)}\right\}^2 + y^2 = \left\{\frac{sK}{(1 - K^2)}\right\}^2. \tag{1.36}$$

Now the equation of a circle can be written in the general form

$$x^2 + y^2 + 2gx + 2fy = -c$$

or

$$(x + g)^2 + (y + f)^2 = g^2 + f^2 - c.$$

The centre of the circle is at $x = -g$, $y = -f$, and the radius is equal to $(g^2 + f^2 - c)^{1/2}$. By comparison it can be seen that eqn (1.36) represents a family of circles with K as parameter. For a particular value of K the circle is centred at

$$x = \frac{s(1 + K^2)}{2(1 - K^2)};\ y = 0, \text{ and has radius } \frac{sK}{(1 - K^2)}.$$

Now consider that P lies on the surface of the left-hand member of a pair of identical cylindrical conductors of radius a and with the separation between their axes equal to b (see Fig. 1.15). For the equipotential defined by the surface of the left-hand conductor,

$$a = \frac{sK}{(1 - K^2)}, \qquad \frac{b}{2} = \frac{s(1 + K^2)}{2(1 - K^2)}.$$

On eliminating s between these two equations,

$$\frac{b}{a} = \frac{1}{K} + K.$$

From eqns (1.34) and (1.35), $K = \exp(2\pi\varepsilon_0 V_P/Q)$, and so

$$\frac{b}{a} = \exp(-2\pi\varepsilon_0 V_P/Q) + \exp(2\pi\varepsilon_0 V_P/Q)$$

$$= 2\cosh(2\pi\varepsilon_0 V_P/Q), \tag{1.37}$$

where $\cosh\theta \equiv (e^\theta + e^{-\theta})/2$ is one of the so-called **hyperbolic functions.**

Since the potential difference between the two conductors is $2V_P$, and the charge per unit length is $\pm Q$, it follows that the capacitance per unit length C is given by $C = Q/2V_P$, i.e.

$$C = \frac{\pi\varepsilon_0}{\cosh^{-1}(b/2a)} \text{ F m}^{-1}, \tag{1.38}$$

since $\cosh(-\theta) = \cosh(\theta)$.

Since $\cosh^{-1}(b/2a) = \ln\{(b/2a) + \sqrt{[(b^2/4a^2) - 1]}\}$, this means that $C = \pi\varepsilon_0/\ln(b/a)$ to an accuracy better than 0.5 per cent for $b/a > 10$. For $b/a = 5$, for example, $C = 17$ pF m^{-1}.

It should be noticed that the charge on each conductor is not distributed uniformly over the cross-section, owing to the influence of the charge on the neighbouring conductor; this feature is known as the **proximity effect** and is exemplified by the fact that $b \neq s$.

The capacitance per unit length between a long cylindrical conductor of radius a and an infinite conducting plane (e.g. the ground plane in an integrated circuit) can be obtained easily from the result of the previous analysis. The physical situation of a line charge Q per unit length at a distance $s/2$ above a conducting plane of infinite extent can be represented in terms of a hypothetical line charge, or 'image' charge, $-Q$ per unit length situated at a distance $s/2$ below the position of the conducting plane, the plane itself being imagined to have been removed. The equipotential surfaces for the line charge and its image are identical to those illustrated in Fig. 1.15; the conducting plane is an equipotential surface and its position would coincide with the plane equipotential surface containing the line $x = 0$ and lying perpendicular to the plane of the diagram.

From eqn (1.37),

$$2\pi\varepsilon_0 V_P/Q = \cosh^{-1}(b/2a).$$

Now in the present case the potential difference between the two conductors (i.e. the line and the plane) is V_P in terms of the definition used in the previous analysis, so

$$C = \frac{Q}{V_P} = \frac{2\pi\varepsilon_0}{\cosh^{-1}(b/2a)},$$

or, putting $b/2 = h$, where h is the distance from the axis of the conductor to the plane,

$$C = \frac{2\pi\varepsilon_0}{\cosh^{-1}(h/a)}\ \text{F m}^{-1}. \tag{1.39}$$

For $(h/a) = 2$, then $C \approx 40\ \text{pF m}^{-1}$, which could have a very significant effect on the characteristics of a circuit operating at a high frequency. Equation (1.39) also gives a rough but useful guide to the capacitance per unit length of a metal strip above a conducting plane, a configuration that occurs commonly on printed-circuit boards and in integrated circuits. For instance, for square-section strip with the width of the strip equal to h, the (sophisticated) computations yield $C \approx 50\ \text{pF m}^{-1}$; the particular value obtained depends on the theoretical model used.

In practice the electric field, and hence the capacitance, between two conductors A, B will be influenced by the presence of at least one other conductor D, say; for instance the metal box which contains the circuit in question (see Fig. 1.16(a)). The potential of conductor D must be intermediate between the potential of A and that of B and so there will be capacitance between A and D and between D and B. If C_{AB} is the desired capacitance in a circuit, then C_{AD}, C_{DB} are called **stray capacitances**. The effects of stray capacitances can be very significant in high-frequency circuits and also in relation to electromagnetic interference (see Section 9.4).

In electrical circuits, capacitors may be connected together in parallel or in series (see Fig. 1.17(a) and (b)). A common practical situation in a circuit is where the stray capacitance between connecting wires, or between a connecting wire and a metal ground plane, is in parallel with a capacitor; this can be especially significant in high-frequency circuits where the capacitor may be of small value ($\sim$100 pF or less). If two

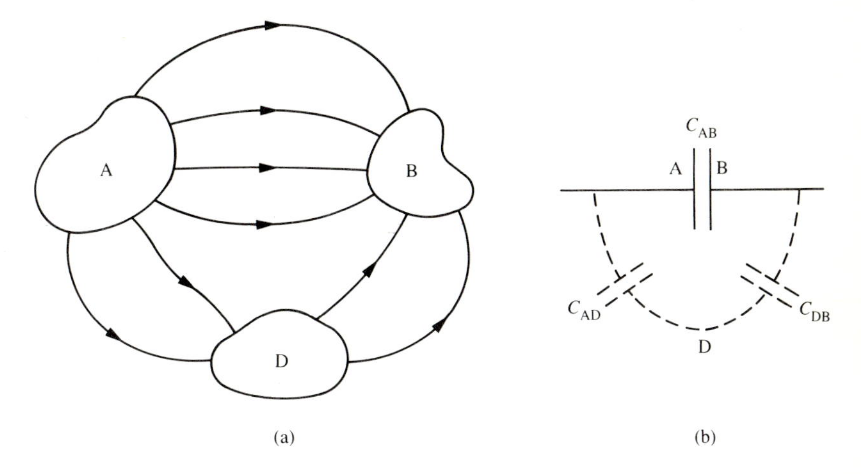

Fig. 1.16 (a) The electric field between three conductors A, B, D. (b) The capacitance between A and B and the stray capacitances C_{AD}, C_{DB}.

capacitances C_1 and C_2 are connected in parallel, then the potential difference (V, say) will be the same across both of them, so the charges on the capacitors are $Q_1 = C_1 V$ and $Q_2 = C_2 V$ respectively. So the effective capacitance (C) of the parallel combination is given by the quotient—see eqn. (1.25)—(total charge)/(potential difference), i.e.

$$C = \frac{Q_1 + Q_2}{V}$$

or

$$C = C_1 + C_2 \qquad \textit{(capacitors in parallel)}. \tag{1.40}$$

Consider two capacitors C_1, C_2 connected in series; initially the system will be electrically neutral. Now imagine that a source of e.m.f. V is connected in series with the capacitors (see Fig. 1.17(b)). Charge will flow round the circuit (with conventional positive charges flowing in the sense shown) until the total potential difference across the two capacitors is equal and opposite to V. Note that the positive charge which accumulates on the 'positive plate' of C_1 induces a negative charge of equal magnitude on the 'negative plate'; this is accomplished by a net flow of positive charge to the 'positive plate' of C_2. The final, equilibrium, situation is shown in Fig. 1.17(c) and in this case the potential differences across the capacitances C_1 and C_2 are given by $V_1 = Q/C_1$ and $V_2 = Q/C_2$ respectively.

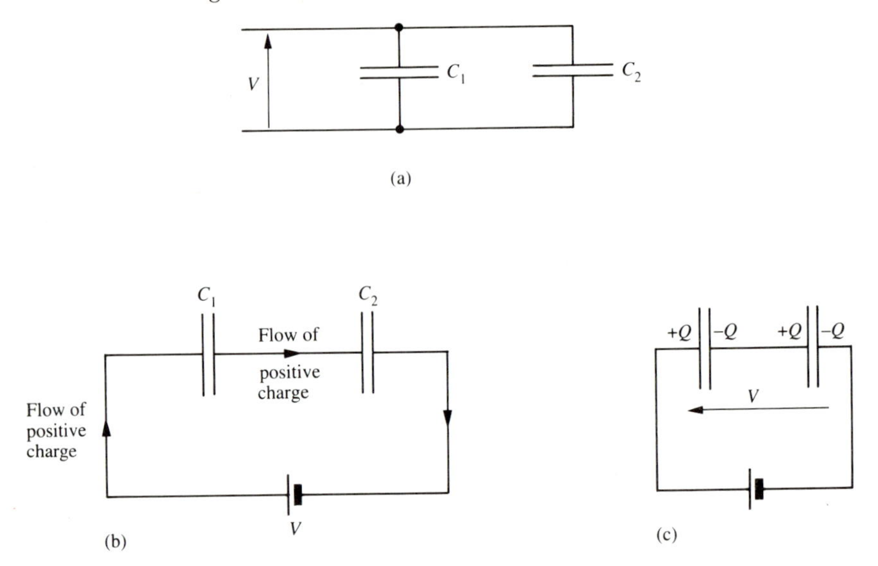

Fig. 1.17 Two capacitors (a) in parallel and (b) in series; (b) and (c) show the flow of charge that establishes the charges $\pm Q$ on the two capacitors in series.

Hence an expression for the effective capacitance (C) of the series combination can be obtained by once again applying eqn (1.25):

$$C = \frac{Q}{V_1 + V_2}$$

or, on substituting for V_1, V_2:

$$\frac{1}{C} = \frac{1}{C_1} + \frac{1}{C_2} \qquad \textit{(capacitors in series)}. \tag{1.41}$$

Note that for capacitors in series, if one of the capacitors is of much smaller capacitance than the other, then it will dominate the combination.

Equations (1.40) and (1.41) can be simply extended for more than two capacitors.

Example 1.5 Two capacitors having capacitances C_1, C_2 are each charged to a voltage V. The capacitors are now connected in parallel but with the positive terminal of each capacitor connected to the negative terminal of the other. Obtain expressions for the energy stored in the two capacitors before and after they are connected together.

Using eqn (1.32), the initial energy W_i of the two capacitors is given by

$$W_i = (C_1 + C_2)V^2/2.$$

After connecting the capacitors together in the specified manner, the total charge Q on the combination is given by

$$Q = (C_1 - C_2)V.$$

Since the capacitance of the two capacitors in parallel is $(C_1 + C_2)$, the final energy W_f of the combination is (using $Q^2/2C$)

$$W_f = \frac{(C_1 - C_2)^2 V^2}{2(C_1 + C_2)}.$$

So

$$\frac{W_f}{W_i} = \frac{(C_1 - C_2)^2}{(C_1 + C_2)^2}.$$

Clearly $W_f < W_i$, and this raises the question: what has happened to the missing energy?

One source of energy loss is the heating ('ohmic' loss) due to the (transient) current flowing through the resistance of the connecting wire. Also, a time-dependent current generates electromagnetic waves (see Section 5.4.3) which transport energy away from the system. In making the connection between the capacitors there may be a spark (due to the potential difference between the conductors) and again electromagnetic waves will be generated, not to mention sound waves! These sources of energy loss, and any other unaccounted for mechanisms, must make up the difference between W_i and W_f. The generation of electromagnetic waves by sparks is a common source of **electromagnetic interference** (see Chapter 9). □

1.6 Magnetic effects of steady electric currents

If a moving electric charge, either in vacuum or in a metal wire, say, experiences a velocity-dependent force, then the charge is moving through a region in which a magnetic field is said to exist.

It has been demonstrated *experimentally* that a magnetic field exists in conjunction with an electric current. The measurements are consistent with an expression for dB, the element of magnetic field B at a point P

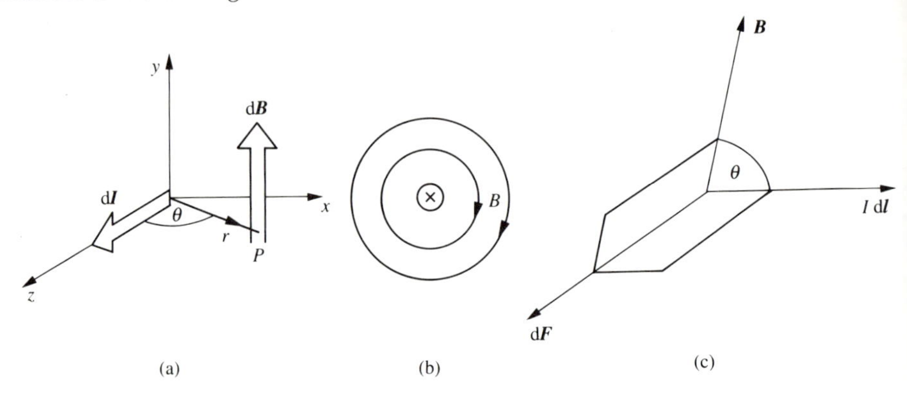

Fig. 1.18 (a) The magnetic field d***B*** due to a current I flowing in an element of circuit d***l***. (b) The pattern of the magnetic field B around a long, straight wire carrying a current directed into the plane of the diagram [a 'right-hand screw' law]. (c) The force d***F*** on a 'current element' I d***l*** situated in a magnetic field ***B***.

due to an element dl of a circuit carrying a current I, which is of the form

$$dB = \frac{\mu_0}{4\pi} \frac{I \sin\theta \, dl}{r^2}, \tag{1.42}$$

where r is the distance from the circuit element to the point in question and θ is the angle between the element and the line joining the element to P (see Fig. 1.18(a)).

The quantity $\mu_0/4\pi$ (equal to 10^{-7} henrys per metre) is the constant of proportionality in SI units.*

Equation (1.42) is known as the *Biot–Savart law*; note that the magnetic field ***B*** is a vector quantity.

The total magnetic field at P due to a complete circuit is given by

$$B = \frac{\mu_0 I}{4\pi} \oint \frac{\sin\theta \, dl}{r^2}, \tag{1.43}$$

where $\oint$ indicates integration over all the elements constituting the complete circuit.

* The fundamental constants of electromagnetism ε_0, μ_0 are not independent. This follows from the fact that ε_0 determines the unit for charge (Q) (eqn (1.10)) and μ_0 determines the unit for current (I) (see eqn (1.47)), and Q and I are related through $I = dQ/dt$.

Historically the subject of magnetism developed from studies of the properties of permanent magnets ('lodestone', bar magnets made of iron, navigational compasses); eventually the experimental work of Oersted drew magnetism and electricity together by demonstrating the magnetic effects of electric currents. The magnetism of the (ferromagnetic) materials of which permanent magnets are composed is now ascribed to microscopic electric currents at the atomic and molecular level. In an electromagnet the magnetic field is due to a macroscopic current flowing in a conductor which is usually wound as a coil; the coil is often wound on a core of ferromagnetic material in order to enhance the magnetic field.

Example 1.6 Derive an expression for the magnetic field B at a distance a from a long straight wire carrying a current I. Calculate the value of B if $I = 1$ A and $a = 1$ cm.

From eqn (1.42),

$$B = \frac{\mu_0 I}{4\pi} \int_{\theta=0}^{\theta=\pi} \frac{\sin\theta \, \mathrm{d}l}{r^2},$$

where the approximation has been made that the wire is infinitely long so that the range of θ is from 0 to π (see the diagram below).

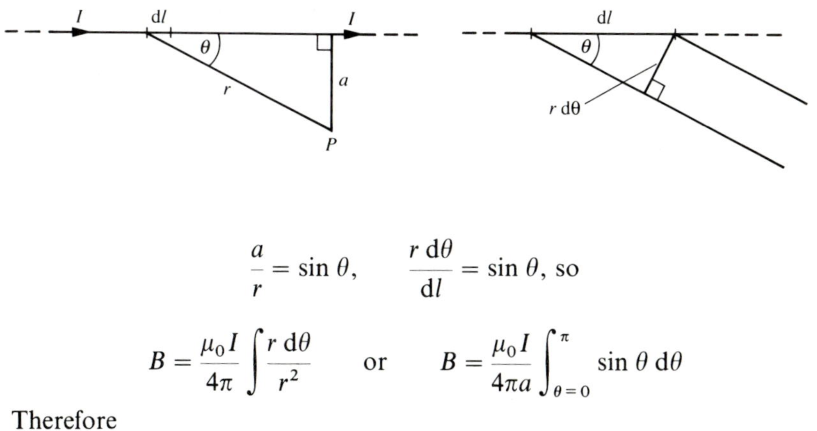

$$\frac{a}{r} = \sin\theta, \qquad \frac{r\,\mathrm{d}\theta}{\mathrm{d}l} = \sin\theta, \text{ so}$$

$$B = \frac{\mu_0 I}{4\pi} \int \frac{r\,\mathrm{d}\theta}{r^2} \qquad \text{or} \qquad B = \frac{\mu_0 I}{4\pi a} \int_{\theta=0}^{\pi} \sin\theta\,\mathrm{d}\theta$$

Therefore

$$B = \frac{\mu_0 I}{4\pi a} \{-\cos\theta\}_{\theta=0}^{\pi}$$

and, finally,

$$B = \frac{\mu_0}{4\pi}\frac{2I}{a}. \tag{1.44}$$

For the given data, $B = 2 \times 10^{-5}$ tesla (T).

Note that for r constant, the magnitude of B is constant, i.e. the magnetic field has cylindrical symmetry with the wire as axis (see Fig. 1.18(b)).

It is left as a problem to show that at a point on the axis of a circular loop of wire of radius a carrying current I, the magnetic field is axially directed and given by

$$B = \frac{\mu_0}{4\pi}\cdot 2\pi I a^2(a^2 + x^2)^{-3/2}, \tag{1.45}$$

where x is the distance of the point in question from the plane of the loop.

In the case of a circular loop of wire carrying a current of 1 A, the magnetic field at a point on the axis distant 1 cm from the plane of the coil is given by eqn (1.45) as 2.2×10^{-5} T; this field increases in proportion to the number of turns on the coil, at least as long as it remains a 'thin' coil (see Section 5.2.2).

The following expression (eqn (1.46)) for the force $\mathrm{d}F$ acting on an *element* of circuit $\mathrm{d}l$ carrying current I and situated in a magnetic field B has been found to be in accordance with experimental measurements (which, of course, had to be made using *closed* circuits):

$$\mathrm{d}F = BI\,\mathrm{d}l \sin\theta. \tag{1.46}$$

The relative directions of B, $\mathrm{d}l$ and $\mathrm{d}F$ are shown in Fig. 1.18(c). Notice that the force is proportional to the component of the element of circuit $\mathrm{d}l \sin\theta$ that is *perpendicular* to B. This relation is fundamental to the operation of the common types of electric motor. Equation (1.46) yields $\mathrm{d}F/\mathrm{d}l = 10^{-2}\ \mathrm{N\,m^{-1}}$ for $I = 1$ A and $B = 10^{-2}$ T.

Equations (1.42) and (1.46) can be combined to yield a general expression for the force F between two circuits, but a particular result that is of practical interest is that which applies to two long, straight, parallel wires separated by a distance x and carrying currents I_1 and I_2, namely

$$F = \frac{\mu_0}{4\pi} \cdot \frac{2I_1 I_2}{x} \tag{1.47}$$

The ampere (A) is so defined that for $I_1 = I_2 = 1$ A and $x = 1$ m, then $F = 2 \times 10^{-7}$ N m^{-1}, which fixes the value of the fundamental magnetic constant (or 'permeability of free space') μ_0 as $4\pi \times 10^{-7}$ henrys per metre (H m^{-1}).

The equivalent expression to eqn (1.46) for a single particle of charge Q moving at speed v is

$$F = BQv \sin \theta. \tag{1.48}$$

The equivalence may be seen by writing $I = \mathrm{d}Q/\mathrm{d}t$ and $v = \mathrm{d}l/\mathrm{d}t$.

1.7 Electromagnetic induction

Historically, Oersted discovered that a magnetic field is produced by an electric current, and in the early nineteenth century Faraday investigated the proposition that a current could be produced by a magnetic field.

As far as closed circuits are concerned, Faraday's important experimental discovery was that an electromotive force (e.m.f.) was induced in a circuit for which the magnetic flux linked with the circuit was *changing with respect to time.* This is probably the best-known of *Faraday's laws of electromagnetic induction.*

If a circuit consists of a plane loop of area A and there is a uniform magnetic field B directed normal to the plane of the loop (see Fig. 1.19(a)), then the magnetic flux ϕ 'linked' with the loop is given by

$$\phi = BA. \tag{1.49(a)}$$

If B is uniform but makes an angle θ with the normal to the plane of the loop (see Fig. 1.19(b)), then

$$\phi = BA \cos \theta \qquad \text{or} \qquad \phi = \boldsymbol{B} \cdot \boldsymbol{A} \tag{1.49(b)}$$

(the scalar product of $\boldsymbol{B}$ and $\boldsymbol{A}$).

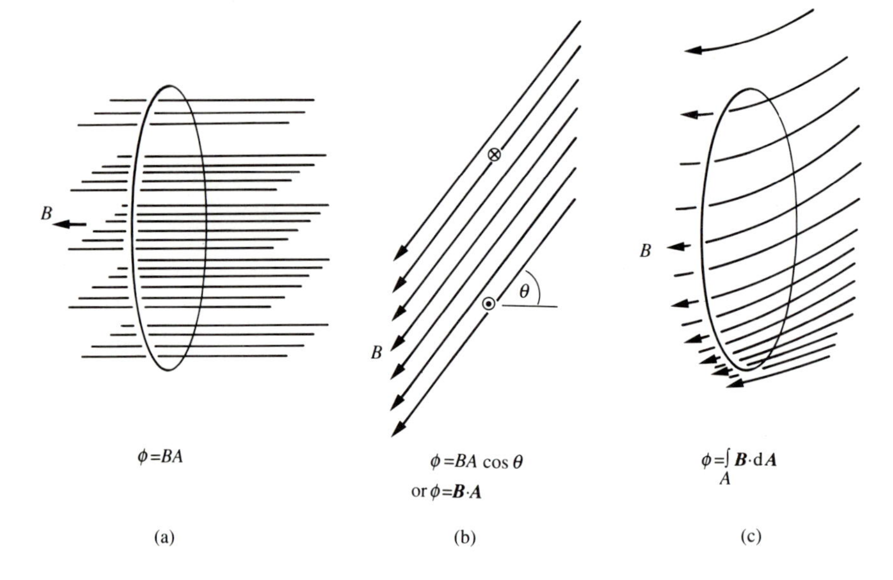

Fig. 1.19 Magnetic flux ϕ linked with a closed circuit in the form of a plane loop. (a) Uniform magnetic field B normal to the plane of the loop. (b) Uniform magnetic field B at an angle θ to the normal to the loop. (c) Non-uniform magnetic field B.

Finally, if B varies in magnitude and direction over the plane of the loop (see Fig. 1.19(c)), then

$$\phi = \int_A \boldsymbol{B} \cdot \mathrm{d}\boldsymbol{A} \tag{1.49(c)}$$

Here d$\boldsymbol{A}$ is a vector of magnitude dA directed perpendicular to the plane of an elementary area of magnitude dA, and the integration is carried out over the total area of the loop.

If the loop or circuit in question has N turns, then the above expressions for ϕ are multiplied by N.

Neumann put the matter on a more quantitative footing by assuming that the induced e.m.f. E is proportional to the rate of change of flux linkage with respect to time, $\mathrm{d}\phi/\mathrm{d}t$, an assumption which is supported by the experimental evidence.

In SI units (E in volts, t in seconds, ϕ in webers (Wb)),

$$E = d\phi/dt. \tag{1.50}$$

What is the direction of the induced e.m.f. in a circuit?

Lenz's law states, in effect, that whenever a change in flux linkage causes an induced current to flow, as is the case with a closed circuit, then the sense of the current flow is such as to produce physical effects **opposing the original change**.

This statement is rather cumbersome, and its physical significance can be better illustrated by two simple examples.

First, consider a closed circuit in the form of a rectangular loop for which the magnetic field B is normal to the plane of the loop and spatially uniform over the area of the loop but **decreasing with time** (see Fig. 1.20(a)). The sense of the induced e.m.f. is such as to produce an induced current I that in turn generates a magnetic field (refer to Fig. 1.20(b)) in a sense so as to try to **maintain** the field through the loop. Conversely, if the primary magnetic field B were **increasing with time**, then I would be in the opposite sense to the previous case and would generate an **opposing** magnetic field so as to reduce the total magnetic field towards its earlier value. Electromagnetic induction of this kind, where the magnetic field B is due to a current flowing in another circuit, is the basis of transformer action (see Fig. 1.21 and Section 3.7).

Second, consider a rectangular loop lying in the x–y plane, moving in the y-direction through a **time-independent** magnetic field that is in the z-direction but whose magnitude varies with y, i.e. $B \to B_z(y)$ (see Fig. 1.20(b)). If $B_z(y)$ decreases with increasing y so that the magnetic flux ϕ linked with the loop is decreasing with time, then the direction of the induced current I will be such as to try to maintain the value of ϕ.

As mentioned above, the magnetic flux linked with a circuit may be due to the magnetic field generated by a current flowing in another circuit, and if this flux is time-dependent, then an e.m.f. will be induced (see Fig. 1.21). However, it is very important to realize that there is a 'self-flux' linked with a circuit due to its own current, and if this current is time-dependent, then again an e.m.f. will be induced; the sense of this e.m.f. will be such as to oppose the change that is itself producing the e.m.f. (see Fig. 1.22). The relation between an induced e.m.f. E and the rate of change of current dI/dt, either in another circuit or in the same circuit, is expressed

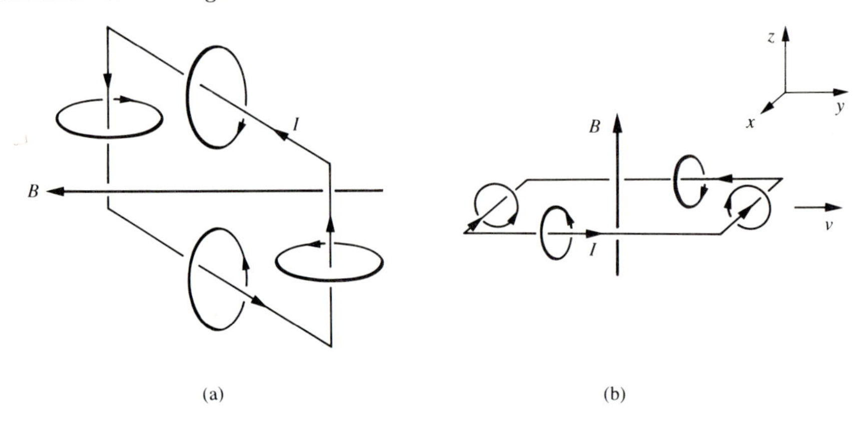

Fig. 1.20 Electromagnetic induction: (a) The sense of the induced current I when the magnetic flux due to the field B is decreasing with time. (b) The sense of the induced current I when B is time-independent but decreases in magnitude in the y-direction and the loop is moving in the y-direction. In both (a) and (b) the magnetic flux due to the induced current is in the same direction as B, i.e. the induced flux is in a sense so as to try to maintain the value of B.

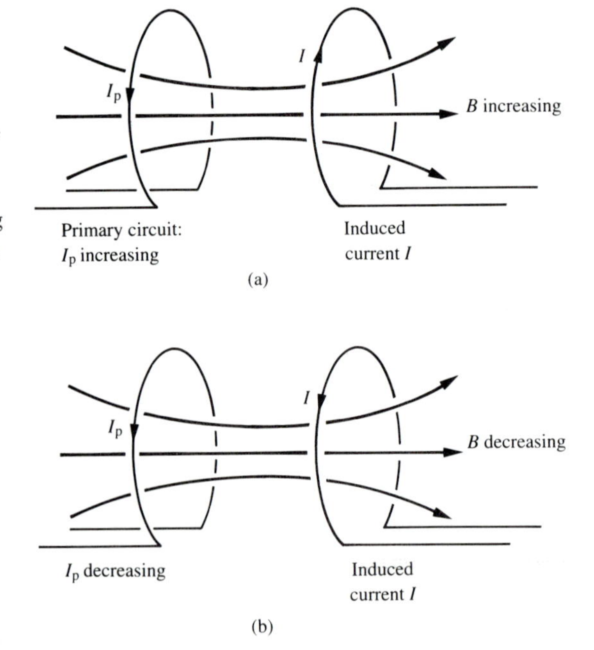

Fig. 1.21 Mutual induction between two circuits (transformer action). Note that although I_P is in the same sense in (a) and (b), the induced current I is in opposite senses depending on whether I_P is increasing or decreasing.

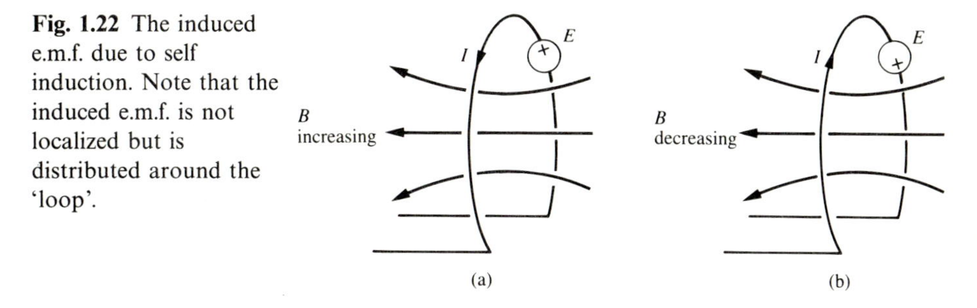

Fig. 1.22 The induced e.m.f. due to self induction. Note that the induced e.m.f. is not localized but is distributed around the 'loop'.

in terms of a characteristic property called the **coefficient of mutual inductance** M or the **coefficient of self-inductance** L:

$$M \quad \text{or} \quad L \equiv \frac{E}{\mathrm{d}I/\mathrm{d}t} \tag{1.51}$$

The SI unit of M and of L is the henry (H).

Using eqn (1.50), eqn (1.51) can be written also as

$$M \quad \text{or} \quad L = \frac{\mathrm{d}\phi}{\mathrm{d}I} \tag{1.52}$$

Every component of a circuit has a coefficient of self-inductance, however small, but a component designed to have a specific value of L is usually in the form of a coil consisting of many turns of wire (except for operation at very high frequencies: $\geqslant 1000$ MHz).

An important point to notice is that an induced e.m.f. is not localized in any particular part of a circuit, since the flux linkage is associated with the circuit as a whole.

In the calculation of the self-inductance and the mutual inductance of circuit elements, it is necessary to know the magnetic field produced by currents flowing in the elements. A law of electromagnetism that is very useful in this context is *Ampère's circuital law*. This states that the line integral of the magnetic field B around a closed path is equal to the total current I linked with the path, multiplied by the fundamental magnetic constant μ_0:

$$\oint \boldsymbol{B}\cdot \mathrm{d}\boldsymbol{l} = \mu_0 I \tag{1.53}$$

Strictly speaking the law in this form applies only in vacuum; in material media it has to be modified (see Section 5.3).

An example of the application of this law is the calculation of the magnetic field B inside a 'long' cylindrical solenoid, i.e. a solenoid whose length is much greater than its diameter (see Fig. 1.23). If the solenoid has n turns per metre and the current in each turn is I, then, for the closed path ABCDA situated far from either end of the solenoid,

$$\oint \boldsymbol{B}\cdot \mathrm{d}\boldsymbol{l} = \oint_{\mathrm{A}}^{\mathrm{B}} \boldsymbol{B}\cdot \mathrm{d}\boldsymbol{l} + \oint_{\mathrm{B}}^{\mathrm{C}} \boldsymbol{B}\cdot \mathrm{d}\boldsymbol{l} + \oint_{\mathrm{C}}^{\mathrm{D}} \boldsymbol{B}\cdot \mathrm{d}\boldsymbol{l} + \oint_{\mathrm{D}}^{\mathrm{A}} \boldsymbol{B}\cdot \mathrm{d}\boldsymbol{l} = Bz.$$

Here it has been assumed that the magnetic field is uniform inside this part of the solenoid and is negligibly small outside the solenoid in the region containing the element of path CD. Also, for those portions of BC, and DA lying within the solenoid the magnetic field is perpendicular to the path and so the contributions to the scalar product $\boldsymbol{B}\cdot \mathrm{d}\boldsymbol{l}$ are zero. Now the total current linked with the closed path ABCDA is nzI, so, using Ampere's circuital law,

$$Bz = \mu_0 nzI$$

or

$$B = \frac{\mu_0}{4\pi} 4\pi nI. \tag{1.54}$$

If the area of cross-section of the solenoid is A and its length is l (so that the total number of turns is nl), then the magnetic flux ϕ linked with the solenoid due to the current in its own windings is given by $\phi = BA \cdot nl$ if end-effects are neglected. Thus

$$\phi = \frac{\mu_0}{4\pi} 4\pi A\, l\, n^2 I$$

and, using eqn (1.52) the coefficient of inductance L_0 is given by

$$L_0 = \frac{\mu_0}{4\pi} 4\pi A\, l\, n^2. \tag{1.55}$$

If a toroidal coil (see Fig. 1.24) is wound around a material medium, then the **relative permeability** μ_r of the medium is defined through $\mu_r \equiv L/L_0$,

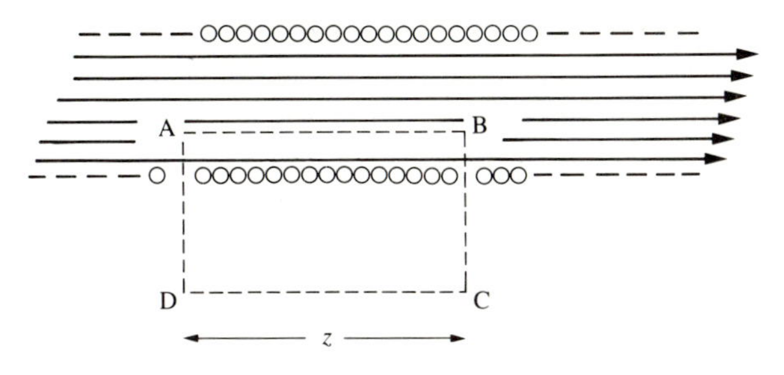

Fig. 1.23 The path of integration ABCDA in the calculation of the magnetic field inside a 'long' cylindrical solenoid.

where L, L_0 are the self-inductances of the solenoid with and without the medium respectively (cp. eqn (1.28), which defines the relative permittivity ε_r of a medium). For ferromagnetic media μ_r may be as large as several thousand (see Section 5.3). Values of self-inductance encountered in practice range from ~ 1 henry (H) (in inductors for use at low frequencies, e.g. the mains frequency) down to ~ 1 microhenry (μH) in high-frequency circuits.

A transformer is an example of a mutual inductor. Two coils electrically insulated from each other are wound on a ferromagnetic core and, as a result, a changing current in the 'primary' coil induces an e.m.f. in the 'secondary' coil (see Fig. 1.21) whose magnitude depends on the number of turns in the two coils and on the 'tightness' of the coupling between them. If the primary current I_P varies sinusoidally with time ($I_P = \hat{I}_P \sin \omega t$, say), then the resulting magnetic flux will have the same time-dependence, and the amplitude of the induced e.m.f., and hence of the induced current, will be proportional to ω, since it is equal to the derivative of the flux with respect to time (eqn (1.50)). Since ferromagnetic materials such as iron and steel alloys are electrical conductors, currents will also be induced (**eddy currents**) in a ferromagnetic core. Because of its electrical resistance there will be power dissipation in the core associated with the eddy

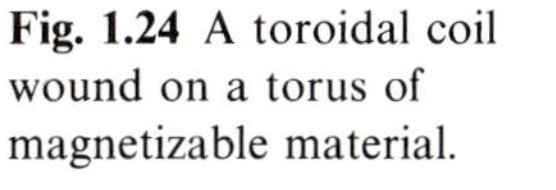
Fig. 1.24 A toroidal coil wound on a torus of magnetizable material.

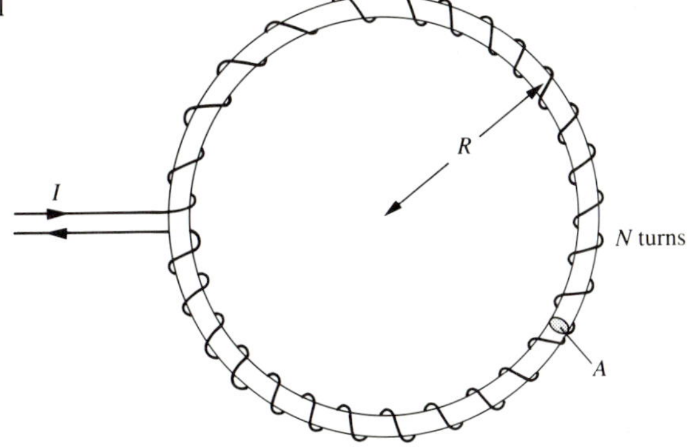

currents, so cores are made up of laminations oriented so as to block these currents and thus reduce the effective conductivity. Since power is proportional to (current)^2, these eddy current losses will be proportional to ω^2, and in practice this sets an upper limit of about 20 kHz to the range of frequency over which such transformers will operate efficiently. At higher frequencies, cores made from ferrites (materials with fairly high values of μ_r but very low electrical conductivity) are used. Transformers are discussed in more detail, from the point of view of their properties as circuit elements, in Section 3.7.

Now consider a circuit that can be represented by a resistance R and an inductance L in series with a source of e.m.f. E. Remembering that the induced e.m.f. in the inductor opposes any change in the current I through it, and noting also that the total potential difference around the closed circuit must be zero (*Kirchhoff's 'voltage law'*), it follows that

$$E - L\frac{dI}{dt} - RI = 0.$$

The source of e.m.f. is supplying energy at the rate EI and so

$$EI = LI\frac{dI}{dt} + RI^2.$$

For I increasing with time, the induced e.m.f. is in the opposite sense to I and the energy supplied by the source of e.m.f. can be considered as being stored in the magnetic field generated in and around the inductor by the current I.

The energy W stored in the inductor is given by

$$W = L\int_{I=0}^{I} I\frac{dI}{dt}\,dt \qquad \text{or} \qquad W = L\int_{I=0}^{I} I\,dI.$$

Hence

$$W = \frac{LI^2}{2}. \tag{1.56}$$

The expression for the stored energy density associated with a magnetic field can be obtained by considering the magnetic field inside an ideal toroidal coil (see Fig. 1.24). If the toroid is uniformly wound, with N turns, then the magnetic field within the coil will be concentric and the application of Ampère's circuital law to a circular path of radius R inside the coil gives the magnetic field B. In this case,

$$\oint \boldsymbol{B}\cdot \mathrm{d}\boldsymbol{l} = B\cdot 2\pi R \qquad \text{and} \qquad B\cdot 2\pi R = \mu_0 I.$$

Hence

$$B = \frac{\mu_0 NI}{2\pi R}.$$

The magnetic flux ϕ over the cross-section (area A) of the coil is $B\cdot A$ and the total flux linkage with the toroidal coil is NBA, i.e.

$$\phi = \frac{\mu_0 N^2 AI}{2\pi R}.$$

Hence, using eqn (1.52), the self-inductance L of the coil is

$$L = \frac{\mu_0 N^2 A}{2\pi R}.$$

Using eqn (1.56) for the stored energy in an inductance and noting that the volume of the toroidal coil is equal to $2\pi RA$, it follows that the energy density W is given by

$$W = \frac{B^2}{2\mu_0}\ \mathrm{J\,m^{-3}}. \tag{1.57}$$

It should be noted that this expression has the same general mathematical form as that for the stored energy associated with an electric field (see eqn (1.33)); both expressions involve the square of the field quantity (E or B) and the respective fundamental constants (ε_0 or μ_0). These expressions apply also to time-dependent electric and magnetic fields and, in particular, to electromagnetic waves.

It is useful to acquire a feeling for the electromagnetic fields associated with the voltages and currents in circuits. On the one hand there are devices in which electric and magnetic fields have obvious roles, such as electrostatic generators, magnetic actuators, and motors. On the other hand there are a great many effects in which the role of electromagnetic fields is more subtle but which are nevertheless extremely important; the following list is illustrative but by no means exhaustive:

electromagnetic coupling between elements of a circuit and between circuits screening of electromagnetic fields
antennas
transmission lines
semiconductor devices, particularly in integrated circuits
transformers
sensors/transducers.

1.8 Summary

Electric field E is defined as force per unit charge: $E = \frac{F}{Q}$.

Electric current (I) is the flow of electric charge (Q): $I = \frac{\mathrm{d}Q}{\mathrm{d}t}$.

Conversely, $Q = \int I\,\mathrm{d}t$.

Electrical resistance R is defined through $R \equiv \frac{V}{I}$.

Work done = force × (distance moved by the point of application of the force) (but remember the vector nature of force; see Section 1.3 and eqn (1.17).

Power P dissipated in a resistance R carrying a current I is given by

$$P = VI = I^2R = \frac{V^2}{R}.$$

Rate of working (power P) of a source of e.m.f. $\mathcal{E}$ supplying a current I:

$$P = \mathcal{E}I.$$

Gauss's law: a relation between an electric field and the charges which are its source:

$$\int_S \boldsymbol{E}\cdot\mathrm{d}\boldsymbol{S} = \int_S E\cos\theta\,\mathrm{d}S = \frac{Q}{\varepsilon_0}.$$

The normal electric field 'close to' a surface carrying a surface charge density σ:

$$E = \frac{\sigma}{\varepsilon_0}.$$

Electric potential V as a function of distance R in the field around a point charge Q:

$$V = \frac{Q}{4\pi\varepsilon_0 R}.$$

Definition of capacitance: $C \equiv \dfrac{Q}{V}$.

Parallel-plate capacitor: $C = \dfrac{\varepsilon_0 A}{t}$.

Relative permittivity: $\varepsilon_r \equiv \dfrac{C}{C_0}$.

Coaxial line: $C = \dfrac{2\pi\varepsilon_0}{\ln\left(\dfrac{b}{a}\right)}\ \mathrm{F\ m^{-1}}$.

Capacitors in parallel: $C = C_1 + C_2$.

Capacitors in series: $\dfrac{1}{C} = \dfrac{1}{C_1} + \dfrac{1}{C_2}$.

Biot–Savart law: $\mathrm{d}B = \dfrac{\mu_0}{4\pi} \cdot \dfrac{I \sin\theta\, \mathrm{d}l}{r^2}$.

Force on an element of a circuit in a magnetic field: $\mathrm{d}F = BI\, \mathrm{d}l \sin\theta$.

Force on a charge Q moving at speed v in a magnetic field: $F = BQv \sin\theta$.

Magnetic flux: $\phi = BA\cos\theta$ or, more generally, $\phi = \displaystyle\int_A \boldsymbol{B} \cdot \mathrm{d}\boldsymbol{A}$.

Induced e.m.f.: $E = \mathrm{d}\phi/\mathrm{d}t$.

Mutual inductance, self-inductance: M or $L \equiv \dfrac{E}{\mathrm{d}I/\mathrm{d}t} = \dfrac{\mathrm{d}\phi}{\mathrm{d}I}$.

Ampère's circuital law: $\displaystyle\oint \boldsymbol{B} \cdot \mathrm{d}\boldsymbol{l} = \mu_0 I$.

Energy stored in an inductance: $W = \dfrac{LI^2}{2}$.

Energy stored in a charged capacitor:

$$W = \frac{Q^2}{2C} = \frac{CV^2}{2} = \frac{QV}{2}.$$

Energy density in an electric field: $\dfrac{\varepsilon_0 E^2}{2}$ J m^{-3}.

Energy density in a magnetic field: $\dfrac{B^2}{2\mu_0}$ J m^{-3}.

Further reading

Chapman, S. J. (1985). *Electrical machinery fundamentals.* McGraw-Hill, New York.

Duffin, W. J. (1990). *Electricity and magnetism*, 4th ed. McGraw-Hill, New York.

Parton, J. E. and Owen, S. J. T. (1986). *Applied electromagnetics*, 2nd ed. Macmillan, London.

Problems

1.1 In a van de Graaff generator, charge is delivered to the high-voltage electrode at the rate of 500 microcoulombs per second. If a voltage of 12 kV is produced between the two electrodes, calculate the power required to drive the belt of the generator against electrical forces (assuming negligible power is required to overcome friction).

1.2 A 'chip' of silicon of resistivity 2 Ω m has the following dimensions: length 100 μm; width 10 μm; thickness 5 μm. If a potential difference of 5 V is applied over the long dimension of the chip, calculate the power dissipated and the power density.

1.3 Two small charged objects (i.e. 'point charges') carry charges Q_1, Q_2 and are separated by a distance of 1 cm; the force between them has a magnitude of 27 N. What is the magnitude of the force between the same two charged objects if their distance of separation is 3 cm?

1.4 What is the potential energy W of an electron at a point P_2 relative to its potential energy when it is at point P_1 if the electrostatic potential V at P_2 is +5 kV relative to that at P_1?

1.5 In a region of space the electrostatic field E points in the $+x$-direction and its magnitude is given by $E = 300x$ V m^{-1}. Calculate the change ΔV in the electrostatic potential between two points on the x-axis with coordinates $x = 3$ mm and $x = 5$ mm.

1.6 Four capacitors are connected as shown in the diagram (Fig. 1.25). Find the effective capacitance between the terminals A and B and the charge on each capacitor (in terms of C and V).

Fig. 1.25

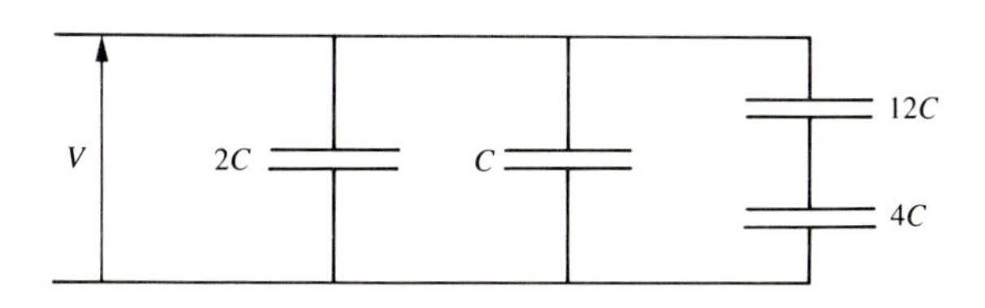

1.7 Show that the magnetic field B at a point on the axis of a circular loop of wire of radius a is axially directed and given by

$$B = \mu_0 2\pi I a^2/\{4\pi(a^2 + x^2)^{3/2}\},$$

where I is the current in the loop and x is the distance of the point in question from the plane of the loop.

1.8 A long wire is bent into a 'hairpin' shape as shown in Fig. 1.26. Show that the magnetic field B at the point P lying at the centre of the half-circle is given by

$$B = \frac{\mu_0 I}{4\pi a}\{2 + \pi\}$$

Fig. 1.26

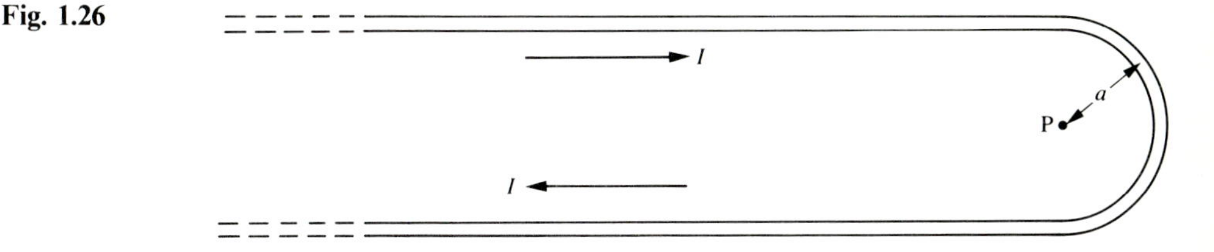

1.9 Derive the expression given below for the coefficient of mutual inductance M between a long straight wire and another wire bent into the shape of a closed rectangle of length a and breadth b (see Fig. 1.27 below). The long wire is coplanar with the rectangle, is parallel to the sides of length a, and is a distance d from the nearer such side.

$$M = (\mu_0/4\pi)2a \ln\{1 + (b/d)\}$$

Fig. 1.27

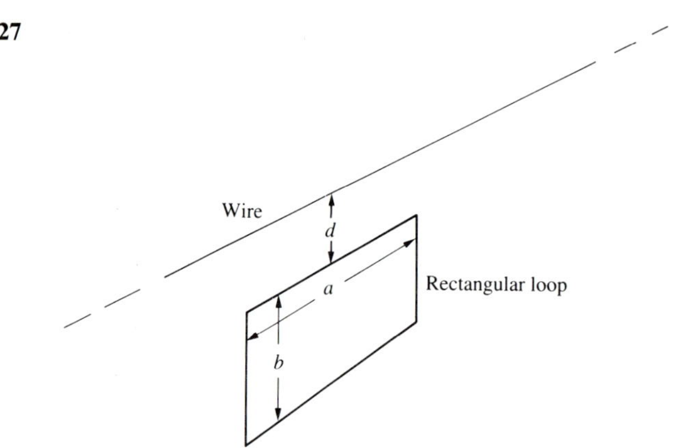

Direct current circuits

2

2.1 Circuit elements

A simple closed circuit consisting of an ideal **active element** (a source of electromotive force (e.m.f. E) and a **passive element** (a resistor R) connected by wires of negligible resistance was discussed in Section 1.3. All the energy supplied by the source to the mobile electric charges in the circuit is dissipated as the charges flow through the resistor and appears as heat and other forms of energy. As was seen in Chapter 1, there are other circuit elements, namely capacitors and inductors, for which the energy supplied to them is partly stored and partly dissipated (see Sections 1.5 and 1.7 respectively). It should be remembered that for time-dependent currents, energy is also lost from a circuit through the radiation of energy as electromagnetic waves. However, this process has practical significance only at relatively high frequencies and/or in circuit elements ('antennas') specifically designed to be efficient radiators.

In the analysis of direct current (d.c.) and alternating current (a.c.) circuits an idealization will be made in which the actual physical components will be represented by circuit elements such that each element has only the property conveyed by its name, i.e. a resistance has resistance only, a capacitance has capacitance only and an inductance has inductance only. The wire of the coil of a real inductor has resistance, so the inductor can be represented by a resistance in series with an inductance. For a capacitor there is a small leakage current between its plates (see Section 5.1), so a real capacitor can be represented by a capacitance having in parallel a resistance that accounts for the high-resistance leakage path. Real resistors have small values of capacitance and inductance associated with them; a non-inductive resistor can be made by using 'doubled' wire so that adjacent equal and oppositely flowing currents produce cancelling magnetic fields. Also, practical sources of e.m.f. always have internal resistance, inductance and capacitance, although in many situations their effects are of little practical significance. In this chapter, d.c. circuits are considered in which the only circuit elements are pure resistances and active elements having internal resistance only.

Now consider a source of e.m.f. E having internal resistance R_0 (see Fig. 2.1(a)); if a load current I is drawn, then the potential difference (p.d.) V_0 between its terminals (or 'terminal voltage') is given by

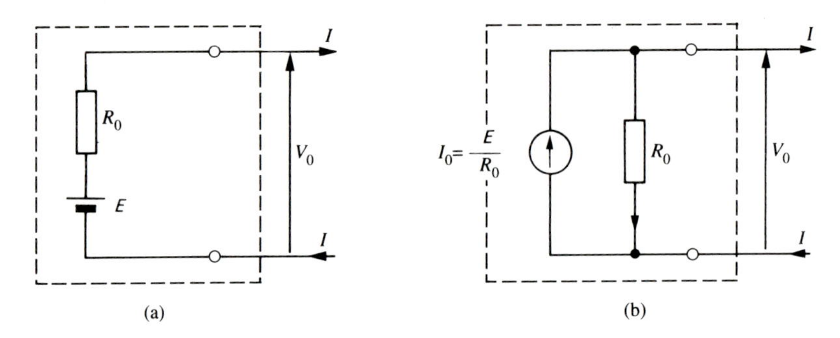

Fig. 2.1 Representations of a source of e.m.f. of internal resistance R_0 in terms of (a) an ideal constant voltage source E and (b) an ideal constant current source I_0 $(= E/R_0)$.

Table 2.1. Characteristics of some common sources of e.m.f.

Source	E_0 (volts)	R_0 (ohms)
Weston-Cadmium cell	1.01864	500
Lead acid accumulator	1.95	0.01
Nickel–iron alkaline accumulator 'Nife' cell	1.2	0.1
'Dry' cell	1.5	30

$$V_0 = E - IR_0 \tag{2.1}$$

(note that for the open-circuit situation $I = 0$ and $V_0 = E$ as discussed in Section 1.3). Nominal values for E and R_0 for some commonly used sources of e.m.f. are given in Table 2.1.

In the situation depicted in Fig. 2.1(a) the source of e.m.f. is represented as an ideal source (with *zero internal resistance*) in series with a resistance representing all the dissipative processes present in the real source. For a source with $R_0 = 0$, from eqn (2.1) the terminal voltage V_0 is independent of the magnitude of the current drawn; such a source is called an (ideal) **constant voltage source**.

It turns out to be useful to represent an active element in another way. Imagine a source that supplies the same value of current irrespective of the value of the resistance connected between its terminals (from which it can be deduced that this imagined source must have an *infinite internal resistance*). If current I is drawn from the terminals of the real source, which is represented as an ideal current source with a resistance R_0 in parallel (see Fig. 2.1(b)), then the terminal voltage V_0 is

$$V_0 = (I_0 - I)R_0, \tag{2.2}$$

since I_0 divides between R_0 and the external circuit. Now, for the open-circuit condition ($I = 0$), $V_{oc} = I_0 R_0$, which by definition is equal to E. Hence $I_0 = E/R_0$ and it follows that eqn (2.2) can be written as

$$V_0 = E - IR_0,$$

which is identical to eqn (2.1).

The resistance R_0 is said to be in 'shunt' (i.e. in parallel) with the ideal current source I_0. Notice also that $1/R_0$ is called the conductance G_0 of the source (NB the reciprocal ohm is called the siemen (S)).

The concept of an (ideal) **constant current source** shunted by a conductance G_0 is very useful in the analysis of many forms of circuit (see Chapters 3 and 7 particularly).

Imagine that a load resistor R_L is connected between the terminals of the source of e.m.f. E in the circuit of Fig. 2.1(a). From the definition of resistance (see Section 1.2),

$$I(R_0 + R_L) = E. \tag{2.3}$$

The power P dissipated in R_L (see eqn (1.9)) is

$$P = I^2 R_L$$

or

$$P = E^2 R_L/(R_0 + R_L)^2. \tag{2.4}$$

It is of practical interest to find the condition for maximum power dissipation (P_{max}) in the load. The first step of the standard mathematical procedure to find the maximum value of P is to find the value of R_L for which $dP/dR_L = 0$. Using the rule for the differentiation of a quotient,

$$\frac{dP}{dR_L} = \frac{E^2(R_0 + R_L)^2 - 2E^2 R_L(R_0 + R_L)}{(R_0 + R_L)^4}$$

or

$$\frac{dP}{dR_L} = \frac{E^2(R_0^2 - R_L^2)}{(R_0 + R_L)}.$$

Clearly, $dP/dR_L = 0$ for $R_L = R_0$, but, formally, the question remains as to whether this is the condition for a maximum or a minimum value of P (or even for a point of inflection on the curve of P vs. R_L). To resolve this question it is necessary to obtain an expression for d^2P/dR_L^2 and to see whether it is >0, <0, or $=0$ when $R_L = R_0$:

$$\frac{d^2P}{dR_L^2} = \frac{E^2\{(R_0 + R_L)^4(-2R_L) - 4(R_0^2 - R_L^2)(R_0 + R_L)^3\}}{(R_0 + R_L)}.$$

Setting $R_L = R_0$, then $d^2P/dR_L^2 < 0$, which means that $R_L = R_0$ is the condition for maximum power dissipation in the load resistance:

$$P_{max} = \frac{E^2}{4R_L} = \frac{E^2}{4R_0}. \tag{2.5}$$

So, in order to obtain the maximum available power from a source it is necessary to know the source resistance R_0 so that the load resistance can be matched to that of the source.

It has been shown already that E is equal to the open-circuit voltage between the terminals of a source. If these terminals are short-circuited ($R_L = 0$), then the short-circuit current I_{sc} is given by

$$I_{sc} = E/R_0 \qquad \text{or} \qquad R_0 = E/I_{sc},$$

i.e.

$$R_0 = \text{(open-circuit voltage)/(short-circuit current)}. \tag{2.6}$$

This relationship emphasizes the very important point that for a real source, R_0 can be determined from *measurements* made at the output terminals of the source. This may be a very much simpler procedure than modelling the source and analysing the associated circuit. Of course you should not thoughtlessly short-circuit any source, or irreparable damage may be done!

If it is not practicable actually to short-circuit the output terminals of a source, an alternative procedure is to measure the load current I as a function of R_L down to the smallest 'safe' value of R_L. From eqn (2.3),

$$I = E/(R_0 + R_L).$$

By extrapolating the graph of I vs. R_L to $R_L = 0$, I_{sc} may be estimated (see Fig. 2.2) and hence R_0 may be determined using eqn (2.6).

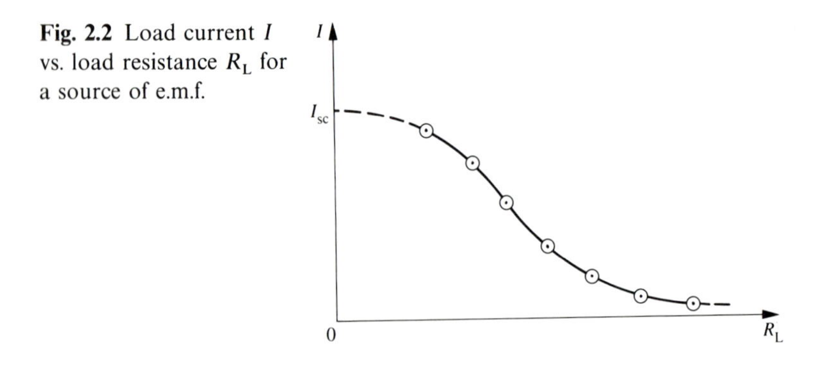

Fig. 2.2 Load current I vs. load resistance R_L for a source of e.m.f.

Fig. 2.3

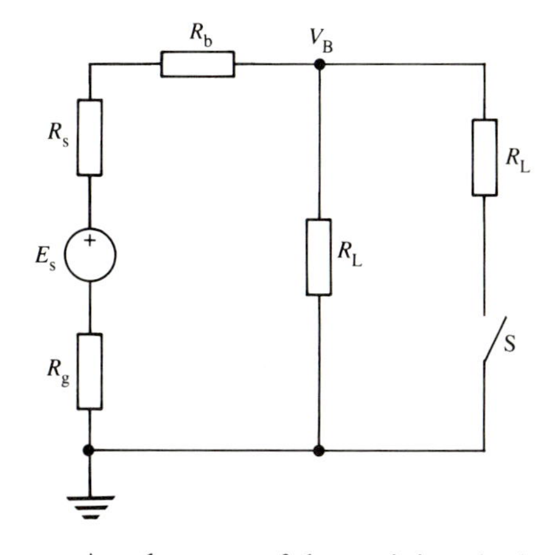

Another way of determining the internal resistance of a source is to find that value of R_L for which the terminal voltage V_0 is equal to $E_s/2$; this value of R_L is equal to R_0 (this follows simply from eqns (2.1) and (2.3)).

Example 2.1

In electronic circuits sharing the same power supply line (or power 'bus'), load-induced interference can be a serious problem (see Chapter 9). In Fig. 2.3 E_s is the e.m.f. of a power supply having an internal resistance R_s. A number of loads are connected in parallel across the supply; for simplicity only two such loads, each of resistance R_L, are shown in the figure. R_b is the resistance of the power bus and R_g is the resistance of the 'ground' (or 'common') connection.

The voltage V_B of the power bus will vary depending on whether or not the second load is switched into the circuit. Incidentally, the switch S could represent a drastic change in the resistance of a device from 'high' to 'low' due to a step change in the signal voltage level, as in the case of a logic gate in a digital system.

If switch S is open, then

$$V_B = \frac{E_s R_L}{(R_T + R_L)}$$

where $R_T \equiv R_g + R_s + R_b$.

On closing S, more current is drawn from the supply and V_B decreases owing to the additional potential difference across the resistance R_T; let ΔV denote this additional difference:

$$V_B - \Delta V = \frac{E_s(R_L/2)}{\left\{R_T + \dfrac{R_L}{2}\right\}}.$$

It follows that

$$\Delta V = E_s R_L\{(R_T + R_L)^{-1} - (2R_T + R_L)^{-1}\}.$$

In a typical situation R_s, R_g and R_b will each be much smaller in value than R_L (e.g. $R_s \approx 10^{-2}\,\Omega$, R_b, $R_g \approx 1\,\Omega$, $R_L \approx 100\,\Omega$), so the expression for ΔV can be written as

$$\Delta V = E_s\left\{\left(1 + \frac{R_T}{R_L}\right)^{-1} - \left(1 + \frac{2R_T}{R_L}\right)^{-1}\right\}.$$

For $R_T/R_L \ll 1$, $\left(1 + \frac{R_T}{R_L}\right)^{-1} \approx 1 - \frac{R_T}{R_L}$ *(using the binomial expansion for* $(1 + x)^{-1}$), so

$$\Delta V \approx E_s\left\{\left(1 - \frac{R_T}{R_L}\right) - \left(1 - \frac{2R_T}{R_L}\right)\right\}$$

or

$$\Delta V \approx \frac{E_s R_T}{R_L}, \qquad \text{i.e.} \quad \Delta V \approx \frac{E_s(R_g + R_s + R_b)}{R_l}.$$

For the values of the resistances given above, $\Delta V/E_s \approx 1$ per cent.

In situations where the operation of the switch S represents a transition between a very high and a very low resistance state of a logic gate, say, then ΔV has the form of a 'step' (see Chapter 4) and there is a possibility of spurious triggering of other logic gates supplied via the same bus.

The drop in line voltage ΔV can be reduced by using separate buses and return leads for each load. It is left as a problem to show that in this circumstance, and for two loads R_L,

$$\Delta V \approx \frac{E_s(R_g + R_s)}{R_L}.$$

This could represent a significant improvement over the previous situation, provided that there is a good earth connection (i.e. R_g not greater than R_b). Of course there are cost and space implications in using separate power lines to each load. □

2.2 Thévenin's theorem

An important feature of circuits more complicated than that shown in Fig. 2.1 is expressed in *Thévenin's theorem*:

Any linear* circuit can be represented at its output terminals by a single source of e.m.f. with a resistance in series.

This means that a complicated circuit which may be, or may be imagined to be, contained within a 'black box' behaves, as far as an external circuit connected to its accessible output terminals is concerned, just like the simple battery in Fig. 2.1. The values of E, R_0 for the simple **Thévenin equivalent source** can be calculated from a knowledge of the circuit by methods to be described in the following sections. Alternatively, E and R_0 may be determined experimentally by open- and short-circuit measurements as just described in Section 2.1.

So far, the sources considered have been **independent sources**: the e.m.f. and internal resistance of an independent source do not depend on voltages or currents elsewhere in the circuit in question. A voltage amplifier is an example of a **dependent source**: 'looking in' at its output terminals, the amplifier may be represented by a single equivalent Thévenin source of e.m.f. E together with an equivalent series resistance R_0. However, the values of E and R_0 depend on voltages and currents elsewhere in the amplifier circuit; e.g. the value of E will depend directly on the value of the input voltage of the amplifier.

An illustration of Thévenin's theorem is provided by the circuit shown in Fig. 2.4(a). This apparently complicated network in fact consists of series and parallel branches only, together with a single source of e.m.f. By employing the rules for combining resistances in series and parallel, namely

$$R = R_1 + R_2 + R_3 + \cdots \qquad \text{and} \qquad \frac{1}{R} = \frac{1}{R_1} + \frac{1}{R_2} + \frac{1}{R_3} + \cdots$$

respectively, the circuit can be simplified as shown in Fig. 2.4(b) and (c).

NB for two resistors in parallel,

$$R = \frac{R_1 R_2}{(R_1 + R_2)}.$$

This is a useful little rule to remember.

* The significance of 'linear' in this context will be explained in Section 2.6..

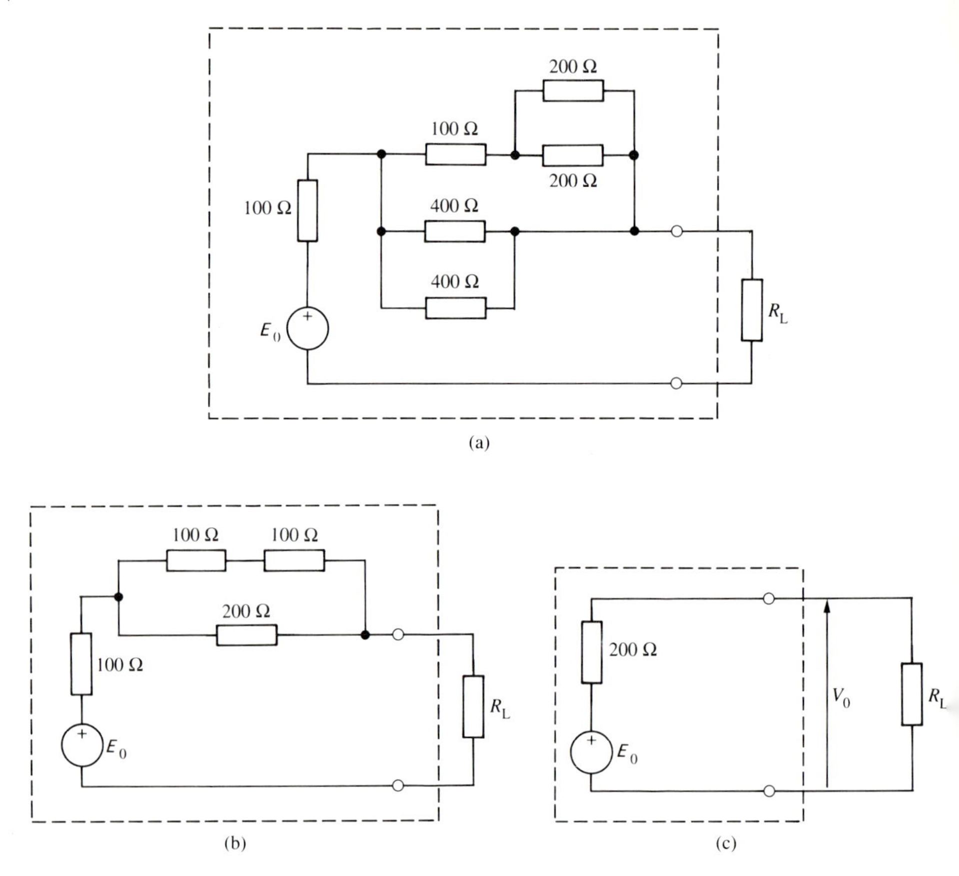

Fig. 2.4 An illustration of Thévenin's theorem for a simple network contained within a 'black box'; (b) and (c) represent simplifications of (a).

However, even a Wheatstone bridge, which is apparently quite a simple circuit, cannot be reduced in this elementary way and a different approach based on Kirchhoff's laws must be adopted; this approach is described in the following section.

Example 2.2 Obtain the Thévenin equivalent of the circuit shown in Fig. 2.5.

The open-circuit output voltage V_{oc} is given by

$$V_{oc} = E_2 + \frac{(E_1 - E_2)}{(R_1 + R_2)} R_2$$

or

$$V_{oc} = \frac{E_2 R_1 + E_1 R_2}{(R_1 + R_2)} \qquad \text{('}E_0\text{' for the equivalent source).}$$

Fig. 2.5

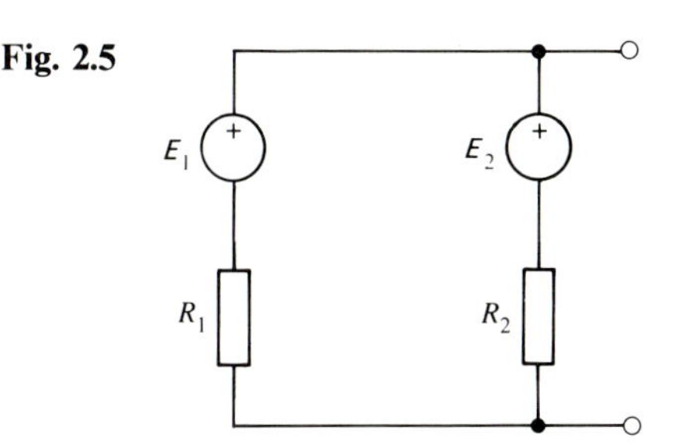

If the output terminals are short-circuited, then the short-circuit current I_{sc} is the sum of the short-circuit currents that would flow if E_1 and E_2, in turn, acted alone. So

$$I_{sc} = \frac{E_1}{R_1} + \frac{E_2}{R_2} = (E_1R_2 + E_2R_1)/R_1R_2.$$

Hence, using eqn (2.6), the series resistance 'R_0' of the equivalent source is given by

$$R_0 = \frac{(E_2R_1 + E_1R_2)R_1R_2}{(R_1 + R_2)(E_1R_2 + E_2R_1)},$$

i.e.

$$R_0 = \frac{R_1R_2}{(R_1 + R_2)} \qquad (R_1 \text{ in parallel with } R_2).$$

A simple check on the expressions for V_{oc} and 'R_0' is to set $E_2 = 0$ and $R_2 = \infty$. In this case,

$$V_{oc} = E_1 \qquad \text{and} \qquad R_0 = R_1, \quad \text{as they should.}$$ □

2.3 Kirchhoff's laws

(i) The **voltage law**: The algebraic sum of the potential differences taken around a loop (or 'mesh') of a circuit is zero.
(ii) The **current law**: The algebraic sum of the currents into a node of a circuit is zero.

The abbreviations KVL and KCL will be used from now on for the voltage and current laws respectively.

Consider the circuit shown in Fig. 2.6 and note that it cannot be resolved simply into series and parallel branches as was the case with the circuit of Fig. 2.4. Note also what is meant by a **mesh** and a **node**: Fig. 2.6(b) and (c) respectively.

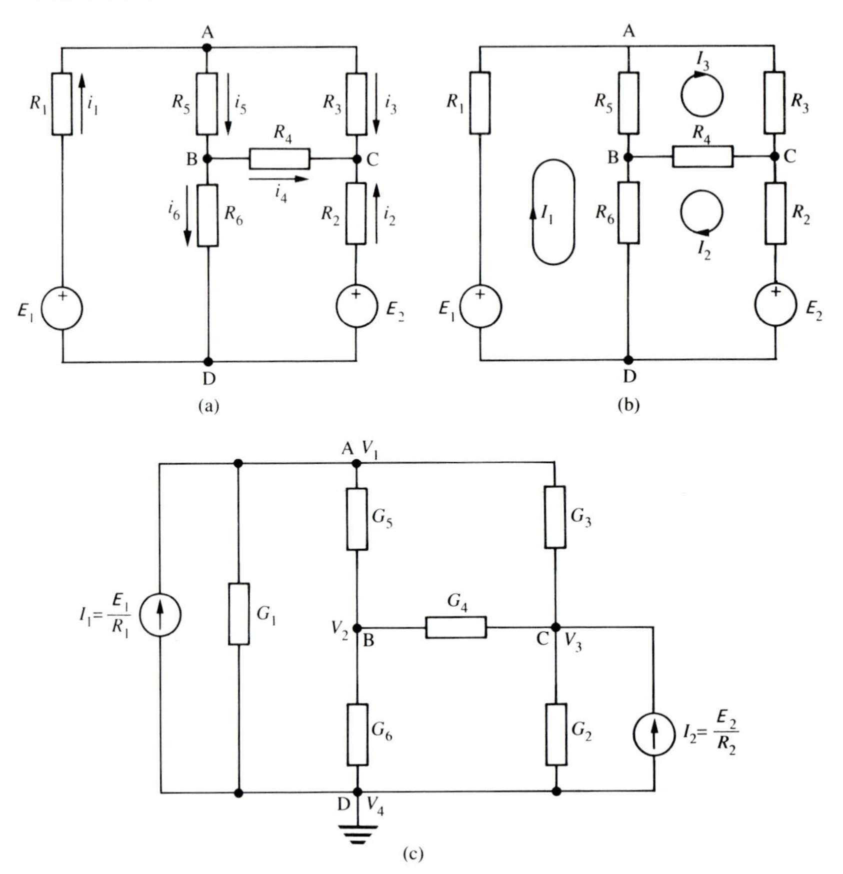

Fig. 2.6 A general network illustrating (a) branch currents and (b) mesh currents. (c) Node voltages V_1, V_2, V_3, V_4; the voltage sources E_1 and E_2 have been replaced by their equivalent current sources I_1 and I_2—see Section 2.1 (NB I_1, I_2 are not the same currents as I_1, I_2 in (b)).

If $E_2 = 0$ then the circuit is a **Wheatstone bridge** (see eqn (2.19) later) with R_4 representing the resistance of the detector that indicates the value of the current i_4.

Alternatively, if $R_3 = \infty$ (open circuit) and $E_2 < E_1$, then the circuit becomes a **potentiometer**, in which E_2 may be determined in terms of E_1 (a 'standard' source of e.m.f.). This is achieved by adjusting the ratio R_5/R_6 until no current flows in the branch containing the detector (R_4) (see eqn (2.17)).

Letting the branch currents be $i_1, i_2, \ldots, i_6$ as shown, the application of KVL to the mesh ABDA yields the equation

$$i_i R_1 + i_5 R_5 + i_6 R_6 = E_1.$$

The application of KVL to the other five possible meshes, and of KCL to the four nodes, yields ten equations in all. However, it transpires that there are only six independent equations. In principle, these six simultaneous equations can be solved to yield the six branch currents; consequently the potential differences in the branches can be calculated by using $v_n = R_n i_n$.

The process of solving two or three simultaneous equations is not too tedious, but for more than three it is necessary to devise systematic and efficient procedures. The first step is to exploit KVL and KCL to reduce the number of equations and of unknowns; two procedures, namely **mesh analysis** and **nodal analysis**, emerge and are introduced in the two following sections. Second, mathematical elimination procedures, such as the Gauss-Jordan elimination method, are used to reduce the labour (and computing time) required in solving the remaining simultaneous equations.

Note that the term 'voltage' will be used in place of the more precise, but ponderous, term 'potential difference'.

2.4 Mesh analysis

Consider a set of **mesh currents** I_1, I_2, I_3 traversing complete loops of the circuit of Fig. 2.6(b). Since each mesh current flows right through any junction (node) in its path, KCL is automatically satisfied. Hence the number of independent simultaneous equations obtained is equal to the number of meshes, in this case three. This number is significantly less than the six equations obtained in the previous section, where branch currents were used. The application of KVL to the meshes indicated in Fig. 2.6(b) yields the equations:

$$I_1R_1 + (I_1 - I_3)R_5 + (I_1 - I_2)R_6 = E_1$$

$$I_2R_2 + (I_2 - I_1)R_6 + (I_3 - I_1)R_4 = -E_2$$

$$I_3R_3 + (I_3 - I_2)R_4 + (I_3 - I_1)R_5 = 0.$$

The terms of these equations can be arranged more systematically as follows:

$$\left\{\begin{aligned} I_1(R_1+R_5+R_6) - I_2R_6 - I_3R_5 &= E_1 \\ -I_1R_6 + I_2(R_2+R_4+R_6) - I_3R_4 &= -E_2 \\ -I_1R_5 - I_2R_4 + I_3(R_3+R_4+R_5) &= 0. \end{aligned}\right. \tag{2.7}$$

The total resistance traversed by I_1 in mesh 1 $(R_1 + R_5 + R_6)$ is called the **self-resistance** R_{11} of mesh 1 and the **mutual resistances** with meshes 2 and 3 are R_6 and R_5, which will be denoted by R_{12} and R_{13} respectively. Note that if the two relevant mesh currents flow through a mutual resistance in opposite directions, then the sign of the mutual resistance is taken as negative (and vice versa). Following this convention, eqns (2.7) take the systematic form

$$I_1R_{11} + I_2R_{12} + I_3R_{13} = E_1$$
$$I_1R_{21} + I_2R_{22} + I_3R_{23} = -E_2$$
$$I_1R_{31} + I_2R_{32} + I_3R_{33} = 0.$$

This process can be developed for a general network, containing say m meshes, to give the set of simultaneous equations

$$\left\{\begin{array}{l} I_1R_{11} + I_2R_{12} + \cdots + I_mR_{1m} = E_1 \\ I_1R_{21} + I_2R_{22} + \cdots + I_mR_{2m} = E_2 \\ \quad\vdots \qquad\quad \vdots \qquad\qquad\quad \vdots \qquad\quad \vdots \\ I_1R_{m1} + I_2R_{m2} + \cdots + I_mR_{mm} = E_m \end{array}\right. . \tag{2.8}$$

To solve such a set of equations for $m = 10$, say, would require hundreds of multiplications and divisions, even using elimination techniques (such as the Gauss–Jordan scheme), so it is not surprising that digital computers are a powerful tool in this field. In the rapidly developing field of electronics computer-aided design (ECAD), information becomes outdated very rapidly, but currently a software package capable of handling a circuit with up to 60 nodes and 180 components is not unusual. Notwithstanding the availability of software packages, it is useful for illustrative purposes and for practice to consider the derivation and solution of eqns (2.8) for some particular circuits where m is not too large.

For the circuit of Fig. 2.6(b), eqns (2.7) can be written in matrix form as

$$\underbrace{\begin{bmatrix} (R_1 + R_5 + R_6) & -R_6 & -R_5 \\ -R_6 & (R_2 + R_4 + R_6) & -R_4 \\ -R_5 & -R_4 & (R_3 + R_4 + R_5) \end{bmatrix}}_{3 \times 3 \text{ matrix}} \begin{bmatrix} I_1 \\ I_2 \\ I_3 \end{bmatrix} = \begin{bmatrix} E_1 \\ -E_2 \\ 0 \end{bmatrix} \tag{2.9}$$

↑ ↑ column vector

or

$$[R][I] = [E]$$

The rules of matrix algebra will not be described and exploited in a formal way here. However, by writing the mesh equations in matrix form, the unknowns and constants in the problem are displayed in a helpful way which, incidentally, can reveal mistakes in the setting-up of the equations. For instance, for a passive network the matrix must be symmetrical about its diagonal, as is the case in the example above.

The **determinant** D of the coefficients of the currents I_1, I_2, I_3 in eqns (2.7) and (2.9) is the array

$$D = \begin{vmatrix} (R_1 + R_5 + R_6) & -R_6 & -R_5 \\ -R_6 & (R_2 + R_4 + R_6) & -R_4 \\ -R_5 & -R_4 & (R_3 + R_4 + R_5) \end{vmatrix} \tag{2.10}$$

and the current I_2, say, is given by

$$I_2 = \begin{vmatrix} (R_1 + R_5 + R_6) & E_1 & -R_5 \\ -R_6 & -E_2 & -R_4 \\ -R_5 & 0 & (R_3 + R_4 + R_6) \end{vmatrix} \Bigg/ D,$$

i.e. the numerator determinant consists of the determinant D but with the second column replaced by the e.m.f.s in question. To determine I_1 and I_3 the first and third columns, respectively, would be replaced by the e.m.f.s: this rule can be extended to matrices (and determinants) of higher dimensions corresponding to networks with greater numbers of meshes.

The simplest way of describing the method for evaluating determinants is to return to the notation of eqns (2.8):

$$D = \begin{vmatrix} R_{11} & R_{12} & R_{13} \\ R_{21} & R_{22} & R_{23} \\ R_{31} & R_{32} & R_{33} \end{vmatrix}.$$

First it is necessary to define the term **cofactor**; e.g. the cofactor C_{11} is

$$C_{11} = (-1)^{1+1} \cdot \begin{vmatrix} R_{22} & R_{23} \\ R_{32} & R_{33} \end{vmatrix},$$

i.e. D with its 1st row and 1st column deleted. Similarly, C_{12} is given by (1st row and 2nd column deleted)

$$C_{12} = (-1)^{1+2} \cdot \begin{vmatrix} R_{21} & R_{23} \\ R_{31} & R_{33} \end{vmatrix}$$

and C_{13} by

$$C_{13} = (-1) \cdot \begin{vmatrix} R_{21} & R_{22} \\ R_{31} & R_{32} \end{vmatrix}$$

The aim in expanding a determinant is to break it down systematically into determinants of smaller and smaller dimensions until only 2×2

arrays remain. For a 2 × 2 determinant,

$$\begin{vmatrix} a_{11} & a_{12} \\ a_{21} & a_{22} \end{vmatrix} = a_{11}a_{22} - a_{12}a_{21}. \tag{2.11}$$

(This can be checked by solving *two* simultaneous equations.)

A determinant can be expanded about any row or column. For instance, the rule for expanding D about the first row is

$$D = R_{11}C_{11} + R_{12}C_{12} + R_{13}C_{13},$$

where the C's are already 2 × 2 in this case and can be evaluated according to eqn (2.11). The determinant in the numerator of eqn (2.10) can be expanded according to the same rule.

Note that if some of the elements of a determinant are zero, then a judicious choice of row or column for the expansion can significantly reduce the labour involved.

So, on drawing together these various points, it follows that

$$I_2 = \frac{E_1C_{12} + (-E_2)C_{22} + (0)C_{32}}{D} \tag{2.12}$$

and, in general, for a network with m meshes,

$$I_1 = \frac{E_1C_{11}{}^{j} + E_2C_{2j} + E_3C_{3j} + \cdots + E_{mj}}{D},$$

or, in a more compact form,

$$I_j = \frac{1}{D}\sum_{n=1}^{n=m} E_nC_{nj}. \tag{2.13}$$

Equation (2.13) is a statement of *Cramer's rule.*

This has been a bit long-winded, necessarily, but some illustrative examples should be illuminating.

Example 2.3 Evaluate the mesh currents I_1, I_2, I_3 in the circuit shown in Fig. 2.7. Following eqn (2.9), the mesh equations can be written as

$$\begin{bmatrix} 3 & -2 & 0 \\ -2 & 9 & -3 \\ 0 & -3 & 8 \end{bmatrix}\begin{bmatrix} I_1 \\ I_2 \\ I_3 \end{bmatrix} = \begin{bmatrix} 10 \\ 0 \\ 0 \end{bmatrix}$$

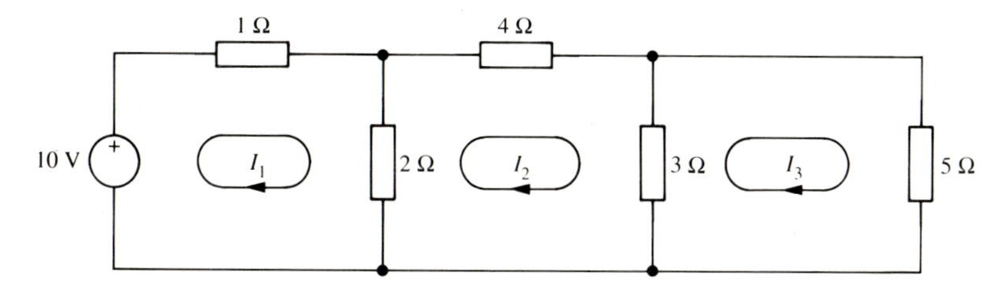

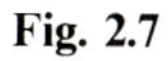

Fig. 2.7

So I_1 is given by

$$I_1 = \begin{vmatrix} 10 & -2 & 0 \\ 0 & 9 & -3 \\ 0 & -3 & 8 \end{vmatrix} \Bigg/ \begin{vmatrix} 3 & -2 & 0 \\ -2 & 9 & -3 \\ 0 & -3 & 8 \end{vmatrix}$$

$$= \frac{10\cdot(-1)^{1+1}(9 \times 8 - 3 \times 3) + (-2)\cdot(-1)^{1+2}(0 \times 8 + 0 \times 3) + 0}{3\cdot(-1)^{1+1}(9 \times 8 - 3 \times 3) + (-2)\cdot(-1)^{1+2}(-2 \times 8) + 0}$$

$$= \frac{10 \times 63}{3 \times 63 - 2 \times 16} = \frac{630}{157}\,\text{A}.$$

Similarly, I_2 is given by

$$I_2 = \begin{vmatrix} 3 & 10 & 0 \\ -2 & 0 & -3 \\ 0 & 0 & 8 \end{vmatrix} \Bigg/ 157 = \frac{160}{157}\,\text{A}.$$

Finally, following the same procedure, $I_3 = 60/157$ A. □

Example 2.4 For the circuit shown in Fig. 2.8, find the power dissipated in the resistor denoted $3R$ if $R = 1\ \Omega$.

Again following eqn (2.9), the mesh equations can be expressed as

$$\begin{bmatrix} 9R & -R & -8R \\ -R & 7R & -2R \\ -8R & -2R & 13R \end{bmatrix} \begin{bmatrix} I_1 \\ I_2 \\ I_2 \end{bmatrix} = \begin{bmatrix} 5 \\ 0 \\ 0 \end{bmatrix}$$

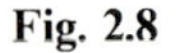

Fig. 2.8

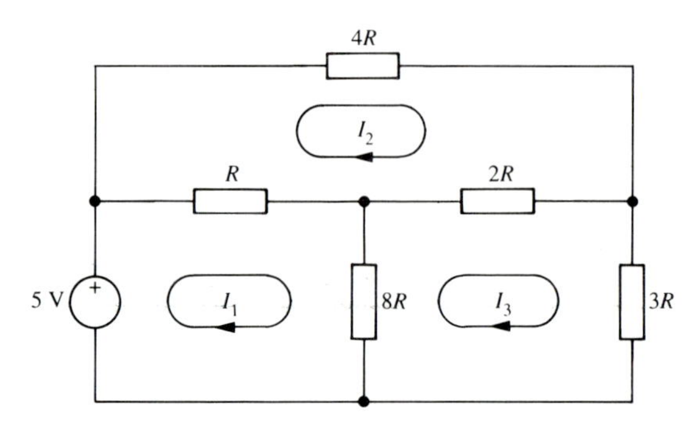

and so

$$I_3 = \begin{vmatrix} 9R & -R & 5 \\ -R & 7R & 0 \\ -8R & -2R & 0 \end{vmatrix} \Bigg/ \begin{vmatrix} 9R & -R & -8R \\ -R & 7R & -2R \\ -8R & -2R & 13R \end{vmatrix}$$

$$= \frac{9R \times 0 + R \times 0 + 5 \times (2R^2 + 56R^2)}{9R \times (91R^2 - 4R^2) + R \times (-13R^2 - 16R^2) - 8R \times (2R^2 + 56R^2)}.$$

Whence $I_3 = 1/R$ A and the power $= I_3^2 R = 1$ W. □

If certain simple modifications are made to the circuit of Fig. 2.6 it can function as a potential-divider, a potentiometer or a Wheatstone bridge. For instance, consider the circuit of Fig. 2.9(a), which has been obtained from that of Fig. 2.6(b) by setting $R_3 = \infty$, $R_4 = 0$, $E_2 = 0$. The independent mesh equations are, in general terms,

$$I_1 R_{11} + I_2 R_{12} = E_1,$$
$$I_1 R_{21} + I_2 R_{22} = E_2. \tag{2.14}$$

Here

$$R_{11} = R_5 + R_6 \qquad R_{12} = -R_6 \qquad E_1 = E$$

$$R_{21} = -R_6 \qquad R_{22} = R_2 + R_6 \qquad E_2 = 0.$$

Hence, using Cramer's rule (eqn (2.13)),

$$I_1 = E_1 C_{11}/D,$$

where

$$D = \begin{vmatrix} R_{11} & R_{12} \\ R_{21} & R_{22} \end{vmatrix} = R_{11}R_{22} - R_{12}R_{21},$$

$$C_{11} = R_{22}.$$

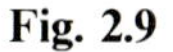
Fig. 2.9

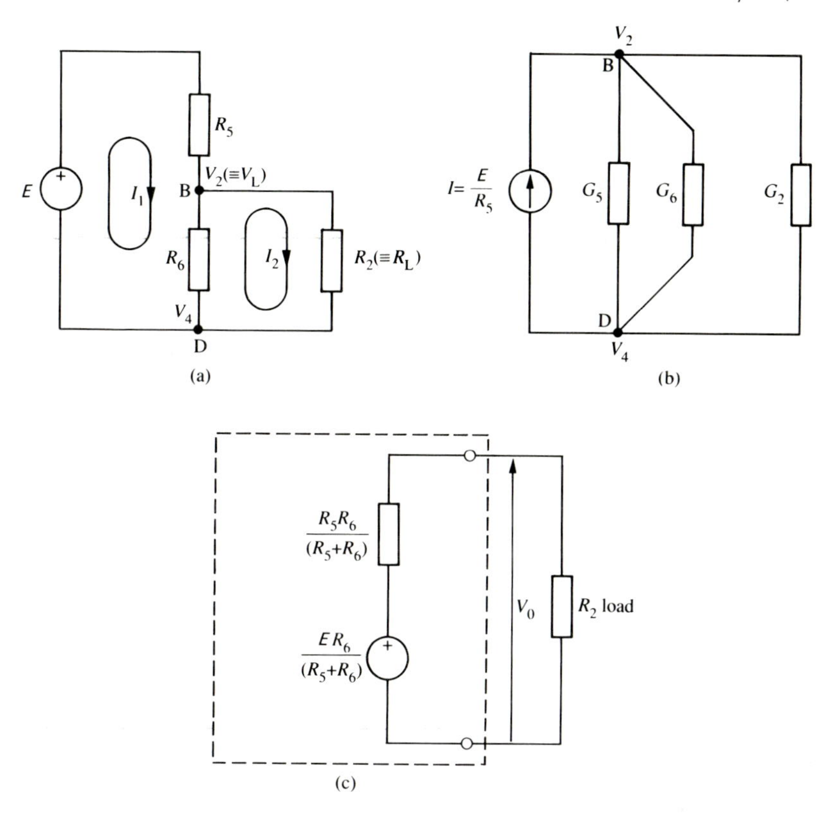

Also, $I_2 = EC_{12}/D$, where $C_{12} = -R_{21} = R_6$. After substitution, the expressions for I_1, I_2 are

(2.15)*
$$I_1 = E \cdot \left\{ R_5 + \frac{R_2 R_6}{R_2 + R_6} \right\}^{-1} \quad \text{and} \quad I_2 = \frac{ER_6}{(R_2R_5 + R_2R_6 + R_5R_6)}.$$

If R_2 is called the load resistor, and is denoted by R_L, then the voltage V_L across it ($= I_2 R_L$) is given by

$$V_L = \frac{ER_6}{(R_5R_6/R_L) + (R_5 + R_6)}$$

* A useful test of the validity of the solution to a problem that can be applied in many situations is to let one or more of the variables or parameters tend to extreme values (e.g. zero or infinity) or to other values which would greatly simplify the problem. In the problem for which eqns (2.15) are solutions, if $R_2 = \infty$, then $I_1 = E/(R_5 + R_6)$, $I_2 = 0$, which is as it should be. Additionally, if $R_2 = 0$, then $I_1 = I_2 = E/R_5$, which again is what is to be expected. Note: such tests are necessary, but not sufficient, conditions for the solution to be correct..

or

$$V_L = \frac{ER_6R_L}{(R_5 + R_6)\left\{\dfrac{R_5R_6}{(R_5 + R_6)} + R_L\right\}}. \tag{2.16}$$

Note that if $R_L \gg R_5R_6/(R_5 + R_6)$, then $V_L \approx ER_6/(R_5 + R_6)$ to a good approximation and the circuit functions as what is known as a **potential-divider** network.

Now refer to the circuit of Fig. 2.6(b) and consider the condition under which the current in the branch containing R_4 is zero, i.e. $I_2 - I_3 = 0$. Equation (2.12) can be used to give I_2, and I_3 can be obtained by again applying Cramer's rule:

$$I_3 = (E_1C_{13} - E_2C_{23})/D.$$

Using the definition of a cofactor to obtain the expressions for C_{13}, C_{23}, it follows that

$$I_2 - I_3 = [E_1(R_3R_6 - R_2R_5) - E_2(R_1R_3 + R_1R_5 + R_3R_5 + R_3R_6)]/D.$$

If it is assumed for simplicity that $R_1 = 0$ so that all the e.m.f. E_1 appears across $(R_5 + R_6)$, and that R_3 is extremely large, then the 'balance condition' $(I_2 - I_3 = 0)$ becomes

$$E_1(R_3R_6 - R_2R_5) - E_2(R_3R_6 + R_3R_5) = 0. \tag{2.17}$$

Now since it has just been assumed that $R_3 \gg R_2, R_5, R_6$, the balance condition becomes

$$E_1R_3R_6 = E_2R_3(R_5 + R_6)$$

or

$$\frac{E_2}{E_1} = \frac{R_6}{(R_5 + R_6)}. \tag{2.18}$$

In a **potentiometer** the node B corresponds to a 'slider', or 'wiper', that is in contact with a continuous resistive wire or strip constituting the series combination $(R_5 + R_6)$, and R_4 represents the resistance of a current-measuring instrument that indicates the value of $(I_2 - I_3)$. There is no physical resistor between the nodes A and C; R_3 represents the leakage paths across and through the insulators that support the circuit components of the potentiometer. Hence R_3 is extremely large in value in practice, which satisfies the assumption made above.

A potentiometer is used to measure an 'unknown' e.m.f. (E_2) in terms of a 'standard' e.m.f. (E_1) by finding the balance condition specified by eqn (2.18). One commonly encountered situation is where E_2 is the e.m.f. of a thermocouple and another is where it is the e.m.f. of a 'secondary' cell (see Table 2.1).

Another circuit that is very widely used is obtained from the circuit of Fig. 2.6 if the source of e.m.f. E_2 is removed but a resistance R_2 is left in the branch. The balance condition of eqn (2.17), with $E_2 = 0$, now becomes

$$R_3R_6 = R_2R_5, \tag{2.19}$$

i.e. the products of the resistances in opposite branches (or 'arms) of the circuit are equal at balance.

If R_2, R_3, R_6 are calibrated resistances, then the value of an 'unknown' resistance R_5 can be determined experimentally by adjusting R_3 and R_2 (the 'ratio arms') to give a convenient value of the ratio R_3/R_2 and then adjusting R_6 to obtain the balanced condition. When used in this way the circuit is known as a **Wheatstone bridge** circuit.

Some further aspects of the Wheatstone bridge circuit will be discussed in Section 2.7, and some related important bridge circuits in the domain of alternating currents and voltages are discussed in Section 3.9.

It should be noted that in analysing a circuit by mesh analysis, the sense of circulation chosen for any particular mesh current is not important; if you happen to have chosen 'wrongly', the expression/value you eventually obtain for the current will have a negative sign. However, it is probably prudent to choose the sense of circulation systematically and to take particular notice of the sense of the current through the 'load' resistance, since this will determine the sense of the voltage across the load.

2.5 Nodal analysis

For many circuits the voltages at the nodes are the quantities of primary interest; nodal analysis leads more directly to these voltages than does the mesh analysis approach. Also, and very importantly, the use of node voltages rather than mesh currents makes for greater efficiency in the analysis of circuits that are predominantly parallel in character, since the number of simultaneous equations that have to be solved is significantly smaller.

A simple but useful example is the potential-divider network illustrated in Fig. 2.9. As far as the mesh containing R_2 and R_6 is concerned, the nodes B and D are the accessible terminals of a voltage source E having a series internal resistance R_5 (i.e. the actual internal resistance R_1 of the voltage source (see Fig. 2.6(b)) can be imagined either to be negligible or to be contained in R_5). Since only differences of potential have physical significance, it is convenient to choose D as a **reference ('common') node** and to refer all potentials to it.

In the circuit being considered, the current I_2 flowing from node B to node D through R_2, for instance, is obtained from

$$V_B - V_D = I_2 R_2 \qquad \text{or} \qquad I_2 = (V_B - V_D)/R_2.$$

The quantity $1/R_2$ is the conductance of R_2 and is denoted by G_2 (the units of conductance being the siemens (S)).

It is convenient to make the network entirely parallel in nature (see Fig. 2.9(b)) by replacing the voltage source E (with R_5 in series) by the equivalent current source, that is, an ideal constant current source E/R_5 shunted by the conductance G_5 $(= 1/R_5)$ (refer to Section 2.1).

By KCL, the total current flowing out of node B is equal to the total current I flowing into B, which is equal to E/R_5. So, on taking account of the three branches connected to node B, containing the resistances R_2, R_6, R_5 respectively, we have

$$(V_2 - V_4)\left(\frac{1}{R_2} + \frac{1}{R_5} + \frac{1}{R_6}\right) = \frac{E}{R_5}$$

or

$$(V_2 - V_4)(G_2 + G_5 + G_6) = I. \tag{2.20}$$

Setting $V_4 = 0$ for convenience,

$$V_2 = IR_2R_5R_6/(R_2R_5 + R_2R_6 + R_5R_6)$$

or

$$V_2 = ER_2R_6/(R_2R_5 + R_2R_6 + R_5R_6), \tag{2.21}$$

which, on putting $V_2 = V_L$, $R_2 = R_L$, becomes identical to eqn (2.16).

Notice that only one nodal equation has to be solved, compared with two mesh equations (eqns (2.14))!

Although the potential-divider is quite simple to analyse, it does serve to emphasize the very important point that, for more complicated circuits, making a judicious choice between mesh and nodal analysis can lead to a significant reduction in the number of simultaneous equations to be solved.

As another illustration of the application of nodal analysis, consider again the circuit of Fig. 2.6. Note that in Fig. 2.6(c) the voltage source E_1 (with series resistance R_1) and the voltage source E_2 (with series resistance R_2) have been replaced by their equivalent current sources I_1 (with shunt conductance $G_1 = 1/R_1$) and I_2 (with shunt conductance $G_2 = 1/R_2$) respectively. On applying KCL to nodes A, B, C, respectively, and setting $V_4 = 0$ ('common'), the following equations are obtained:

$$V_1G_1 + (V_1 - V_2)G_5 + (V_1 - V_3)G_3 = I_1$$
$$(V_2 - V_3)G_4 + (V_2 - V_1)G_5 + V_2G_6 = 0$$
$$V_3G_2 + (V_3 - V_1)G_3 + (V_3 - V_2)G_4 = I_2$$

Remember that an ideal constant current source has infinite internal resistance.

or, more systematically,

$$\begin{aligned} V_1(G_1 + G_3 + G_5) \quad & - V_2G_5 & - V_3G_3 \quad & = I_1 \\ - V_1G_5 \quad & + V_2(G_4 + G_5 + G_6) & - V_3G_4 \quad & = 0 \\ - V_1G_3 \quad & - V_2G_4 & + V_3(G_2 + G_3 + G_4) & = I_2 \end{aligned} \tag{2.22}$$

The formal similarity between these nodal equations and the mesh equations (eqns (2.7)) should be noted; the mathematical method for solving them is the same.

The procedure by which eqns (2.22) were obtained can be extended to more complex circuits, but the general rules have already emerged:

(a) Choose a convenient reference (usually common or earth) and assume a voltage for every other node.
(b) In the equation for a particular node, the coefficient of the voltage of that node is the sum of the conductances connected to it; the coefficients of the other voltages are the conductances joining their respective nodes to the node in question, always taken negative.

For a circuit containing m nodes,

$$V_j = \frac{1}{D} \sum_{n=1}^{n=m} I_n C_{nj} \tag{2.23}$$

(Compare with eqn (2.13), or use the analogue of eqn (2.10).)

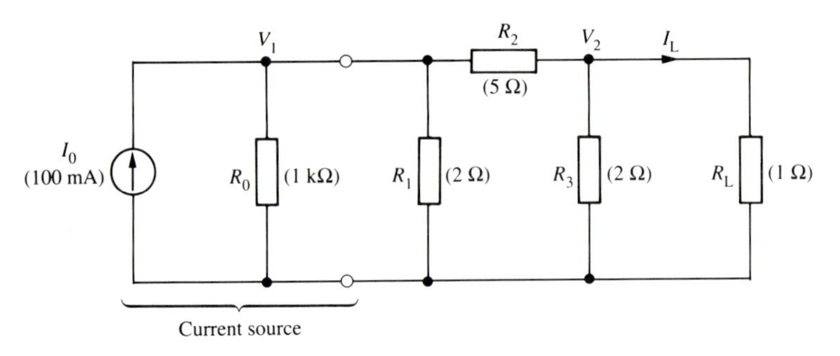

Fig. 2.10

Example 2.5 Determine the value of the current I_L flowing through the load resistance R_L in the circuit shown in Fig. 2.10.

Since there are fewer nodes than meshes (the two nodes are designated by the voltages V_1, V_2 in Fig. 2.10), it is convenient to use nodal analysis.

Applying KCL at the nodes yields the two simultaneous equations

$$V_1\left(\frac{1}{R_0}+\frac{1}{R_1}+\frac{1}{R_2}\right)-\frac{V_2}{R_2}=I_0$$

$$-\frac{V_1}{R_2}+V_2\left(\frac{1}{R_2}+\frac{1}{R_3}+\frac{1}{R_L}\right)=0.$$

Remember that the ideal constant current source I_0 has infinite internal resistance.

Also

$$I_L=\frac{V_2}{R_L}$$

On substituting values for I_0, R_0, R_1, R_2, R_3, R_L and then solving the two simultaneous equations, it follows that

$$V_2 = 17.4\ \text{mV} \qquad I_L = 17.4\ \text{mA}.$$ □

Numerous software packages are available in the realm of electronics computer-aided design (ECAD), of which a well-known example is **SPICE** (Simulation Program with Integrated Circuit Emphasis). SPICE is based on nodal analysis; a circuit is specified by numbering the nodes and identifying the type and value of the circuit element connected between each node.

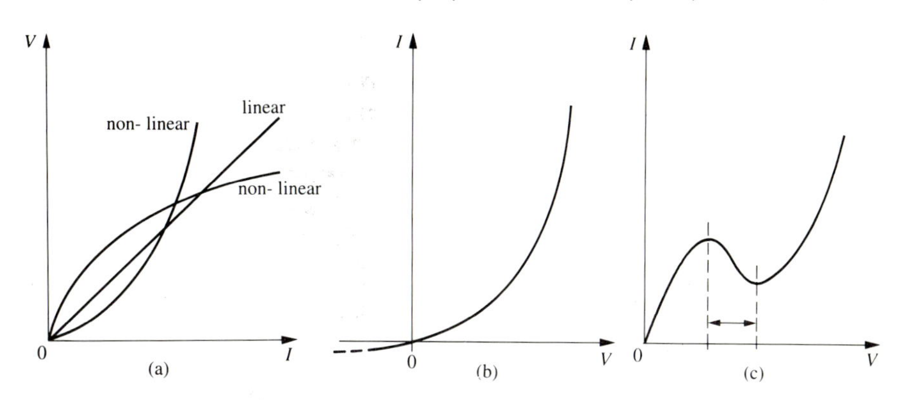

Fig. 2.11 (a) Linear and non-linear current–voltage characteristics (schematic). (b) The I–V characteristic for a semiconductor p-n junction rectifier. (c) An example of a differential resistance ($\equiv \mathrm{d}V/\mathrm{d}I$) that is negative for a certain range of values of V.

2.6 The superposition and reciprocity theorems

In the analyses leading to eqns (2.13) and (2.23) an implicit assumption was that the resistors were linear, i.e.

$$V \propto I$$

or that the ratio V/I was independent of the value of the current I (e.g. the straight-line relation in Fig. 2.11(a)).

A non-linear relationship (see Fig. 2.11(a)) between V and I for a circuit element can be expressed in the form

$$V = AI + BI^2 + CI^3 + DI^4 + \cdots$$

(2.24) or

$$\frac{V}{I} = A + BI + CI^2 + DI^3 + \cdots$$

Here A is the coefficient of the linear term and $B, C, D \ldots$ are coefficients of non-linear terms; clearly the resistance R ($\equiv V/I$) is a function of I.

A significant dependence of resistance upon current commonly arises from the self-heating effect of the current itself. The tungsten filament of a light bulb is a good example of this, but in copper wires the effect is insignificant for many practical purposes. The resistance of the junction between two dissimilar materials depends on the applied voltage, e.g. in a semiconductor p-n junction rectifier (see Fig. 2.11(b)). In some materials, of which the semiconductor gallium arsenide is the best-known example, the resistance depends on the value of the applied electric field (see Fig.

2.11(c)); the negative differential resistance (due to the *Gunn effect*) is exploited to fabricate very high-frequency oscillators.

Circuit elements for which the resistance is independent of the current to a good enough approximation (so that the value of the resistance is uniquely defined, effectively) are referred to as **linear elements** (or **'ohmic' elements**). Elements that are significantly non-linear in this sense are often described as **'non-ohmic'**.

It is important to notice that the transfer characteristics of active (amplifying) devices are linear only to an approximation and even then only over a limited range of the amplitude of the input variable; this limits the **dynamic range** of the device, since the non-linearity of the transfer characteristic means that the output is a distorted version of the input.

A consequence of the assumed linearity of the circuit elements in mesh and nodal analysis is that the algebraic equations that arise are linear; the practical manifestation of this is the **superposition theorem**:

> The current flowing in a particular branch of a circuit is the algebraic sum of the currents that would flow if each of the sources of e.m.f. were in turn, present alone (the internal resistance of each of the sources must be left in place, of course).

To illustrate this, suppose that there are two sources in a circuit so that in eqn (2.24)

$$I = I_a + I_b.$$

Also assume for simplicity that $C, D, \ldots$ are all zero. In this case

$$V = AI_a + AI_b + B(I_a^2 + I_b^2 + 2I_aI_b).$$

For linearity, $B = 0$ and V is related to the superposition of the currents due to the two sources acting separately.

In Sections 2.4 and 2.5, eqns (2.13) and (2.23) exemplify the principle of superposition.

A source of voltage or current may be an **independent source** or a **dependent source**. In the latter case the voltage or current, as the case may be, depends on voltages and/or currents elsewhere in the circuit. A clear example is an active element such as a transistor, where the current in the 'output mesh' depends on the voltage applied in the 'input mesh': commonly used symbols for voltage and current sources are shown in Fig. 2.12.

If a circuit contains only linear elements, one independent source and no dependent sources, then the circuit exhibits reciprocity. This property

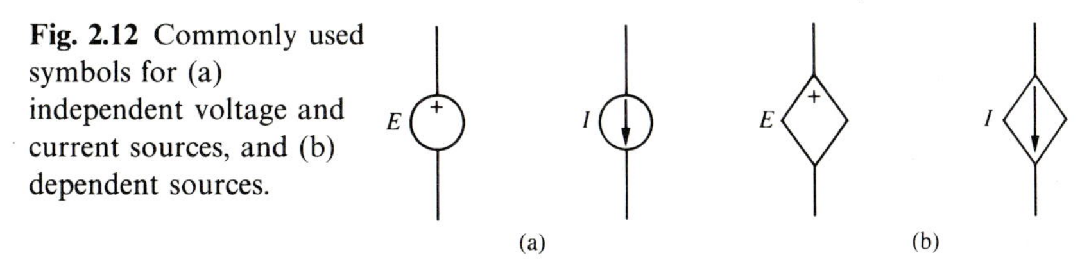

Fig. 2.12 Commonly used symbols for (a) independent voltage and current sources, and (b) dependent sources.

can be illustrated by considering the circuit of Fig. 2.6(b) with $E_2 = 0$. The current in mesh 3 is given by

$$I_3 = \begin{vmatrix} (R_1 + R_5 + R_6) & -R_6 & E_1 \\ -R_6 & (R_2 + R_4 + R_6) & 0 \\ -R_5 & -R_4 & 0 \end{vmatrix} \Big/ D.$$

If now the source is transferred from mesh 1 to mesh 3,

$$I_1 = \begin{vmatrix} 0 & -R_6 & -R_5 \\ 0 & (R_2 + R_4 + R_6) & -R_4 \\ -E_1 & -R_4 & (R_3 + R_4 + R_5) \end{vmatrix} \Big/ D.$$

By inspection of the respective determinants on the numerators of the right-hand sides of these two equations, it can be quickly seen that the mesh current I_1 in the second case is equal to I_3 in the first case.

This argument applies equally well no matter how many meshes there are in a circuit. Hence, if there is a voltage source E in the jth mesh and a consequent current I_n in the nth mesh (see Fig. 2.13), then if E is now

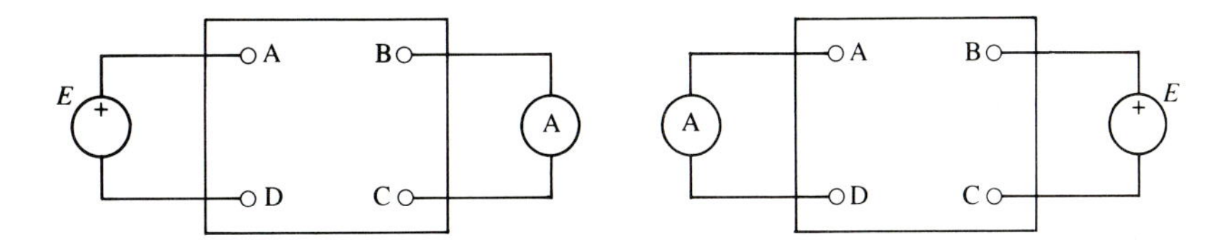

Fig. 2.13 The principle of reciprocity.

placed in the nth mesh the consequent current I_j in the jth mesh is equal to the previous value of I_n.

This is the **principle of reciprocity**.

The interchangeability of the positions of the source and the detector in a Wheatstone bridge circuit (see Fig. 2.15) is an example of reciprocity.

Another example is the interchangeability of the transmitting and receiving antennas in a communications system.

2.7 Equivalent circuits

Frequently, interest is centred on only one particular branch (the 'output' branch) of a circuit. A simple example is the potential-divider circuit shown in Fig. 2.9(a), where the voltage across the load resistor R_2 is of prime interest and R_5, R_6, and the source of e.m.f. E could be contained within a 'black box'. The expression (eqn (2.15)) for the current through the load resistor can be written as

$$I_2\left\{\frac{R_5R_6}{(R_5+R_6)}+R_2\right\}=\frac{ER_6}{(R_5+R_6)} \tag{2.25}$$

If this equation is compared with eqn (2.3), then the internal series resistance 'R_0' and e.m.f. 'E_0' of the Thévenin equivalent source are $R_5R_6/(R_5+R_6)$ and $ER_6/(R_5+R_6)$ respectively. Note that the expression for 'R_0' is that for the parallel combination of the resistances R_5, R_6.

Now since the voltage, 'V_0' say, at the accessible terminals of the imagined black box is equal to I_2R_2, eqn (2.25) can be rearranged as

$$V_0=\underbrace{\frac{ER_6}{(R_5+R_6)}}_{\text{`}E_0\text{'}}-I_2\underbrace{\frac{R_5R_6}{(R_5+R_6)}}_{\text{`}R_0\text{'}} \tag{2.26}$$

It can be seen that eqn (2.26) is in the same form as eqn (2.1) for a simple voltage source of e.m.f. E_0 with series resistance R_0.

So, if it is required to find the **Thévenin equivalent** of a circuit, it is useful to express the relation between the terminal voltage 'V_0' and the load current in the form of either eqn (2.25) or eqn (2.26).

The **Norton equivalent circuit** follows immediately from these considerations, i.e. it consists of an ideal constant current source 'I_0' (= 'E_0'/'R_0') shunted by a conductance 'G_0' (= 1/'R_0').

In the cases of potentiometer circuits and bridge circuits the accuracy with which the balance condition can be judged is of great practical importance. Assuming that the resistance of the variable resistances can be varied smoothly enough (NB R_5, R_6 constitute the 'slide wire' of total resistance (R_5+R_6)), then the sensitivity of the 'detector' becomes the determining factor. Consider, for example, the potentiometer circuit shown in Fig. 2.14(a), which has been derived from the circuit of Fig. 2.6(b) by assuming for simplicity that $R_1=0$, by letting $R_3=\infty$, and by denoting R_4 by R_M, the resistance of the detecting instrument. The balance condition ($I_2=0$) has already been shown to be $E_2/E_1=R_6/(R_5+R_6)$.

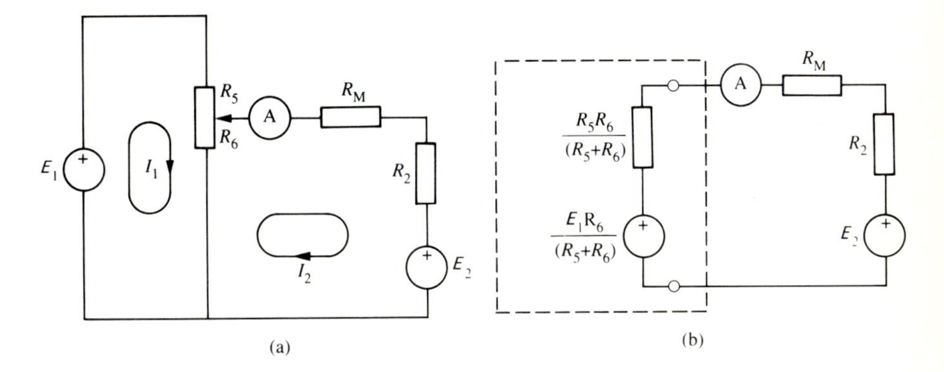

Fig. 2.14

However, the question arises: what is the magnitude of I_2 for a given degree of unbalance of the circuit? The answer will reflect the accuracy with which the balance condition can be determined in practice.

The equivalent circuit in Fig. 2.14(b) can be found by arguments analogous to those leading to eqn (2.25) earlier, and it follows that

$$I_2 = \left\{\frac{E_1R_6}{(R_5 + R_6)} - E_2\right\} \Bigg/ \left\{\frac{R_5R_6}{(R_5 + R_6)} + R_M + R_2\right\}. \tag{2.27}$$

It is convenient, for generality, to define $f \equiv E_2/E_1$ and, if $R \equiv (R_5 + R_6)$, the balance condition becomes

$$f = \frac{R - R_5}{R}$$

and $R_5 = R(1 - f)$ at balance.

A small deviation from balance can be specified through $R_5 = \{R(1 - f) - r\}$, where r is a small value of resistance compared with the values of R_5, R_6 in the balanced situation. On substituting in eqn (2.27) it follows, after a little manipulation, that

$$I_2 = \frac{E_1\left\{\dfrac{(fR + r)}{R} - f\right\}}{fR^2(1 - f)\left\{1 - \dfrac{r}{R(1 - f)}\right\}\left\{1 + \dfrac{r}{fR}\right\} + R_M + R_2}.$$

On multiplying out the terms in the denominator, the product of $r/R(1 - f)$ with r/fR is neglected, since r/R has already been assumed to be $\ll 1$ and $(r/R)^2$ will be even smaller.

So

$$I_2 \approx \frac{rE_1/R}{f(1-f)R\left\{1+\dfrac{r}{fR}-\dfrac{r}{R(1-f)}\right\}+R_M+R_2}$$

and, finally,

$$I_2 \approx \frac{rE_1/R}{fR(1-f)+r(1-2f)+R_M+R_2}. \tag{2.28}$$

For an unbalance of $x\%$ then $r/R_{5_{\text{balance}}} = x/100$, or $r/R(1-f) = x/100$. If f is taken as $\frac{3}{4}$, say, $R = 1\ \text{k}\Omega$, $R_M = 1\ \text{k}\Omega$, and $R_2 = 50\ \Omega$, then substitution in eqn (2.28) yields $I_2 \approx 4\ \mu\text{A}$, which can be detected with commonly available current meters, e.g. a digital multimeter.

Incidentally, if in eqn (2.28) the term $r(1-2f)$ in the denominator can be assumed negligible, then $I_2 \propto r$, approximately; i.e. the detector current is a linear function of the out-of-balance for small deviations from the balance condition.

Another illustration of the utility of equivalent circuits is the problem of obtaining an expression for the out-of-balance current in the detector branch of a Wheatstone bridge. In an approach using mesh analysis, the mesh currents are chosen so that only one such current traverses the branch containing the detector, since the detector current is the quantity of particular interest (see Fig. 2.15). It is a useful but somewhat laborious exercise (see Problem 2.4) in setting up mesh equations and solving the resulting determinants to show that $I_2(R_0 + R_4) = E_0$, where

$$R_0 = \frac{R_5R_6}{(R_5+R_6)} + \frac{R_2R_3}{(R_2+R_3)}, \qquad E_0 = \frac{E(R_2R_5 - R_3R_6)}{(R_5+R_6)(R_2+R_3)}. \tag{2.29}$$

As the first step of an alternative approach to determining R_0 and E_0, consider the simple source of Fig. 2.1; R_0 is equal to the resistance between the terminals with the ideal source of e.m.f. E_0 removed. This statement is also true for more complex circuits; to calculate the equivalent resistance R_0, remove all sources of e.m.f. and replace them by their respective internal resistances, or by short-circuits if their internal resistances are negligible.

For the Wheatstone bridge circuit of Fig. 2.15(a) the network as 'seen' from terminals A, B is as shown in Fig. 2.15(b) (where the source has been assumed to have negligible internal resistance), i.e. R_5 in parallel with R_6 and R_3 with R_2. Hence again,

$$R_0 = R_5R_6/(R_5+R_6) + R_2R_3/(R_2+R_3).$$

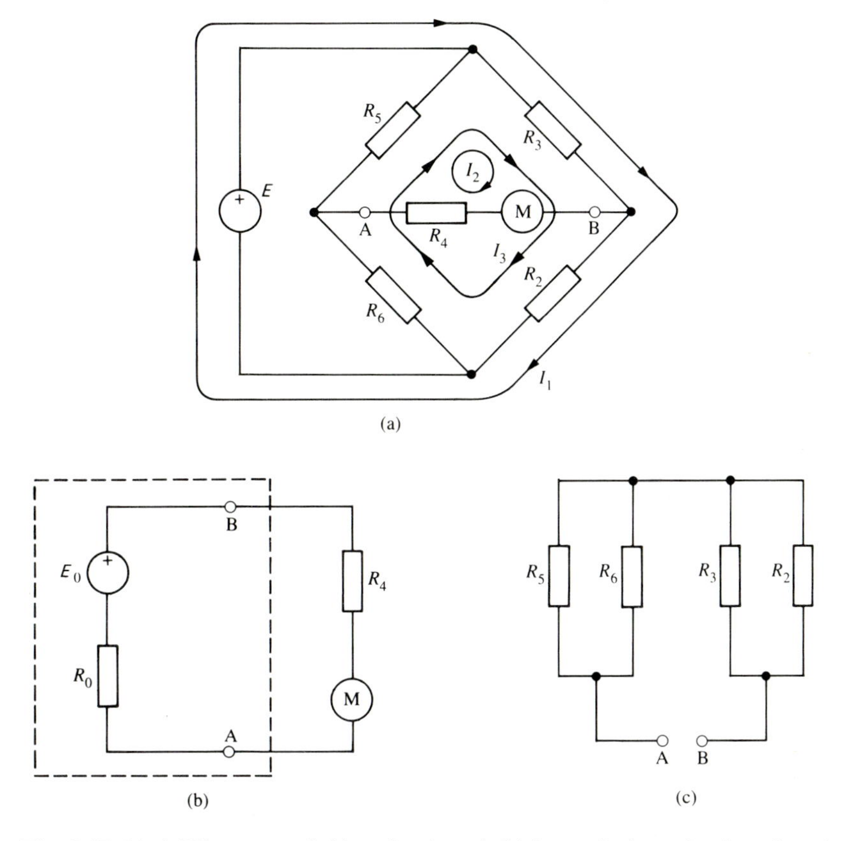

Fig. 2.15 (a) A Wheatstone bridge circuit and (b) its equivalent circuit as 'seen' from the detector. The method for obtaining the expression for R_0 is shown in (c).

Furthermore, the open-circuit voltage 'E_0' between A and B is just

$$E_0 = E\frac{R_2}{(R_2 + R_3)} - E\frac{R_6}{(R_5 + R_6)},$$

which reduces to the same expression for E_0 as in eqns (2.29).

As with the potentiometer, the accuracy with which the unknown quantity (the resistance R_5 in this case) can be determined is related to the smallest out-of-balance signal (current or voltage, depending on the type of detector) that can be detected. In order to obtain an expression for the out-of-balance current, assume in the circuit of Fig. 2.15 that $R_2 = R_3 = R_6 = R$ (this merely simplifies the consequent algebra; no matter of principle is involved), $R_5 \equiv R_x$, and $R_4 \equiv R_M$ (the resistance of the detector). It follows that the expressions for the out-of-balance current

I_2 and the out-of-balance e.m.f. E_0 (eqns (2.29)) take the forms

$$I_2 = \frac{E(R - R_x)}{2R_M(R + R_M) + R(R + 3R_x)}$$

(2.30) and

$$E_0 = \frac{E(R - R_x)}{2(R + R_x)}.$$

Near to balance, $R \approx R_x$ and these expressions can be written as

$$I_2 \approx \frac{E(R - R_x)}{4R(R + R_M)} \quad \text{and} \quad E_0 \approx \frac{E(R - R_x)}{4R}, \tag{2.31}$$

i.e. I_2 and E_0 are linearly dependent on $(R - R_x)$, to an approximation.

Assuming that $(R - R_x)/R = 10^{-3}$, then consider two extreme practical situations, namely (a) where $R = 10\ \Omega$ and $R_M = 100\ \Omega$, and (b) where $R = 100\ \text{k}\Omega$ and $R_M = 1\ \text{M}\Omega$. In situation (a), if $E = 2$ V, then $I_M \approx 5\ \mu\text{A}$, and in (b), if $E = 200$ V, then $E_0 \approx 50$ MV. Hence in (a) a low-resistance current-measuring meter would be used and in (b) a high-resistance voltmeter.

It is appropriate to point out at this stage that the analogue in nodal analysis to Thévenin's theorem is: Any linear circuit having two terminals can be represented by an ideal constant current source in parallel with a conductance. This is *Norton's theorem.*

Strain gauge circuits provide a practical example of the use of bridge circuits. A typical modern type of strain gauge consists of a metal foil strip (of the order of micrometres in thickness) on a plastic backing sheet that can be bonded to the material to be put under mechanical strain; the component of strain in the direction of the length of the strip causes a change in the resistivity of the metal foil. If the strip is incorporated as the 'unknown' resistance R_x in a Wheatstone bridge circuit, then, from eqn (2.31), the change δE_0 in E_0 due to a change δR_x in R_x is given by

$$\delta E_0 = \frac{\partial E_0}{\partial R_x}\,\delta R_x = -\frac{E}{4R_x}\,\delta R_x.$$

For the maximum strains existing in practice, $\delta R_x/R_x \approx 1$ per cent, so for $E \approx 10$ V, $\delta E_0 \approx 25$ mV.

A useful gain in sensitivity can be gained by using (nominally) identical strain gauge elements in the four arms of the bridge network. By arranging the strain gauges at right angles on the member under strain and in a differential configuration in the bridge circuit so that, for example, for a given value of strain R_2 and R_5 decrease by δR_x, and R_3 and R_6 increase by δR_x, then it follows immediately from eqn (2.29) that the out-of-balance voltage δE_0 is given by

$$\delta E_0 \approx -\frac{\delta R_x}{R_x},$$

i.e. there is a fourfold increase in sensitivity.

A very useful additional feature of this differential configuration is that resistance changes in the strain gauge elements, and in the connecting leads, due to temperature changes cancel out.

2.8 D.c. measurements

The most important operational features of a measuring instrument as far as the indication is concerned (e.g. pointer-and-scale, recorder chart, digital display) are the **accuracy**, the **sensitivity**, the **resolution** and the **rated operating conditions**.

The accuracy* of a measurement is the closeness of agreement between the result of the measurement and the true value of the measurand (i.e. the voltage, current, or resistance in this context). Obviously the accuracy of an instrument depends on the quality of the calibration process. The sensitivity is the change in response divided by the change in the stimulus, and the resolution is a quantitative expression of the ability of an indicating device to distinguish meaningfully between closely adjacent values of the quantity indicated. The rated operating conditions refer, fairly obviously, to the range of the measurand for which the metrological characteristics of the instrument in question are intended to lie within specified limits, but they also apply to other operating requirements such as ambient temperature commonly, and for example vibration, static electric field strength, and radiofrequency field strength.

From a circuit standpoint, a current-measuring instrument should have a relatively small internal resistance, and a voltage-measuring instrument should have a relatively high internal resistance (since it will be connected across a circuit element, i.e. in 'shunt'). A modern digital multimeter typically has voltage-measuring ranges from ~ 100 mV to ~ 1 kV with an

* The statements concerning accuracy, sensitivity, resolution, and rated operating conditions have been adapted closely from the definitions of these quantities given in the *International vocabulary of basic and general terms in metrology* prepared by a joint working group appointed by the International Bureau of Weights and Measures (BIPM), the International Electrotechnical Commission (IEC), the International Organization for Standardization (ISO), and the International Organization of Legal Metrology (OIML), and published by the ISO in 1984; it was also published by the British Standards Institution as PD 6461: Part 1: 1985.

internal resistance of at least 10 MΩ, current ranges from ~100 μA (series resistance ~1 kΩ) to ~10 A (series resistance ~0.1 Ω), and resistance-measuring ranges from ~100 Ω to ~10 MΩ.

Suppose that in the potential-divider circuit of Fig. 2.9(a), R_2 represents the resistance of a voltmeter that is to be used to measure the voltage across R_6. Further suppose that a condition of the problem is that $I_2 \leqslant I_1/100$. Then, from eqns (2.15), it is required that $R_2 \geqslant 99R_6$. If $E = 1.5$ V and $I_1 = 10$ mA, then $R_5 + R_6 = 150\ \Omega$, very closely, and if $R_6 = 100\ \Omega$, say, then R_2 must be at least 9967 Ω. An implication is that the meter employed must give a significant reading for $I_2 = 100\ \mu$A.

Although moving-coil meters are still quite widely used, they have been replaced to a large extent by 'digital' meters. In a digital voltmeter (DVM) an **analogue-to-digital converter** (ADC) converts a d.c. voltage (the measurand) into a count proportional to the input voltage; this count is processed and displayed digitally. Perhaps the most important feature from an operational point of view is the very high input resistance of DVMs, commonly to 10 MΩ. The output display may be '$3\frac{1}{2}$ digits'; this means that the most significant digit can take the values 0 and 1 only For instance, the voltage-measuring ranges of an instrument could be

0–199.9 mV

0–1.999 V

0–19.99 V

and so on. Obviously the resolution of a digital instrument is related to the number of digits displayed.

Often, current- and voltage-measuring capabilities are incorporated in a single unit, which is then called a **digital multimeter** (see Fig. 2.16). For current measurements the voltage drop across a small internal resistance (~1 Ω) is measured. In the resistance-measuring mode an internal constant current source drives a small current through the 'unknown' resistance and the voltage drop across a small internal resistance is measured. A cautionary note! It has been known for users to change the mode selection switch of a multimeter from 'voltage' to 'current' without changing the mode of connection of the instrument. Remember that voltage is measured 'across' a circuit element, whereas a current-measuring instrument must be connected 'in series' with the circuit element.

A figure-of-merit for voltmeters that is sometimes used is the 'ohm per volt'. For instance, an instrument of resistance 1 kΩ for which the full-scale deflection (f.s.d.) current is 1 mA has a figure of merit of 1 volt/10^{-3} ampere, or 10^3 ohm per volt. For a 10^4 ohm per volt voltmeter with a 10 V range the meter's resistance is $10 \times 10 = 10^5\ \Omega$ and the f.s.d. current sensitivity is 10 volt/$10^5\ \Omega = 100\ \mu$A.

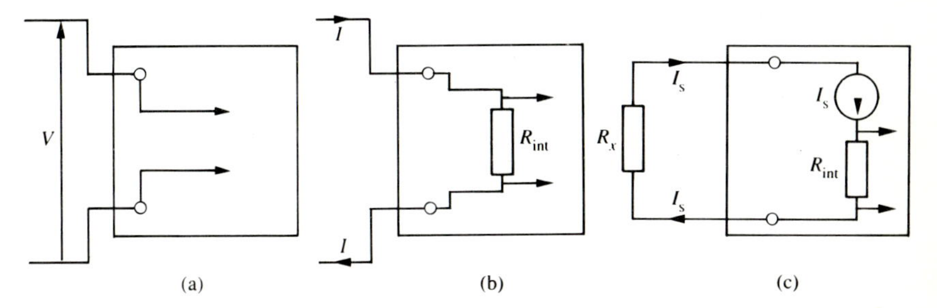

Fig. 2.16 The modes of connection of a digital multimeter: (a) as a voltmeter; (b) as a current-measuring instrument; (c) as an 'ohmmeter' to measure the resistance R_x.

Problems

2.1 A power supply has terminal voltages 999 V and 990 V when the load currents are 10 mA and 100 mA respectively. If the load resistances in the two situations are 99.9 kΩ and 9.9 kΩ respectively, calculate the e.m.f. and internal resistance of the power supply.

2.2 Find the voltage V_0 and the power delivered by the source in the circuit shown in Fig. 2.17.

Fig. 2.17

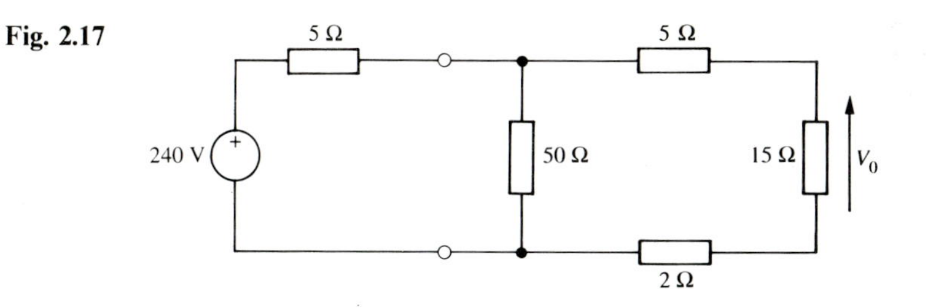

2.3 In the circuit of Fig. 2.9(a), for what range of values of R_2 is the voltage across R_2 equal to $ER_6/(R_5 + R_6)$ to at least 1 per cent if $R_5 = 6$ kΩ, $R_6 = 4$ kΩ?

2.4 Show that the resistance between the input terminals of the network shown in Fig. 2.18 is equal to $4R$.

2.5 Obtain eqn (2.29) for the out-of-balance current in the Wheatstone bridge circuit of Fig. 2.15 using mesh analysis.

Fig. 2.18

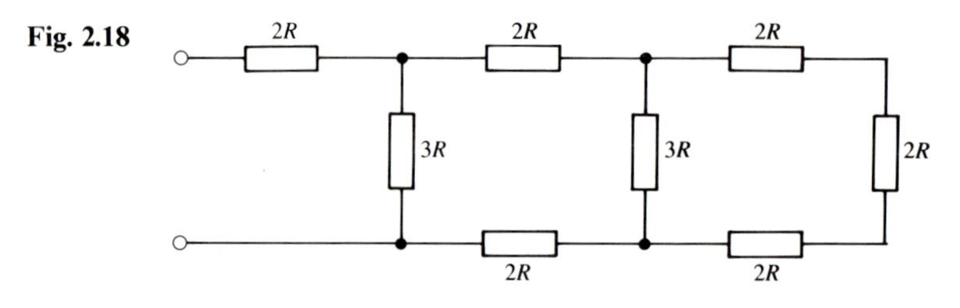

2.6 Find the Norton equivalent of the network shown in Fig. 2.19.

Fig. 2.19

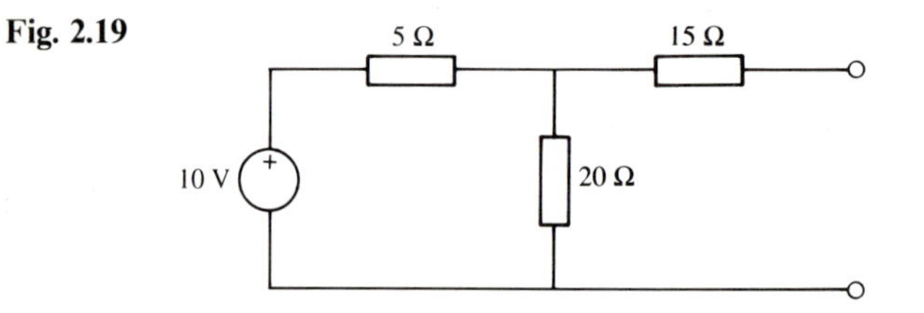

2.7 Find the out-of-balance current through the microammeter (of resistance 100 Ω) in the Wheatstone bridge circuit of Fig. 2.20 if the value of the resistance R_x is changed from 3000 Ω to 3010 Ω.

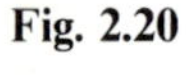
Fig. 2.20

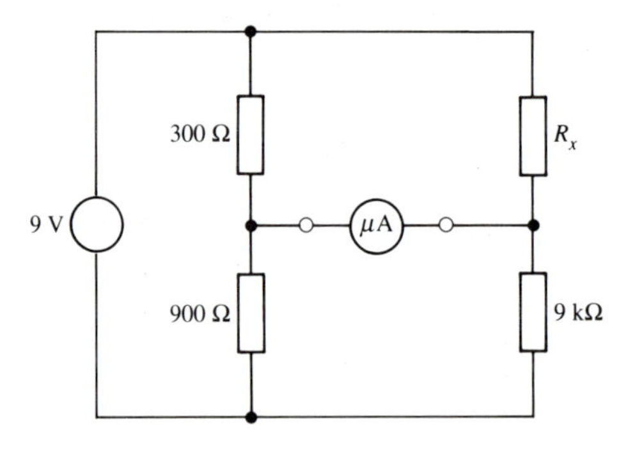

2.8 For a two-terminal 'black-box' containing resistances and current or voltage sources only, the voltage and current at the terminals are related by $I = (-V/4 + 3)$ A. Find (a) the Thévenin equivalent circuit, and (b) the Norton equivalent circuit, of the 'black box'.

2.9 Calculate the value of the ratio I_L/I_s for the circuit shown in Fig. 2.21 using the methods of both mesh and nodal analysis.

2.10 Find the Thévenin and Norton equivalents of the network shown in Fig. 2.22.

Fig. 2.21

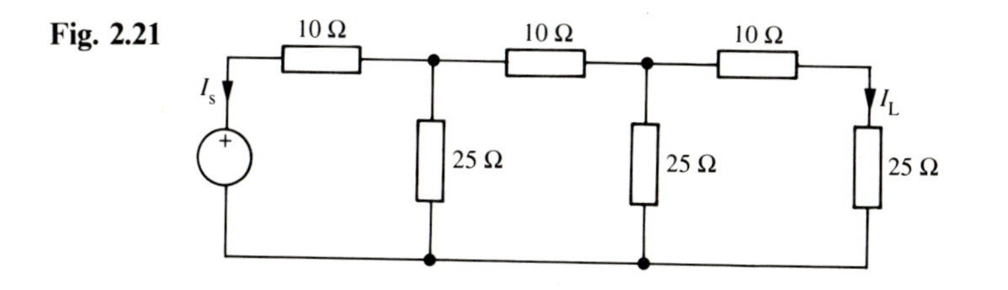

Fig. 2.22

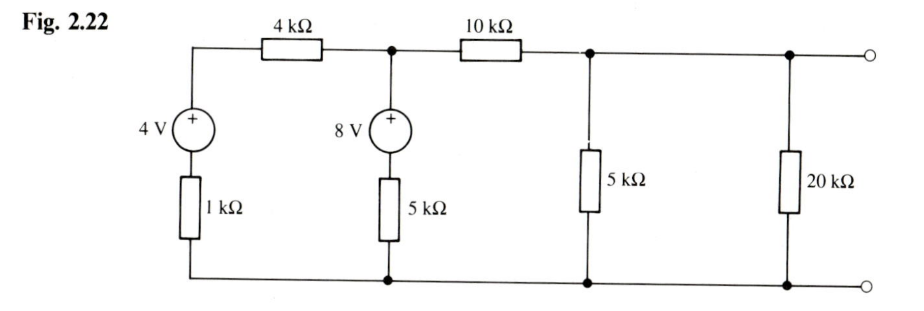

2.11 Two batteries A and B act in the same circuit, as shown in Fig. 2.23. What power does each battery deliver?

Fig. 2.23

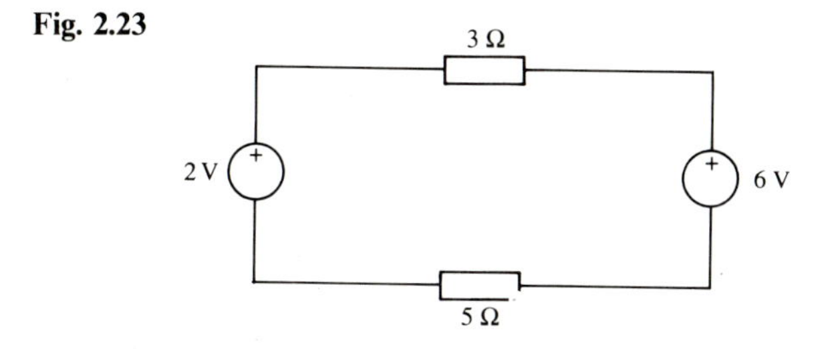

Alternating currents

3

In order to provide a basis for a thorough understanding of alternating current (a.c.) circuits it is useful first to study the relations between transient time-dependent currents and voltages in circuits containing inductors and capacitors as well as resistors. This study (Section 3.1) should lead to a clear understanding of the physical reasons why the sinusoidally varying current and voltage in an a.c. circuit are generally not 'in step' with one another. The subsequent sections of this chapter describe the basis of a.c. circuit theory, using complex numbers, and apply the theory to a number of basic types of circuit.

3.1 Transient currents in series *RC* and *RL* circuits

The applicability of Kirchhoff's laws to circuits containing inductors and capacitors as well as resistors, and in which the voltages and currents are time-dependent, will now be examined.

The idea of a current flowing continuously in a closed circuit, with no local accumulations of charge, will be retained (KCL). That is, for any element of the circuit, the rate at which charge is carried into the element is equal to the rate at which charge is leaving it; this concept is very important. In the case of a current flowing 'through' a capacitor, for example, charges accumulate on the two plates but these two accumulations are of opposite sign and so the statement of the previous sentence remains valid.

KVL is related to the principle of the conservation of energy: the rate of working of the source of e.m.f. is equal to the rate at which energy is being dissipated in the circuit.

So it will be assumed that Kirchhoff's laws can be applied to the instantaneous values of the currents and voltages in a circuit.

For a circuit containing resistive elements only, the drainage of energy from the source of e.m.f. occurs through transfer to the moving electrons constituting the current, as outlined in Sections 1.2 and 1.3. However, in a complete circuit containing inductive and/or capacitive elements there is potential energy associated with the electric and magnetic fields in these circuit elements. Hence:

$$\text{rate of supply of energy by the source} = \text{rate of dissipation of energy in resistances} + \text{rate of increase of energy stored in inductances and capacitances.}$$

A particular electron making a complete traversal of a closed loop (or 'mesh') can have no net gain or loss of energy. This energy balance is exemplified by

e.m.f. = sum of all the potential differences around the loop,

where the potential differences can be due to

(a) current (I) flowing through resistances (R): IR
(b) induced e.m.f.s in inductances (L): $L\,\mathrm{d}I/\mathrm{d}t$ (eqn (1.51))
(c) charge accumulations (Q) on capacitances (C): Q/C (eqn (1.25))

If, in a particular closed loop where the current is I, there is one resistance R, one inductor L, and one capacitor C, then

$$E = RI + L\frac{\mathrm{d}I}{\mathrm{d}t} + \frac{Q}{C}. \tag{3.1}$$

Consider the series RC circuit shown in Fig. 3.1(a), in which the switch S is closed at a time (t) that can be designated, for convenience, as $t = 0$. In this case eqn (3.1) becomes

$$E = RI + \frac{Q}{C} \quad \text{or} \quad E = R\frac{\mathrm{d}Q}{\mathrm{d}t} + \frac{Q}{C}, \tag{3.2}$$

since, for current I flowing through a capacitance C, $I = \mathrm{d}Q/\mathrm{d}t$ (or, $Q = \int I\,\mathrm{d}t$). Equation (3.2) can be rewritten as

$$EC - Q = CR\frac{\mathrm{d}Q}{\mathrm{d}t} \quad \text{or} \quad \frac{1}{CR}\int \mathrm{d}t = \int \frac{\mathrm{d}Q}{(EC - Q)}.$$

Now $\mathrm{d}(EC - Q) = -\mathrm{d}Q$, since E, C are constants, and so

$$\frac{1}{CR}\int \mathrm{d}t = -\int \frac{\mathrm{d}(EC - Q)}{(EC - Q)}.$$

The integral on the right-hand side is now in the standard form $\int \mathrm{d}x/x = \ln x$, so it follows that

$$\frac{t}{CR} = -\ln(EC - Q) + (\text{a 'constant of integration'}),$$

or

$$EC - Q = (\text{a constant}) \times \mathrm{e}^{-t/CR}. \tag{3.3}$$

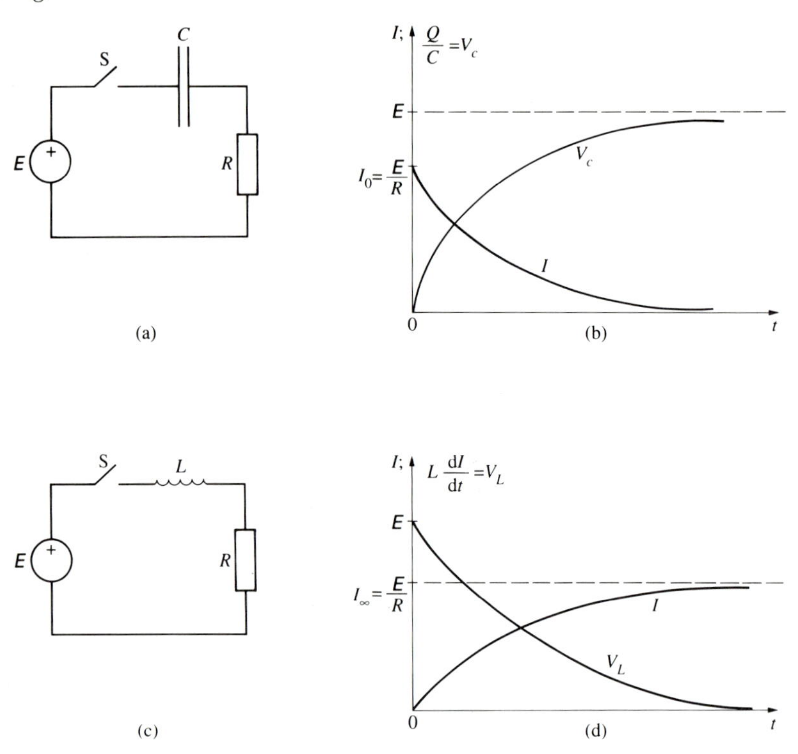

Fig. 3.1 The time-dependence of the voltage V_c across the capacitor C, and the current I, on closing the switch S in the circuit (a) are illustrated in (b). The time-dependence of the voltage across the inductor L and the current I in the circuit (c) are illustrated in (d).

The constant of integration is determined by a boundary condition, namely that the charge Q on the capacitance is zero at the time ($t = 0$) when the switch S is closed and the charging of the capacitance begins. Setting $Q = 0$ at $t = 0$ in eqn (3.3) means that the constant takes the value EC, and so

$$Q = EC(1 - e^{-t/CR}). \tag{3.4}$$

Also,

$$I = \frac{dQ}{dt} = \frac{E}{R} e^{-t/CR} \quad \text{or} \quad I = I_0 e^{-t/CR}. \tag{3.5}$$

Here I_0, the current at $t = 0$, is equal to the steady current that would flow if the capacitance were short-circuited (bypassed). The quantity CR, which has the dimensions of time, is called the **time constant** of the circuit and is the time taken for the current to decrease to a fraction $1/e$ of the initial current. The time-dependence of I, and of the capacitor voltage V_c, are sketched in Fig. 3.1(b).

For the series *RL* circuit of Fig. 3.1(c), eqn (3.1) becomes

$$E = RI + L\frac{\mathrm{d}I}{\mathrm{d}t} \quad \text{or} \quad \frac{E}{R} - I = \frac{L}{R}\cdot\frac{\mathrm{d}I}{\mathrm{d}t}, \tag{3.6}$$

and it follows that

$$\frac{R}{L}\int \mathrm{d}t = \int \frac{\mathrm{d}I}{(E/R - I)}.$$

From analogous considerations to those for the *RC* circuit,

$$\frac{Rt}{L} = -\ln(E/R - I) + (\text{a constant})$$

or

$$\frac{E}{R} - I = (\text{a constant}) \times \mathrm{e}^{-Rt/L}.$$

Since $I = 0$ at $t = 0$, then the constant must be equal to E/R, and so

$$I = \frac{E}{R}(1 - \mathrm{e}^{-Rt/L})$$

or

$$I = I_\infty(1 - \mathrm{e}^{-Rt/L}), \tag{3.7}$$

where I_∞ $(=E/R)$ is the current that would be established after an infinite lapse of time. The quantity L/R is called the time constant of the circuit, i.e. the time taken for the value of the quantiy $(I_\infty - I)$ to decrease to I_∞/e.

From Fig. 3.1(b) it is now clear that in the case of a time-dependent current (I) through a capacitor, the voltage (V_c) across the capacitor is 'out of step' with the current: the voltage rises as the current falls. In the case of an inductor (see Fig. 3.1(d)) the converse occurs: the voltage (V_L) across the inductor falls as the current rises.

Similarly, sinusoidal currents and voltages are 'out of step' with each other in capacitors and inductors. As will be seen in Section 3.2, there is said to be a **phase difference** between the voltage and the current.

A useful fact to remember from a practical point of view is that in an *RC* circuit, $I/I_0 = 0.007$ at $t = 5RC$, i.e. the current has fallen to what could be considered to be a negligible level in many circumstances (but not necessarily all circumstances!). A similar argument applies to RL circuits, i.e. the current rises to within 0.7 per cent of I_∞ in a time equal to $5L/R$.

It is left as an exercise (Problem 3.1) to show that if a capacitor (C) is charged to a voltage (E) and then discharged through a series resistance (R), then the charge (Q) on the capacitor is given by

$$Q = Q_0 \, e^{-t/CR},$$

where Q_0 ($=CE$) is its initial charge.

A potentially hazardous situation may arise in circuits where large-value capacitors are charged to a high voltage, e.g. in a high-voltage power supply unit. If, after such a supply is switched off, the series resistance through which the capacitors are able to discharge has a large value (possibly the input resistance of the circuit or device to which the supply is connected), then the voltages on the capacitors will decay with a very large time constant and may remain dangerously high for a long time. For instance, in an extreme case the series resistance may be that of a leakage path between the terminals of the capacitor (leakage via a thin film of dirt and/or moisture on the surface of insulators); this may be $10^{10}\ \Omega$ or higher, so, if $C = 100\ \mu\text{F}$, then $CR = 10^6$ s and, if $E = 1$ kV, the voltage remaining between the plates of the capacitor after approximately 12 days would be $1000/\text{e} \approx 370$ V. Provision should always be made for the safe discharge of high-value capacitors that have been charged to high voltages.

Example 3.1 Consider the circuit of Fig. 3.2. What is the time constant for the charging of the capacitor C when the switch S is closed?

By applying KVL,

$$(I_c + I_R)R_0 + I_R R = E.$$

Also, $I_c = \text{d}Q/\text{d}t$ and $I_R R = Q/C$, and so, after substitution,

$$R_0 \frac{\text{d}Q}{\text{d}t} + \frac{Q}{CR}(R + R_0) = E,$$

whence

$$-\frac{CRR_0}{(R + R_0)} \frac{\text{d}\left\{\dfrac{E}{R_0} - \dfrac{Q(R + R_0)}{CRR_0}\right\}}{\left\{\dfrac{E}{R_0} - \dfrac{Q(R + R_0)}{CRR_0}\right\}} = \text{d}t.$$

This may look complicated, but the consequent integral is in exactly the same general mathematical form as that which led to eqn (3.3). So, on

Fig. 3.2

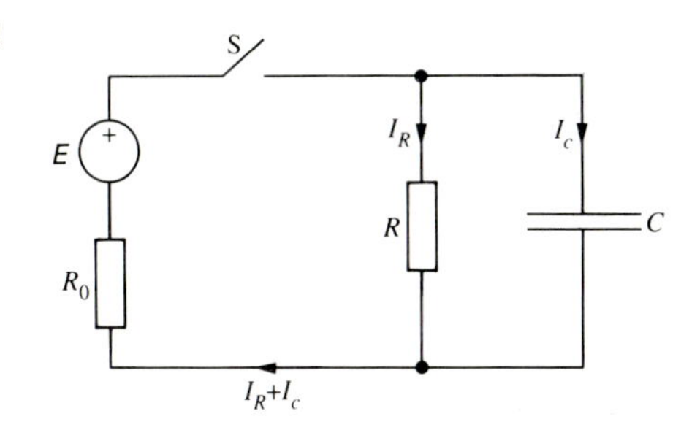

integrating,

$$\ln\left\{\frac{E}{R_0} - \frac{Q(R + R_0)}{CRR_0}\right\} = -\frac{(R + R_0)}{CRR_0}t + A,$$

where A is a constant. Hence

$$\frac{E}{R_0} - \frac{Q(R + R_0)}{CRR_0} = A\exp\left\{-\frac{(R + R_0)t}{CRR_0}\right\}.$$

(Note that e^x is often written as $\exp x$ for convenience.)

Boundary condition: $Q = 0$ at $t = 0$. Hence $A = E/R_0$.

So, finally,

$$Q = \frac{CR}{(R + R_0)}\cdot E\cdot\left[1 - \exp\left\{-\frac{(R + R_0)t}{CRR_0}\right\}\right].$$

Hence, from the exponential term, the time constant is $\dfrac{CRR_0}{(R + R_0)}$. □

If an *RL* circuit is 'broken', by opening a switch for example, then the induced e.m.f. in the inductor tries to sustain the current that was flowing initially. The voltage may be large enough to generate a spark across the contacts of the switch (or, if the inductor consists of a tightly-wound coil of insulated wire, the large induced e.m.f. may cause a breakdown of the insulation). The breakdown voltage for air is about $30\,\mathrm{kV\,cm^{-1}}$. For instance, if a current of 10 A in the coil of an electromagnet of inductance 5 H were to be reduced to zero in a time of 1 ms, the induced e.m.f. would be of magniude $10 \times (5/10^{-3}) = 50\,\mathrm{kV}$! Thus the power supplies to large inductive loads are designed so that the current cannot be altered rapidly, and additionally it is common practice to connect a rectifier (such as a silicon diode) in parallel with the coil of the inductor. This rectifier is

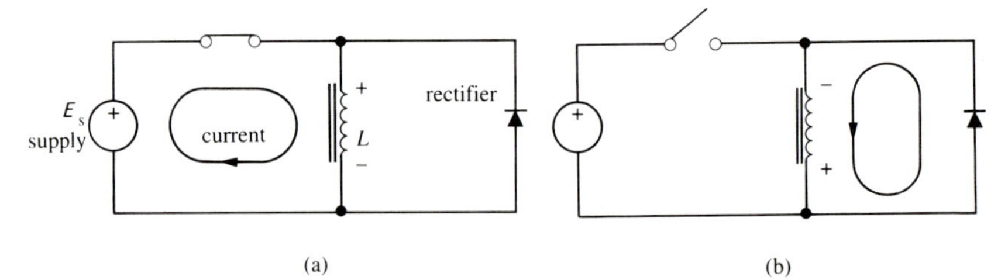

Fig. 3.3 The use of a rectifier to protect an inductive load from large induced e.m.f.s (a) The normal situation. (b) If the circuit is broken, then the current flow due to the induced e.m.f. in the inductor is through the rectifier (note that the polarity of the induced e.m.f. is in such a sense as to forward-bias the tectifier).

connected in the sense that it presents a very high resistance (reverse-biased) to the normal flow of current from the supply (see Fig. 3.3) but a very low resistance (forward-biased) to the current that would be generated by the induced e.m.f. that would arise if the circuit were to be sharply interrupted. Because of this very low resistance there is only a small voltage across the rectifier despite the (transiently) large current that would flow. Hence the resulting maximum voltage across the open contacts of the switch would be little more than the supply voltage and damaging sparking would be unlikely to occur.

3.2 The graphical and mathematical representation of alternating currents and voltages

3.2.1 Graphical representation

Imagine two plane rectangular coils, each consisting of a single turn of wire of area A, situated in a magnetic field B. Further, assume that each coil is rotating with the same constant angular velocity ω radians per second (rad s^{-1}) about a common axis as sketched in Fig. 3.4(a) and that each coil has a resistor R connected to its terminals via a suitable commutator.

At an instant of time t, the magnetic fluxes ϕ_1 and ϕ_2 linked with coils 1 and 2 are given by $\phi_1 = BA \sin \omega t$ and $\phi_2 = BA \sin(\omega t + \theta)$ respectively; here θ is the angle (in radians) between the planes of the two coils. The magnitudes of the respective induced e.m.f.s are (see eqn (1.50))

$$E_1 = \omega BA \cos \omega t \qquad \text{and} \qquad E_2 = \omega BA \cos(\omega t + \theta).$$

The resultant sinusoidal currents are

$$I_1 = \hat{I} \cos \omega t \qquad \text{and} \qquad I_2 = \hat{I} \cos(\omega t + \theta), \tag{3.8}$$

where $\hat{I} = \omega BA/R$ is the **amplitude**, or **peak value**, of the current (see Fig. 3.4(c)). For the sense of rotation chosen for the illustration, coil 2 'leads' coil 1 by the angle θ.

a.c. circuit theory is concerned with the analysis and design of circuits where the currents and voltages have sinusoidal time-dependence.

The period of revolution T of a coil, and hence the period T of the induced sinusoidal current, equals $2\pi/\omega$.

The number of periods (or 'cycles') per second is called the **frequency** (f) of the alternating current and is equal to $1/T$: $f \equiv 1/T = \omega/2\pi$. The SI unit of frequency is the **hertz** (Hz); 1 Hz corresponds to one cycle per second.

Warning: It should be noted that the term 'frequency' is often rather loosely used to indicate both frequency as just defined and also 'angular' frequency ω. The units of angular frequency are radians per second.

Although a.c. generators on an engineering scale are much more complicated in design and construction than the simple illustration of Fig. 3.4, their operation depends on the same basic principles: a mechanical source of power causes the coils to rotate in a magnetic field. In the United Kingdom the speed of rotation determines the frequency (50 Hz) of the national 'mains' supply; in the USA the frequency is 60 Hz.

An expression in the form $I = \hat{I}\cos(\omega t + \theta)$ is called a **phasor**, i.e. a quantity that varies sinusoidally with time and for which the amplitude, (angular) frequency and phase angle are specified. It is important to note that the phase angle θ is a relative quantity; in this case the phase of the current in coil 2 is specified relative to that of the current in coil 1 (whose phase is taken as zero, for convenience).

I_1 and I_2 can be represented on a 'circle' diagram as shown in Fig. 3.4(d) where, for generality, they are now taken to have different amplitudes (owing to the coils having different areas, say). The lines OM, ON, which are both taken to rotate in an anticlockwise sense with angular velocity ω, have lengths proportional to $\hat{I}_1$ and $\hat{I}_2$ (i.e. proportional to the *amplitude*/*magnitude*/'*peak value*'). The *instantaneous* values of I_1 and I_2 of eqns (3.8) are given by the projections of OM and ON, respectively, on to the 'horizontal' diameter, or base line X′OX.

If the two coils were connected in series to an external circuit of resistance R, then the total current would be the sum of I_1 and I_2 of eqns

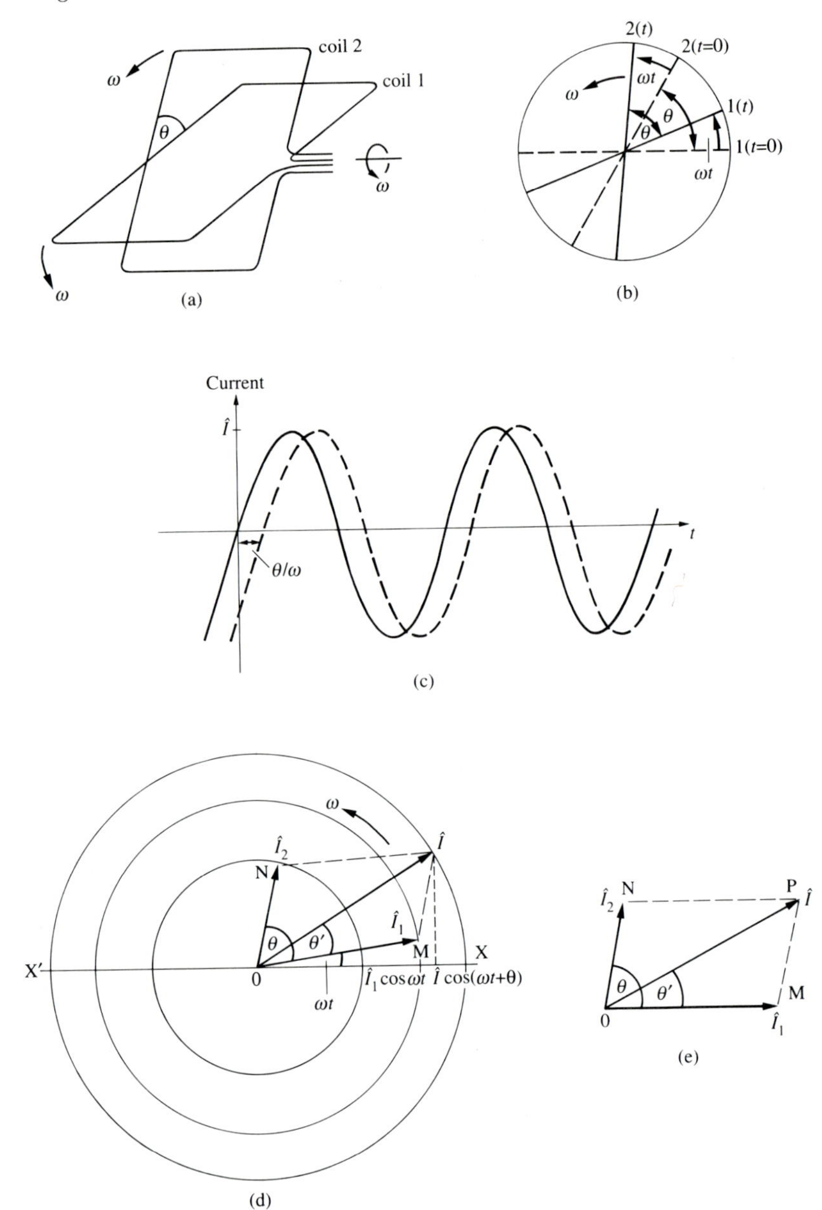

Fig. 3.4 The addition of two phasors of the same frequency. (a) Two coaxial rotating coils. (b) 'End' view of the coils at time $t = 0$ and at time t. (c) The time-dependent induced currents. (d) The phasor diagram. (e) The 'frozen' phasor diagram.

(3.8). How is the sum (or, more strictly, the resultant) of two phasors obtained? The mathematical expression in answer to this question is of crucial importance to the theory of a.c. circuits.

Consider Fig. 3.4(d) and apply the 'cosine rule' of trigonometry to the triangle OMP:

$$\mathrm{OP}^2 = \mathrm{OM}^2 + \mathrm{MC}^2 - 2\cdot\mathrm{OM}\cdot\mathrm{MC}\cdot\cos(180^\circ - \theta)$$

or

$$\hat{I}^2 = \hat{I}_1^2 + \hat{I}_2^2 + 2\hat{I}_1\hat{I}_2 \cos\theta.$$

Since the resultant I is also a phasor of the same frequency, it can be written in the form

$$I = \hat{I}\cos(\omega t + \theta').$$

What is the expression for the phase angle θ' of the resultant phasor?

The expression can be obtained by trigonometry, or otherwise, but the result will simply be stated here. The resultant I of two phasors

$$I = \hat{I}_1 \cos\omega t + \hat{I}_2\cos(\omega t + \theta)$$

can be written as a third phasor of the form

$$I = \hat{I}\cos(\omega t + \theta'),$$

where

$$\hat{I}^2 = \hat{I}_1^2 + \hat{I}_2^2 + 2\hat{I}_1\hat{I}_2\cos\theta \tag{3.9}$$

and

$$\tan\theta' = (\hat{I}_2 \sin\theta)/(\hat{I}_1 + \hat{I}_2\cos\theta). \tag{3.10}$$

So it can be seen that the amplitude (magnitude or peak value) of the resultant is found by completing the 'parallelogram of vectors' OMPN in Fig. 3.4(d), the resultant phasor being given, in magnitude and 'direction' (i.e. phase), by the diagonal OP. Remember that the instantaneous value of the resultant phasor is given by the projection of the rotating line OP on to the reference line X′OX.

Since all of the current 'vectors' are rotating at the same angular velocity, the diagram can be 'frozen' as shown in Fig. 3.4(e). Only differences of phase are of interest, so any direction on the diagram can be chosen for convenience to be the base line or reference phase. Usually the phase differences are specified relative to a reference phasor chosen for convenience: I_1 in this illustration (or a reference sinusoidal voltage if it is voltages that are being combined).

Example 3.2 Find the phasor resultant of the two voltage phasors

$$V_1 = 6\cos\omega t \qquad \text{and} \qquad V_2 = 8\cos(\omega t + 60^\circ).$$

Using eqn (3.9), with $\theta = 60^\circ$, the resultant voltage V is given by

$$V^2 = 36 + 64 + 2 \times 6 \times 8 \times \cos 60^\circ$$

or

$$\hat{V} = 12.2\ V.$$

From eqn (3.10) the phase angle of V is given by

$$\tan \theta' = (8 \sin 60^\circ)/(6 + 8 \cos 60^\circ).$$

Hence

$$\tan \theta' = 0.693 \qquad \text{and} \qquad \theta' = 34.7^\circ. \qquad \square$$

The fact that phasors can be represented by vectors on a two-dimensional plane as shown in Fig. 3.4(e) leads naturally to the representation of phasors by 'complex' quantities (in the mathematical sense of complex numbers, that is); this will be described in the next section. While phasor diagrams such as Fig. 3.4(d) and (e) are not without interest and utility, their use would be tedious in many situations, particularly where it is required to combine more than two phasors. The algebra of complex numbers can be exploited to enable such problems to be solved systematically and in a relatively simple way.

3.2.2 The use of complex numbers

Consider an alternating current $I = \hat{I} \cos \omega t$ flowing through an inductance L; the induced (or 'back') e.m.f. V (see Section 1.7) is given by

$$V = L \frac{\mathrm{d}I}{\mathrm{d}t}$$

or

$$V = -\omega L \hat{I} \sin \omega t.$$

So

$$V = (\omega L \hat{I}) \cos\left(\omega t + \frac{\pi}{2}\right), \tag{3.11}$$

i.e. the induced e.m.f. has amplitude $(\omega L \hat{I})$ and a phase difference of $+\pi/2$ with respect to the reference phasor (in this case the current I); V 'leads' I by $\pi/2$ radians.

If a current $I = \hat{I} \cos \omega t$ is flowing through a resistance R, then the voltage across the resistance is

$$V = RI \qquad \text{or} \qquad V = R\hat{I} \cos \omega t, \tag{3.12}$$

i.e. V is 'in phase' with I (or, in Fig. 3.4(d) and (e), $\theta = 0^\circ$). Since resistance

$R \equiv V/I,$

$$R = \frac{R\hat{I}\cos\omega t}{\hat{I}\cos\omega t}$$

$$= \frac{R\hat{I}}{\hat{I}} = \frac{\text{amplitude of the voltage across the resistance}}{\text{amplitude of the current through the resistance}}.$$

The question to be answered now is: in the case of an inductance in an a.c. circuit, what is the quantity analogous to the resistance of a resistor?

If a current $I = \hat{I}\cos\omega t$ is flowing through an inductance L, then, using eqn (3.11), the voltage across the inductance is given by

$$\frac{V}{I} = \frac{\omega L\hat{I}\cos(\omega t + \pi/2)}{\hat{I}\cos\omega t}.$$

The problem is: how does one handle a ratio such as $\cos(\omega t + \pi/2)/\cos\omega t$, that is, the ratio of two sinusoidal quantities having a phase difference between them?

It can be seen, in the case of an inductance, that the ratio

$$\frac{\hat{V}}{\hat{I}} = \frac{\omega L\hat{I}}{\hat{I}} = \omega L,$$

but *there remains the question of the phase shift of* $\pi/2$.

A brief introduction to the properties of complex numbers will reveal their utility in this context.

Mathematically speaking, the square of any **real number** x is positive, whether the number itself is positive or negative, i.e.

$$(\pm x)^2 \text{ is always positive (and real).}$$

It follows that the square root of a negative real number cannot be real and, in fact, it is termed an **imaginary number**. In particular $\sqrt{-1}$ is denoted by j (i is used in some contexts, but j will always be used here):

$$\mathrm{j} \equiv \sqrt{-1}.$$

The corollary is that $$j^2 = -1.$$

A **complex number** (w say) has a real and an imaginary part, in general:

(3.13)
$$w = x + \mathrm{j}y$$

complex — real part $\mathrm{Re}(w)$ — imaginary part $\mathrm{Im}(w)$.

(Note that both x and y are themselves real numbers.)

It is conveneint to plot the real and the imaginary parts of a complex number on two axes at right angles (see Fig. 3.5(a)) so that they can be considered to be the two components of a vector OP on the **complex plane**. This is the so-called **Cartesian representation** of a complex number and this type of diagram is known as an **Argand diagram**. The real part and the imaginary part of a complex number can be either positive or negative, so the vector representing a complex number may lie in any one of the four quadrants of the complex plane:

Real part	Imaginary part	Quadrant
+	+	1st
−	+	2nd
−	−	3rd
+	−	4th

(See Problem 3.1).

By applying Pythagoras' theorem to triangle OPM in Fig. 3.5(a),

(3.14)
$$w^2 = x^2 + y^2.$$

The **modulus of the complex number** w is denoted by $|w|$, where

(3.15)
$$|w|^2 = (x^2 + y^2),$$

and the **argument of the complex number** w (i.e. the angle ϕ) by arg w, where

(3.16)
$$\arg w = \tan^{-1}(y/x).$$

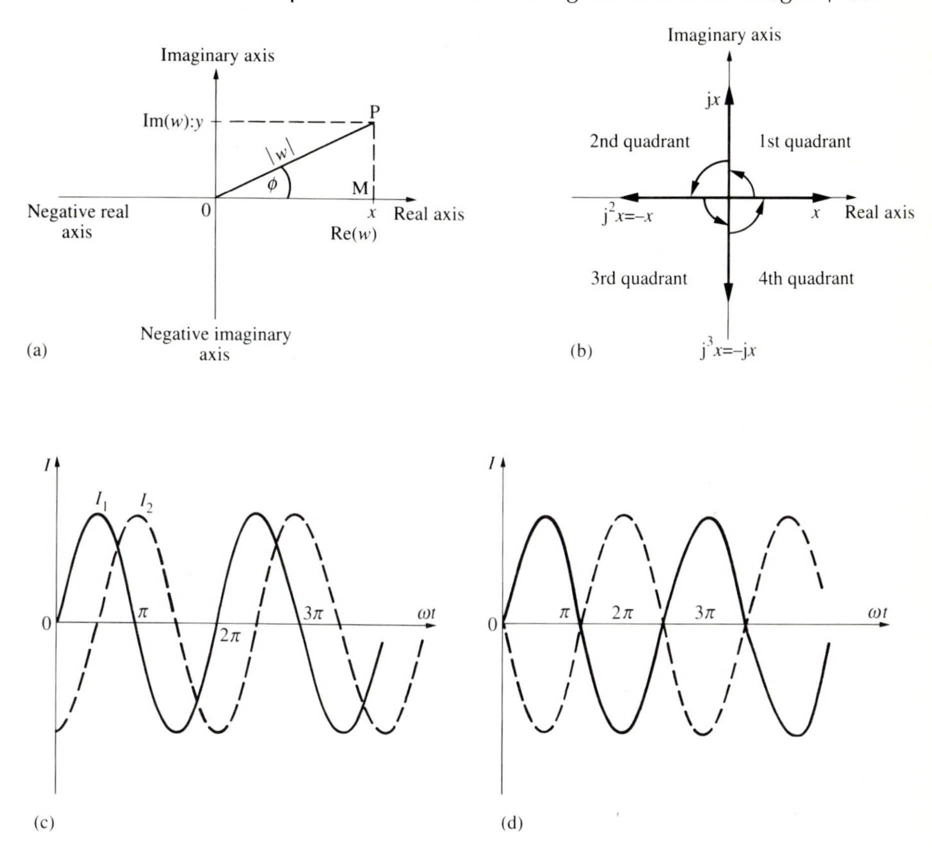

Fig. 3.5 (a) Argand diagram: the representation of a complex number w as a point on the two-dimensional 'complex plane'. (b) j as a rotation operator. (c) Two sinusoidal currents in quadrature: I_2 lags I_2 by $\pi/2$ radians. (d) Two phasors in **anti-phase**; the phase difference between V_1 and V_2 is $\pm\pi$ radians (i.e. $V_1 = -V_2$).

An alternative view of the role of j is as an operator. Consider Fig. 3.5(b). A real number x is represented by a vector of magnitude x pointing in the positive sense of the real axis. As $(j^2)x = -x$, j^2 can be thought of as an operator which rotates x through 180° (or π radians) in the complex plane, as its effect is to reverse the sign of x. On this basis (j^4) is the 'identity operator', since $(j^4) = (j^2)(j^2) = 1$. Hence (j^4) can be interpreted as an operator that rotates x through 360° (or 2π radians).

To summarize:

$$j^2 = j \times j \rightarrow (90^\circ + 90^\circ) \text{ rotation}$$

$$j^4 = j \times j \times j \times j \rightarrow 90^\circ + 90^\circ + 90^\circ + 90^\circ) \text{ rotation.}$$

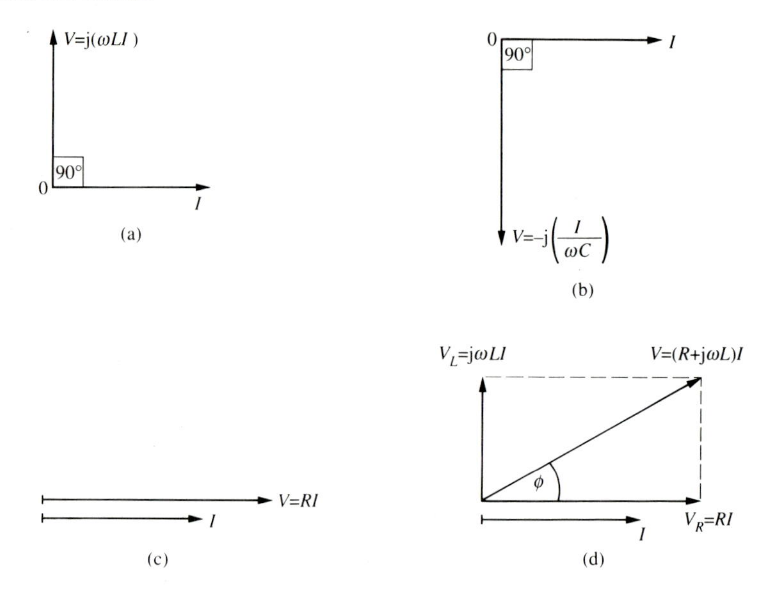

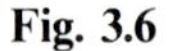
Fig. 3.6

By interpolation it is asserted that

$$\mathrm{j} \rightarrow 90^\circ \text{ rotation}$$

$$\mathrm{j}^3 = \mathrm{j} \times \mathrm{j} \times \mathrm{j} \rightarrow (90^\circ + 90^\circ + 90^\circ) \text{ rotation}$$
$$(270^\circ)$$

$$\text{NB} \quad 270^\circ = 3\pi/2 \text{ radians} \qquad \text{or } -\pi/2 \text{ radians}$$

So, using 'j-notation', eqn (3.11) can be rewritten as

$$V = \mathrm{j}(\omega L I). \tag{3.17}$$

This equation is to be interpreted as saying that the voltage V is a phasor of amplitude $\omega L\hat{I}$ and that V leads I by a phase angle of $\pi/2$ radians (see Fig. 3.6(a)).

Note that phasors that are $\pm\pi/2$ radians out of phase with one another are said to be **in quadrature**.

For a capacitance C,

$$V = \frac{Q}{C} \qquad \text{or} \qquad V = \frac{1}{C}\int I \, \mathrm{d}t.$$

If, as before, $I = \hat{I} \cos \omega t$, then

$$V = \frac{\hat{I}}{\omega C} \sin \omega t$$

or

$$V = \frac{\hat{I}}{\omega C} \cos(\omega t - \pi/2), \tag{3.18}$$

i.e. the voltage across the capacitance C has amplitude $(\hat{I}/\omega C)$ and 'lags' the current I by $\pi/2$ radians (or, equivalently, 'leads' by $3\pi/2$ radians). In j-notation (see Fig. 3.6(b)),

$$V = -\mathrm{j}\left(\frac{I}{\omega C}\right). \tag{3.19}$$

Equations (3.17) and (3.19) can be written in the general form

$$V = \mathrm{j}XI \tag{3.20}$$

where

$$X = \omega L \qquad \text{(inductance)}$$

$$X = -\frac{1}{\omega C} \qquad \text{(capacitance).} \tag{3.21}$$

X is called the **reactance** and L, C are called **reactive circuit elements** (in contrast to resistive elements denoted by R).

For a circuit element consisting of a resistance R and an inductance L in series, and with a current $I = \hat{I} \cos \omega t$ flowing through it, the voltage V across it is the *vector resultant* of the voltage across R and the voltage across L (see Fig. 3.6(c) and (d)). Using eqn (3.17), together with $V = RI$ for a resistance,

$$V = RI + \mathrm{j}(\omega LI), \tag{3.22}$$

Note that eqn (3.22) is in the same form as eqn (3.13), so, following eqn (3.15),

$$|V|^2 = (RI)^2 + (\omega LI)^2$$

and

$$\phi = \arg V = \tan^{-1}\left(\frac{\omega LI}{RI}\right).$$

Since in the vector diagrams, Fig. 3.4(d) and (e), the lengths of the vectors are proportional to the amplitudes (peak values) of the respective phasors, the previous two equations can be interpreted as

$$\hat{V}^2 = \hat{I}^2 R^2 + \omega^2 L^2 \hat{I}^2 \tag{3.23}$$

and

$$\phi = \tan^{-1}\left(\frac{\omega L}{R}\right). \tag{3.24}$$

It is very important to notice that eqns (3.23) and (3.24) are of the same form as eqns (3.9) and (3.10) with θ taken as 90°. This justifies the use of the j-notation.

Two very important relationships that should be noted at this stage are:

(i) *De Moivre's theorem.* A special case of this theorem that is of special interest is

$$e^{j\phi} = \cos\phi + j\sin\phi \tag{3.25}$$

(Memorize this relationship!).

(ii) *The polar form of a complex number.* From Fig. 3.5(a), the real and imaginary parts of the complex number w $(= x + jy)$ are

$$x = |w|\cos\phi \qquad \text{and} \qquad y = |w|\sin\phi. \tag{3.26}$$

So, as a vector in the complex plane, $w = |w|\cos\phi + j|w|\sin\phi$ or

$$w = |w|\,e^{j\phi}. \tag{3.27}$$

This is the so-called **polar form** of a complex number: the polar coordinates of the point P in Fig. 3.5(a) are the length ($|w|$) of the vector OP together with the polar angle ϕ.

As corroboration it can be seen from eqns (3.26) that, since

$$\sin^2\phi + \cos^2\phi = 1,$$

it follows that

$$x^2 + y^2 = |w|^2,$$

which is as it should be.

The product and the quotient of two complex numbers w_1, w_2 can be expressed simply using their polar forms. Writing

$$w_1 = |w_1|\,e^{j\phi_1}, \qquad w_2 = |w_2|\,e^{j\phi_2},$$

it follows that

$$w_1 w_2 = |w_1| \cdot |w_2| \, e^{j(\phi_1 + \phi_2)} \tag{3.28}$$

$$\frac{w_1}{w_2} = \frac{|w_1|}{|w_2|} \cdot e^{j(\phi_1 - \phi_2)} \tag{3.29}$$

The particular relevance of this is that a phasor such as $\cos(\omega t + \phi_1)$ can be expressed as the real part of $e^{j(\omega t + \phi_1)}$ (written $\text{Re}\, e^{j(\omega t + \phi_1)}$). Another phasor of interest, $\cos(\omega t + \phi_2)$ say, can be written as $\text{Re}\, e^{j(\omega t + \phi_2)}$. So the rules for manipulating complex numbers can be applied to the phasors arising in a.c. circuit theory. The end result of an analysis to determine a voltage or a current will be a complex number in polar form that will yield the actual voltage or current by taking its real part. The application of this idea to circuit analysis is described in Section 3.3.

Note that $\sin(\omega t + \phi)$ is the imaginary part of $e^{j(\omega t + \phi)}$, written $\text{Im}\, e^{j(\omega t + \phi)}$.

3.3 Impedance

Consider again the case of a phasor current $I = \hat{I} \cos \omega t$ flowing through a circuit element consisting of a resistance R and an inductance L in series. The ratio of the phasor instantaneous voltage across the element to the instantaneous current through it is called the **impedance** (Z) of the element:

$$Z \equiv \frac{V}{I} \tag{3.30}$$

Now it has just been seen in Section 3.2 that for such a circuit element, V is shifted in phase relative to I and this phase shift is denoted by ϕ. So

$$Z = \frac{\hat{V} e^{j(\omega t + \phi)}}{\hat{I} e^{j\omega t}}$$

or, using eqn (3.29),

$$Z = \frac{\hat{V} e^{j\phi}}{\hat{I}} \tag{3.31}$$

This tells us that Z is in fact a complex number (here expressed in polar form: see eqn (3.27)). The corresponding cartesian form for Z is $Z = (x + \mathrm{j}y)$, where

$$(x^2 + y^2) = |Z|^2 = \frac{\hat{V}^2}{\hat{I}^2}$$

and

$$\tan^{-1}(y/x) = \arg Z = \phi.$$

From a comparison of these two relationships with eqns (3.23) and (3.24) it can be seen that, for a circuit element consisting of a resistance R in series with an inductance L, the real part of the complex impedance is identified with R and the imaginary part with L:

$$Z = R + \mathrm{j}\omega L$$

(3.32) or

$$Z = R + \mathrm{j}X_L.$$

From the discussion leading to eqn (3.21) it follows that for a circuit element consisting of a resistance R in series with a capacitance C, the impedance is

$$Z = R + \mathrm{j}X_C$$

(3.33) or

$$Z = R - \frac{\mathrm{j}}{\omega C}.$$

Summary

For a circuit element of impedance Z, the phasor voltage across it and the phasor current I through it are related by

$$V = ZI, \qquad \text{where } Z = R + \mathrm{j}X.$$

The significance of these relationships is

$$\hat{V} = \hat{I}|Z|, \qquad \text{i.e. } \hat{V} = \hat{I}\sqrt{(R^2 + X^2)}$$

(amplitude of V)

and

$$\arg Z = \tan^{-1}(X/R), \qquad \text{i.e. } \arg Z = \tan^{-1}(\mathrm{Im}\, Z/\mathrm{Re}\, Z)$$

(phase of V *relative to* I).

In phasor form the instantaneous value of V is given by

$$V = \underbrace{\hat{I}\sqrt{(R^2 + X^2)}}_{\text{amplitude of } V} \cdot \begin{cases} \cos(\omega t + \tan^{-1}(X/R)) \\ \sin(\omega t + \tan^{-1}(X/R)). \end{cases}$$

NB 'cos' or 'sin' arises depending on whether the reference phasor (I in this instance) is expressed in terms of cos ωt or sin ωt.

It should be remembered that a.c. circuit theory as developed here applies strictly only to pure sinusoidal voltages and currents, i.e. the amplitude and the frequency do not change with time.

A good working knowledge of a.c. circuit theory is essential in a very wide range of topics in electronic and electrical engineering. Hence it is vital that the material presented in this chapter is clearly understood, including the in-text Examples: attempt all the problems given at the end of the chapter.

An important feature of the algebra of complex numbers is revealed by considering the inverse of the kind of situation considered so far. If there is a voltage V across a circuit element of impedance Z, what is the current I through the element?

$$Z = \frac{V}{I} \quad \text{and so} \quad I = \frac{V}{Z}.$$

Therefore

$$I = \frac{V}{(R + \mathrm{j}X)}. \tag{3.34}$$

I is obviously a complex number, but to convert it to the standard form (see eqn (3.13)) it is necessary to have a complex numerator and a real denominator. This is accomplished by a process known as **rationalisation**, which can be illustrated as follows.

The conjugate of a complex number is obtained by replacing j by $-$j wherever it occurs in the expression for the number; e.g. the conjugate of $(R + \mathrm{j}X)$ is $(R - \mathrm{j}X)$. Commonly, the conjugate of a complex number is denoted by an asterisk:

$$(R - \mathrm{j}X) \rightarrow Z^*.$$

If the numerator and the denominator of the right-hand side of eqn (3.34) are multiplied by $(R - \mathrm{j}X)$, it follows that

$$I = \frac{V(R - \mathrm{j}X)}{(R + \mathrm{j}X)(R - \mathrm{j}X)}$$

or

$$I = V \cdot \frac{(R - \mathrm{j}X)}{(R^2 + X^2)}, \tag{3.35}$$

where the denominator is now a real number. Hence

$$|I| = \hat{I} = \frac{\hat{V}}{(R^2 + X^2)} \cdot |R - \mathrm{j}X|.$$

Now

$$|R - \mathrm{j}X| = (R^2 + X^2)^{1/2}$$

and so

$$\hat{I} = \frac{\hat{V}}{(R^2 + X^2)^{1/2}}. \tag{3.36}$$

Further, from eqn (3.35),

$$\arg I = \tan^{-1}(-X/R) \tag{3.37}$$

(phase of I relative of V).

Note the negative sign in the expression for arg I. This is just saying the obvious, namely that if V leads I by phase angle ϕ (a situation represented by $V = Z \cdot I$), then, conversely, I lags V by a phase angle of the same magnitude (represented by $I = 1/Z \cdot V$).

As mentioned at the end of Section 3.2.2, phasors can be expressed as complex numbers in polar form.

Consider a current $I = \hat{I} \cos \omega t$ (i.e. $I = \hat{I}\, \mathrm{Re}\, \mathrm{e}^{\mathrm{j}\omega t}$) flowing through a circuit element of impedance Z, where $Z = R + \mathrm{j}X$. The voltage across the element is given by $V = ZI$, so

$$V = (R + \mathrm{j}X)I.$$

Fig. 3.7.

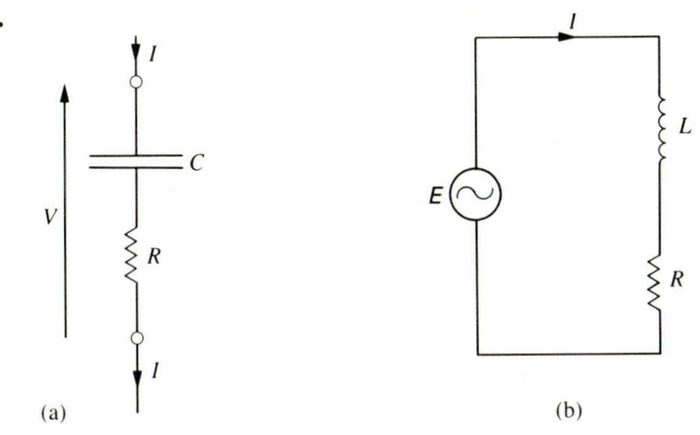

Writing I as $\hat{I}\exp(j\omega t)$ and Z as $|Z|\exp(j\phi)$, where $|Z| = \sqrt{(R^2 + X^2)}$ and $\phi = \tan^{-1}(X/R)$, it follows that

$$V = \hat{I}\sqrt{(R^2 + X^2)}\cdot\exp(j\omega t)\cdot\exp(j\phi)$$

or

$$V = \hat{I}\sqrt{(R^2 + X^2)}\cdot\exp(j\omega t + \phi) \qquad (*)$$

Since the reference phasor (the current I in this case) is the *real* part of $\hat{I}\exp(j\omega t)$, the *real* part of the above expression for V is taken to give the actual voltage in phasor form, i.e.

$$V = \underbrace{\hat{I}\sqrt{(R^2 + X^2)}}_{\text{amplitude of } V}\cdot\cos(\omega t + \phi).$$

It is important to note that, originally, the current could just as well have been represented by $I = \hat{I}\sin\omega t$ (i.e. $I = \hat{I}\,\mathrm{Im}\exp(j\omega t)$). In this event the subsequent analysis would be identical to that leading to equation (*) above, but to obtain the actual voltage it would be necessary to choose the *imaginary* part of the expression for V, namely

$$V = \hat{I}\sqrt{(R^2 + X^2)}\cdot\sin(\omega t + \phi).$$

Example 3.3 Using a cathode ray oscilloscope (CRO), the sinusoidal voltage (V) across a circuit element consisting of a capacitance (C) in series with a resistance (R) (Fig. 3.7(a)) is measured to have an amplitude of 5 V and a frequency of 10^4 radians per second (rad s^{-1}), i.e. 1.6 kHz approximately. If $C = 0.1\ \mu F$ (10^{-7} F) and $R = 3.3\ k\Omega$, calculate the amplitude and phase angle of the current I through the element.

Let $V = \hat{V} \cos \omega t$; then, using eqn (3.34),

$$I = \frac{V}{\left(R - \frac{\mathrm{j}}{\omega C}\right)}$$

or, on rationalizing,

$$I = V \cdot \frac{\left(R + \frac{\mathrm{j}}{\omega C}\right)}{\left(R^2 + \frac{1}{\omega^2 C^2}\right)}.$$

So

$$\hat{I} = \frac{\hat{V}}{\left(R^2 + \frac{1}{\omega^2 C^2}\right)^{1/2}}$$

and

$$\arg I = \tan^{-1}(1/\omega CR).$$

Since in this case $1/\omega C = 10^3\ \Omega$, it follows that

$$\hat{I} = 1.45\ \text{mA} \qquad \text{and} \qquad \arg I = \tan^{-1}(10^3/3300) = 16.9°.$$

So, in phasor form,

$$I = 1.45 \times 10^{-3} \cdot \cos(10^4 t + 16.9°)\ \text{A}.$$

Alternatively, the problem can be solved using the polar form of a complex number to express the phasors. In this case

$$V = \hat{V} \cdot \text{Re} \exp(\mathrm{j}\omega t).$$

Now

$$I = \frac{V}{Z} \qquad \text{or} \qquad I = \frac{V}{\left(R - \frac{\mathrm{j}}{\omega C}\right)}.$$

On rationalizing,

$$I = V \cdot \frac{\left(R + \frac{\mathrm{j}}{\omega C}\right)}{\left(R^2 + \frac{1}{\omega^2 C^2}\right)}.$$

Now

$$\left(R + \frac{\mathrm{j}}{\omega C}\right) = \sqrt{\left(R^2 + \frac{1}{\omega^2 C^2}\right)} \cdot \exp(\mathrm{j}\phi),$$

where $\phi = \tan^{-1}(1/\omega CR)$. Hence, since $V = \hat{V}\cos\omega t = \hat{V} \cdot \mathrm{Re}\exp(\mathrm{j}\omega t)$, then

$$I = \frac{\hat{V}}{\sqrt{\left(R^2 + \frac{1}{\omega^2 C^2}\right)}} \cdot \exp(\mathrm{j}\omega t) \cdot \exp(\mathrm{j}\phi)$$

or

$$I = \frac{\hat{V}}{\sqrt{\left(R^2 + \frac{1}{\omega^2 C^2}\right)}} \cdot \exp[\mathrm{j}(\omega t + \phi)].$$

Since, in this case, the reference phasor (the voltage V) was expressed in terms of the *real* part of $\exp(\mathrm{j}\omega t)$, the actual phasor current is obtained by taking the *real* part of the complex exponential function, i.e.

$$I = \frac{\hat{V}}{\sqrt{\left(R^2 + \frac{1}{\omega^2 C^2}\right)}} \cdot \mathrm{Re}\exp[\mathrm{j}(\omega t + \phi)]$$

or, finally,

$$I = \frac{\hat{V}}{\sqrt{\left(R^2 + \frac{1}{\omega^2 C^2}\right)}} \cdot \cos(\omega t + \phi),$$

which is identical to the result obtained above. □

Example 3.4 A source of e.m.f. (E), of negligible internal impedance, is connected in series with an inductor that can be represented by an inductance L ($=0.1$ H) in series with a resistance R ($=10\ \Omega$) (Fig. 3.7(b)). If the amplitude and frequency of the e.m.f. are 250 V and 50 Hz respectively, calculate the amplitude of the current (I) and the phase angle of I relative to that of the e.m.f.

Using eqn (3.32),

$$I = \frac{E}{Z} = \frac{E}{(R + \mathrm{j}\omega L)}.$$

Rationalizing,

$$I = E\frac{(R - j\omega L)}{(R^2 + \omega^2 L^2)}.$$

Therefore

$$\hat{I} = \hat{E}(R^2 + \omega^2 L^2)^{-1/2}$$

and

$$\arg I = \tan^{-1}(-\omega L/R)$$

(phase of I relative to E).

On substituting for $\hat{E}$, ω $(=2\pi f)$, R, and L, the values obtained are

$$\hat{I} = 7.6\ \text{A} \qquad \text{and} \qquad \arg I = -72^\circ.$$ □

It has been seen that the basic laws governing d.c. circuits are $V = RI$ for a (resistive) circuit element together with $\sum (E + V) = 0$ for a closed loop and $\sum I = 0$ at a node. Analogous relationships are valid in an a.c. circuit, namely $V = ZI$ coupled with $\sum (E + V) = 0$ and $\sum I = 0$, where E, V, I, and Z are now of course represented by complex functions. Thus, for impedances in series and in parallel, respectively,

$$Z = Z_1 + Z_2 + Z_3 + \cdots$$
$$\frac{1}{Z} = \frac{1}{Z_1} + \frac{1}{Z_2} + \frac{1}{Z_3} + \cdots \tag{3.38}$$

Example 3.5 Obtain an expression for the impedance (Z) of a capacitance (C) in parallel with a resistance (R) (Fig. 3.8).

Using eqn (3.38),

$$\frac{1}{Z} = \frac{1}{R} + \frac{1}{jX_C}.$$

From eqn (3.21),

$$X_C = -\frac{1}{\omega C} \qquad \text{or} \qquad jX_C = \frac{1}{j\omega C}. \tag{3.39}$$

So

$$\frac{1}{Z} = \frac{1}{R} + j\omega C$$

Therefore

$$Z = \frac{R}{(1 + j\omega CR)}.$$

Fig. 3.8

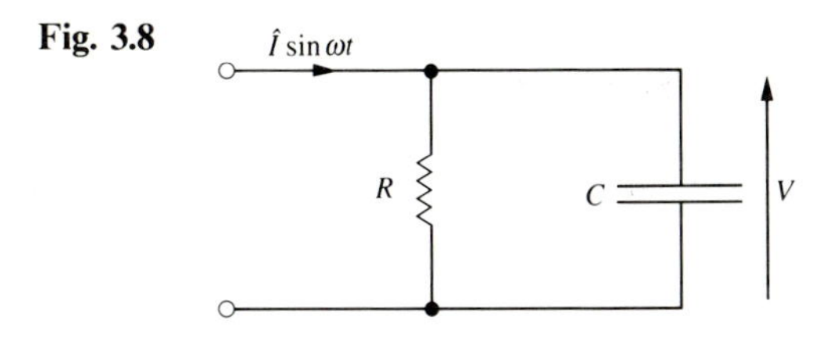

Rationalizing,

$$Z = \frac{R(1 - j\omega CR)}{(1 + \omega^2 C^2 R^2)}. \tag{3.40}$$

Hence

$$|Z| = R(1 + \omega^2 C^2 R^2)^{-1/2} \tag{3.41}$$

and, from eqn (3.40),

$$\arg Z = \tan^{-1}(-\omega CR). \tag{3.42}$$

If a current $I = \hat{I} \sin \omega t$ flows in the circuit, what is the voltage across C?

$$V = ZI \qquad \text{where } I = \hat{I} \,\text{Im} \exp(j\omega t).$$

Now the polar form of the impedance Z (eqn (3.40)) is

$$Z = R(1 + \omega^2 C^2 R^2)^{-1/2} \exp(j\phi) \qquad \text{where } \phi = \tan^{-1}(-\omega CR).$$

Therefore

$$\begin{aligned} V &= \hat{I}R(1 + \omega^2 C^2 R^2)^{-1/2} \exp(j\phi) \cdot \exp(j\omega t) \\ &= \hat{I}R(1 + \omega^2 C^2 R^2)^{-1/2} \exp[j(\omega t + \phi)]. \end{aligned}$$

In this example the phasor voltage is obtained by taking the *imaginary* part of $\exp[j(\omega t + \phi)]$, i.e.

$$V = \hat{I}R(1 + \omega^2 C^2 R^2)^{-1/2} \sin(\omega t + \phi).$$ □

Note

A 'trick of the trade' that may be found useful in practice is that for *two* circuit elements in parallel, the resultant impedance is given by

$$\text{resultant impedance} = \frac{\text{product of the two impedances}}{\text{sum of the two impedances}}$$

Applying this rule in Example 3.5 above,

$$Z = \frac{R(1/\mathrm{j}\omega C)}{\left(R + \dfrac{1}{\mathrm{j}\omega C}\right)},$$

which, on multiplying the numerator and the denominator by $\mathrm{j}\omega C$, again yields

$$Z = \frac{R}{(1 + \mathrm{j}\omega CR)}.$$

In this approach the reciprocals that would otherwise arise are avoided; it is a matter of taste as to which approach is chosen.

This is a good place to point out that if one of the two elements in parallel has a much smaller impedance than the other, then it will dominate the characteristics of the parallel combination; e.g. for two resistances R_1, R_2 in parallel,

$$\text{resultant } R = \frac{R_1 R_2}{(R_1 + R_2)}.$$

If $R_1 \gg R_2$, and the expression for R is written as

$$R = \frac{R_1 R_2}{R_1\left(1 + \dfrac{R_2}{R_1}\right)},$$

then

$$R \approx \frac{R_1 R_2}{R_1} = R_2,$$

since R_2/R_1 is negligible compared with 1.

The circuit theorems introduced in Chapter 2 for d.c. circuits (Kirchhoff's laws, the superposition theorem, Thévenin's and Norton's theorems) apply to the *instantaneous* values of currents, voltages and e.m.f.s in the a.c. case, provided that the impedances are linear. For a capacitor and an inductor, linearity implies that the values of C and L, respectively, are independent of the magnitude of the current passing through them.

3.4 Admittance

For resistances R_1, R_2, $R_3 \ldots$ in parallel, the resistance R of the combination is given by

$$\frac{1}{R} = \frac{1}{R_1} + \frac{1}{R_2} + \frac{1}{R_3} \cdots$$

The conductance G of a resistance R is defined through

$$G \equiv \frac{1}{R}$$

and so for the parallel combination,

$$G = G_1 + G_2 + G_3 \cdots \tag{3.43}$$

In order to handle situations where there are circuit elements in parallel, the concept of **admittance** (Y) has been introduced:

$$Y \equiv \frac{1}{Z}. \tag{3.44}$$

Since $Z = R + \mathrm{j}X$, it follows that

$$Y = \frac{1}{(R + \mathrm{j}X)}$$

or, rationalizing,

$$Y = \frac{(R - \mathrm{j}X)}{(R + \mathrm{j}X)(R - \mathrm{j}X)},$$

i.e.

$$Y = \frac{(R - \mathrm{j}X)}{(R^2 + X^2)},$$

which is written as

$$Y = G + \mathrm{j}B. \tag{3.45}$$

Here the conductance (G) is given by

$$G = \frac{R}{(R^2 + X^2)}, \tag{3.46}$$

and the **susceptance** (B) by

$$B = \frac{-X}{(R^2 + X^2)}. \tag{3.47}$$

So, to summarize, for elements in series the respective impedances add:

$$Z = Z_1 + Z_2 + Z_3 + \cdots;$$

(3.48) and for elements in parallel,

$$Y = Y_1 + Y_2 + Y_3 + \cdots$$

It should be noticed that there are formal similarities between pairs of equations that relate current and voltage in circuit elements. For instance:

	$V = (R) \cdot I$		$I = (G) \cdot V$
	$V = (\mathrm{j}\omega L) \cdot I$		$I = (\mathrm{j}\omega C) \cdot V$
	$V = \left(\dfrac{-\mathrm{j}}{\omega C}\right) \cdot I$		$I = \left(\dfrac{-\mathrm{j}}{\omega L}\right) \cdot V$
elements in series	$V = (Z_1 + Z_2) \cdot I$	elements in parallel	$I = (Y_1 + Y_2) \cdot V.$

In each case the elements in parentheses in each pair of equations are said to be **duals** of one another. Circuits may also be duals of one another, as will be seen in later chapters.

3.5 Power in a.c. circuits

Consider the situation where a source of e.m.f. drives a current $I = \hat{I} \cos \omega t$ through a circuit element of impedance Z. If the voltage across Z is given by $V = \hat{V} \cos(\omega t + \phi)$, then the instantaneous value of the power dissipated in the element is $p = I \cdot V$, or

$$p = \hat{I} \cdot \hat{V} \cos \omega t \cdot \cos(\omega t + \phi)$$

or

$$p = \hat{I}\hat{V}\cos\phi\cdot\cos^2\omega t - \hat{I}\hat{V}\sin\phi\cdot\sin\omega t\cdot\cos\omega t. \tag{3.49}$$

The second term in this expression is oscillatory (since $\sin\omega t\cdot\cos\omega t = (\sin 2\omega t)/2$) and represents a cyclical flow of energy out of and into the source, i.e. the average of this term over a complete number of periods T (where $T = 2\pi/\omega$) is zero. However, there is a net flow of energy from the source, which is represented by the first term in eqn (3.49).

The **average power** P delivered by the source is given by the average of p over a complete cycle of period T:

$$P = \frac{\hat{I}\hat{V}\cos\phi}{T}\int_0^T \cos^2\omega t\cdot \mathrm{d}t - \frac{\hat{I}\hat{V}\sin\phi}{T}\int_0^T \sin\omega t\cdot\cos\omega t\cdot\mathrm{d}t$$

$$= \frac{\hat{I}\hat{V}\cos\phi}{\omega T}\int_{\omega t=0}^{2\omega} \cos^2\omega t\cdot \mathrm{d}(\omega t) - \frac{\hat{I}\hat{V}\sin\phi}{\omega T}\int_{\omega t=0}^{2\pi} \sin\omega t\cdot\cos\omega t\cdot\mathrm{d}(\omega t)$$

or

$$P = \frac{\hat{I}\hat{V}\cos\phi}{2}, \tag{3.50}$$

since

$$\int_0^{2\pi}\cos^2 x\cdot\mathrm{d}x = \int_0^{2\pi}\sin^2 x\cdot\mathrm{d}x = \pi$$

and

$$\int_0^{2\pi}\sin x\cdot\cos x\cdot\mathrm{d}x = 0. \tag{3.51}$$

The **mean square** value of the current, $\overline{I^2}$, is given by

$$\overline{I^2} = \frac{1}{T}\cdot\int_0^T I^2\cdot\mathrm{d}t = \frac{\hat{I}^2}{\omega T}\int_{\omega t=0}^{2\pi}\cos^2\omega t\cdot\mathrm{d}(\omega t)$$

or, using the first of relationships (3.51), $\overline{I^2} = \frac{\hat{I}^2}{2}$ and the **root mean square** value, I_{rms}, is given by

$$I_{rms} = \frac{\hat{I}}{\sqrt{2}}. \tag{3.52}$$

Analogous relationships hold for voltages, so from eqn (3.50),

$$P = I_{rms}V_{rms}\cdot\cos\phi. \tag{3.53}$$

Now since

$$\cos\phi = \frac{R}{\sqrt{(R^2 + X^2)}} \text{ (see Fig. 3.5)} \qquad \text{and} \qquad \hat{V} = \hat{I}\sqrt{(R^2 + X^2)},$$

it follows that eqn (3.50) can also be rewritten as

$$P = \frac{\hat{I}^2 R}{2} = I_{\text{rms}}^2 R$$

(3.54) or

$$P = \frac{\hat{V}^2 \cos^2\phi}{2R} = \frac{V_{\text{rms}}^2 \cos^2\phi}{R}.$$

In the expression for P, $\cos\phi$ is known as the **power factor** of the circuit element and is equal to unity for a pure resistance and zero for a pure reactance (inductive or capacitive). Thus only the real part of the complex impedance (or admittance) is used in calculating the average power dissipated in the circuit element.

Since the power dissipated depends on I^2 or V^2 (or on the product of I and V), P is a non-linear function of I or V. This means that the complex number notation developed for a.c. circuits (Section 3.2.2) cannot be used in calculating power, at least not in a straightforward way.

For instance, consider a situation where a current $\hat{I}\cos\omega t$ flows through a circuit element and the consequent voltage across the element is $\hat{V}\cos(\omega t + \phi)$, i.e. the actual current and voltage are given by

$$I = \hat{I}\,\text{Re}\exp(\text{j}\omega t) \qquad \text{and} \qquad V = \hat{V}\,\text{Re}\exp[\text{j}(\omega t + \phi)].$$

It follows that

$$\begin{aligned} VI^* &= \hat{V}\exp[\text{j}(\omega t + \phi)]\cdot\hat{I}\exp(-\text{j}\omega t) \\ &= \hat{V}\hat{I}\exp(\text{j}\phi) \\ &= \hat{V}\hat{I}(\cos\phi + \text{j}\sin\phi). \end{aligned}$$

The so-called **apparent power** S is defined by

$$S \equiv \tfrac{1}{2}(VI^*), \tag{3.55}$$

Fig. 3.9

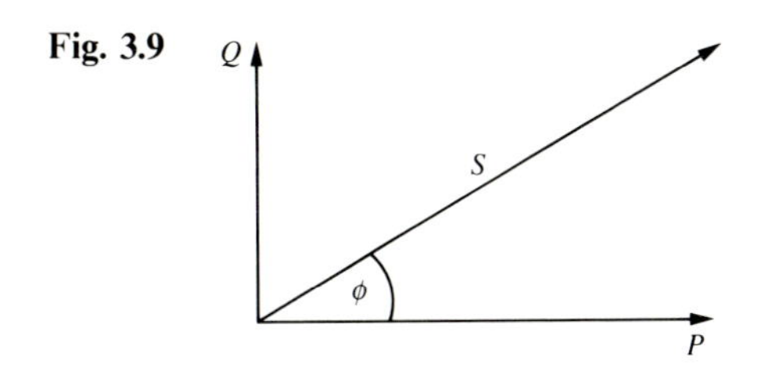

i.e.

$$S = \frac{\hat{V}\hat{I}}{2}\cos\phi + \mathrm{j}\,\frac{\hat{V}\hat{I}}{2}\sin\phi,$$

which is often written as

$$S = P + \mathrm{j}Q$$

where P = (average) real power and Q = 'reactive' power. In an Argand diagram S, P, Q form the sides of a **power triangle** (see Fig. 3.9).

The units of S are the volt-ampere (VA), of Q the volt-ampere reactive (VAR), and of P the watt (W). So the average power is given by

$$P = \tfrac{1}{2}\,\mathrm{Re}\,(VI^*). \tag{3.56}$$

What is the maximum power that an a.c. generator can deliver to a load? Consider the circuit shown in Fig. 3.10. The average power P dissipated in the load is given by

$$P = I_{\mathrm{rms}}^2 R_{\mathrm{L}} = \frac{E_{\mathrm{rms}}^2}{(R_{\mathrm{g}} + R_{\mathrm{L}})^2 + (X_{\mathrm{g}} + X_{\mathrm{L}})^2}\cdot R_{\mathrm{L}}, \tag{3.57}$$

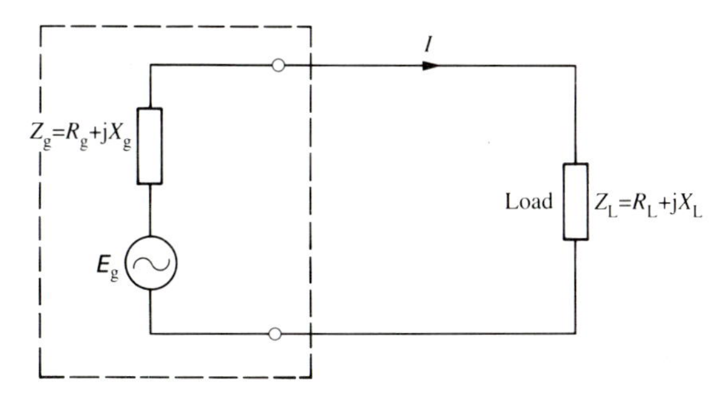

Fig. 3.10

and for P to be a maximum it is required that the denominator of eqn (3.57) be a minimum. In practice the impedance of the generator itself is usually not variable, so P is a maximum if $X_L = -X_g$ and $R_L = R_g$ (see Section 2.1). If these conditions on X_L and R_L are met, then the load is said to be **conjugately matched** to the generator, since $Z_L = -Z_g$. In certain other practical situations it may be preferable to have $Z_L = Z_g$.

There is a possibility of confusion over the use of the word 'load' in this context. The load on a generator is, strictly speaking, the current drawn from it. In the situation of Fig. 3.10, for example, the load on the generator is increased if the magnitude of the load impedance $|Z_L|$ is decreased! So a statement such as 'the generator should not be loaded by more than 1 kΩ' specifies a lower limit to the magnitude of the impedance of the circuit element (the 'load') connected to the generator.

3.6 Real capacitors and inductors

'Pure' resistors, inductors and capacitors cannot be realized in practice: even a straight piece of wire exhibits a small self-capacitance and a small self-inductance (see Section 5.2.2). Also, for example, a metal oxide 1 MΩ resistor may well have a self-capacitance of ~0.3 pF. This may seem to be insignificant, but at 1 MHz the reactance of this value of capacitance is only 500 kΩ approximately, which, since the self-capacitance is effectively in parallel with the resistance, means that the impedance of the resistor as a whole has a magnitude of about 300 kΩ only; obviously the magnitude of the impedance will decrease further with increasing frequency. At frequencies $\geqslant$1 GHz even the capacitance and inductance of the leads of devices may become important. The existence of such self-capacitance, and of 'stray' capacitances in general, means that high-frequency circuits are characterized by low levels of impedance, e.g. '75 Ω', '50 Ω'.

The stray capacitance and mutual inductance between adjacent wires, tracks and components of a circuit may cause significant unwanted coupling even at frequencies as low as the audio range. Thus in general, careful thought must be given to the design of the layout and interconnections of the components of a circuit (see Chapter 9).

3.6.1 Capacitors

Some of the practical aspects of real capacitors will be discussed in Section 5.1, but for the present, attention will be focused on the circuit properties as represented by an equivalent circuit. Energy losses can occur due to leakage currents between the plates of the capacitor and also by loss of energy from the alternating electric field that exists in the space between the plates (**dielectric losses**). As was seen in Section 3.5, the power P dissipated in a resistance R is given by

$$P = I_{\mathrm{rms}}^2 \cdot R \qquad \text{or} \qquad P = V_{\mathrm{rms}}^2/R.$$

It follows that any power dissipation process in a circuit can be represented by an effective resistance R_{eff}, where

$$R_{\mathrm{eff}} = (\text{power dissipated})/I_{\mathrm{rms}}^2$$

(3.58) or

$$R_{\mathrm{eff}} = V_{\mathrm{rms}}^2/(\text{power dissipated})$$

(An important example of this is the effective resistance of an antenna in a radio transmitter circuit, usually called the **radiation resistance**).

For a capacitor the effective resistance may be represented by a series resistance R_s or by a parallel resistance R_p (see Fig. 3.11). The appropriate value of the parallel or series resistance will be frequency-dependent in general, although a constant value may be used over limited ranges of frequency.

For the series equivalent circuit the impedance $(R_s - \mathrm{j}/\omega C_s)$ is represented in the complex plane as shown in Fig. 3.11(c). For most practical purposes capacitors are nearly ideal (i.e. R_s is relatively small), so the angle $(\phi - 270°)$ is very small. Hence it is usual to represent the dissipative processes by the **loss angle** δ, where $\tan\delta = R_s/\omega C_s$, rather than by the power factor $\cos\phi$. Further, since δ is a very small angle (see Table 5.1, page 174, for example), so that $\tan\delta \approx \delta$, it follows that $\delta \approx R_s/\omega C_s$.

It is left as an exercise (Problem 3.23; refer to Example 3.5) to show that for the parallel equivalent circuit of Fig. 3.11(b) the loss angle is given by $\delta \approx (\omega C_P R_P)^{-1}$.

3.6.2 Inductors

Although the resistance of the wire of the coil of an inductor is an obvious source of energy loss (**copper losses**), it is not the sole source, nor in general is it even the most important source. For instance, if the inductor has a ferromagnetic core (made from iron, steel and some of its alloys, or a ferrite) there will be power dissipation associated with both **eddy currents** and with magnetic **hysteresis** in the core material; see Section 5.3). These losses are referred to as **iron losses** or, more generally, as **core losses**. In addition, the coil acts as a radiating antenna, however inefficiently, and this is a further source of power loss; the effect increases with frequency.

As with a capacitor, an inductor can be represented by an equivalent

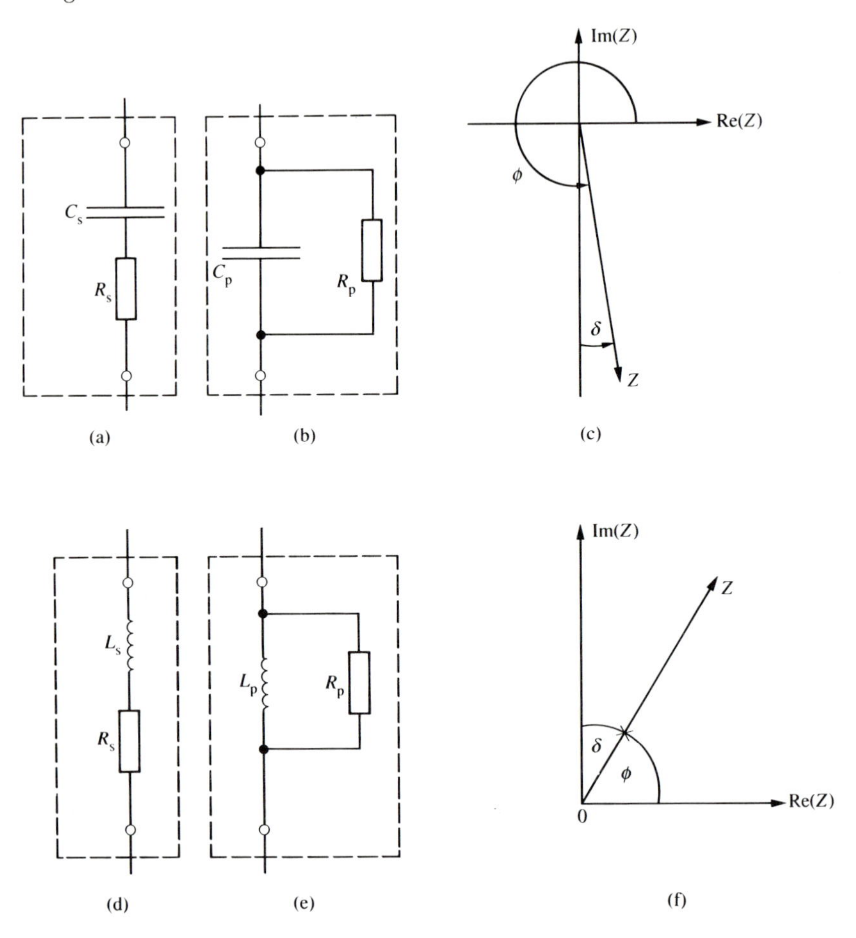

Fig. 3.11 Equivalent circuits and complex impedance vectors for real capacitors and inductors, showing the representations of energy losses. In (c) the loss angle δ is given by $\tan\delta = \omega C_s R_s \approx \delta$, and in (f) by $|\tan\phi| = \omega L_s/R_s$.

circuit containing a series and/or a parallel resistance (see Fig. 3.11). The physical correspondence is between series resistance and copper losses, and between parallel resistance and core losses.

For the series equivalent circuit, $Z = (R_s + j\omega L_s)$ and it is usual to specify the losses through a **quality factor**, Q, defined by $Q = |\tan\phi| = \omega L_s/R_s$. NB $\tan\delta = 1/Q$. In the parallel equivalent, $Q = R_p/\omega L_p$. For the majority of practical cases, $Q \geqslant 10$ and for the high-quality coils used in radiofrequency circuits, $Q \approx 100$ or more.

At frequencies greater than 20 kHz, iron losses become prohibitive, even for laminated cores (see Section 5.3), so ferrite cores are used. At very high frequencies, air-cored coils are used.

3.7 Transformers

As mentioned in Section 1.7, a transformer is an example of a mutual inductor, and, for the purposes of this introductory discussion, may be imagined to be a pair of coils would on a core of ferromagnetic material. Transformers were developed in the late nineteenth century to improve the efficiency of distribution of mains electrical power by 'stepping-up' the a.c. power at the generator end of a system from a low to a higher voltage level for transmission to the consumer. At the consumer end of the system a 'step-down' transformer is used to convert the mains voltage to a more convenient and/or safe level. Some of the other important uses of transformers will be outlined later in this section.

Note that henceforth, lower-case symbols e, v, i will be used to denote *time-dependent* e.m.f.s, voltages, and currents (i.e. sinusoidally varying quantities in the immediate context) and the upper-case symbols E, V, I will be reserved for time-independent (i.e. 'd.c.' quantities). Unless explicitly stated otherwise, the peak value (amplitude) of e.m.f., voltage, or current is given by the modulus of e, v, or i, respectively, and the phase angle, relative to whichever of e, v, i is chosen to define the reference phase, by the argument of e, v, or i.

In considering the transformer shown schematically in Fig. 3.12(a) it will be assumed, initially, that there is no 'leakage' of magnetic flux; i.e. all of the magnetic flux due to currents in the two coils is confined to the material of the core of transformer.

It is necessary to exercise great care in assigning polarities to the induced e.m.f.s. Imagine that the primary current i_p is increasing and generating an increasing magnetic flux ϕ_p linked with the secondary coil. The e.m.f. induced in the secondary coil will be in such a sense that the resultant secondary current generates a flux ϕ_s in the *opposite sense* to ϕ_p (see Section 1.7). In turn, this changing secondary flux will induce an e.m.f. in the primary coil in such a sense as to *reinforce* ϕ_p. Hence, in drawing the equivalent circuit of Fig. 3.12(b), where R_p is the resistance of the primary coil, the mutual induction effect is represented by a source $M\,\mathrm{d}i_s/\mathrm{d}t$ of the same polarity as the primary source v_p. Note that the coefficient of mutual inductance M is defined in eqns (1.51) and (1.52).

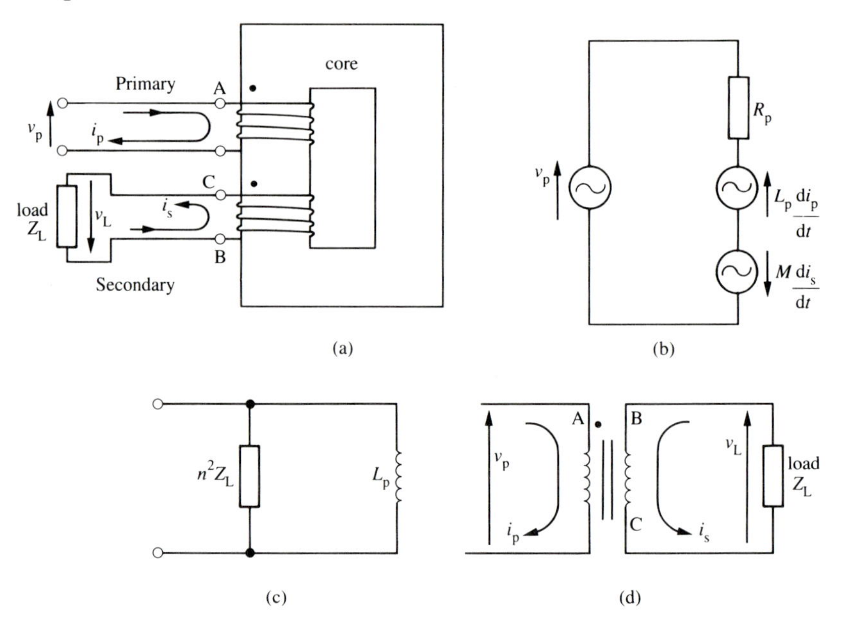

Fig. 3.12

The induced e.m.f. due to the self-inductance of the primary coil opposes the primary source, of course.

On applying KVL to the primary circuit, and assuming that the voltages and currents have an $e^{j\omega t}$ dependence (and remembering that $d/dt(e^{j\omega t}) = j\omega\, e^{j\omega t}$),

$$v_p + j\omega M i_s - j\omega L i_p - i_p R_p = 0$$

(3.59) or

$$i_p R_p + j\omega L_p i_p - j\omega M i_s = v_p.$$

Similarly, for the secondary circuit

$$j\omega M i_p - j\omega L_s i_s - i_s R_s - i_s Z_L = 0, \tag{3.60}$$

where R_s is the resistance of the secondary coil. If $Z_p \equiv (R_p + j\omega L_p)$ and $Z_s \equiv (R_s + j\omega L_s + Z_L)$, then eqn (3.60) can be written as

$$i_s = \frac{j\omega M i_p}{Z_s}, \tag{3.61}$$

and on substituting for i_s in eqn (3.59) it follows that

$$i_p\left(Z_p + \frac{\omega^2 M^2}{Z_s}\right) = v_p,$$

so that

$$Z_{\text{in}}\left(\equiv\frac{v_{\text{p}}}{i_{\text{p}}}\right) = Z_{\text{p}} + \frac{\omega^2 M^2}{Z_{\text{s}}}. \tag{3.62}$$

Also, with the configuration that has been adopted, $v_{\text{L}} = -i_{\text{s}}Z_{\text{L}}$, so

$$\frac{v_{\text{L}}}{v_{\text{p}}} = \frac{-\text{j}\omega M Z_{\text{L}}}{Z_{\text{s}}\left[Z_{\text{p}} + \dfrac{\omega^2 M^2}{Z_{\text{s}}}\right]}$$

or

$$\frac{v_{\text{L}}}{v_{\text{p}}} = \frac{-\text{j}\omega M Z_{\text{L}}}{(Z_{\text{s}}Z_{\text{p}} + \omega^2 M^2)}. \tag{3.63}$$

If the impedance of the secondary circuit is written as $Z_{\text{s}} = R_{\text{s}} + \text{j}X_{\text{s}}$, then eqn (3.62) becomes

$$Z_{\text{in}} = R_{\text{p}} + \text{j}\omega L_{\text{p}} + \frac{\omega^2 M^2 (R_{\text{s}} - \text{j}X_{\text{s}})}{|Z_{\text{s}}|^2}$$

or

$$Z_{\text{in}} = \left[R_{\text{p}} + \frac{\omega^2 M^2 R_{\text{s}}}{|Z_{\text{s}}|^2}\right] + \text{j}\omega\left[L_{\text{p}} - \frac{\omega M^2 X_{\text{s}}}{|Z_{\text{s}}|^2}\right]. \tag{3.64}$$

Thus the effective primary resistance (real part of Z_{in}) is increased by the existence of the secondary circuit and, if X_{s} is inductive, the effective primary inductance is decreased (imaginary part of Z_{in}). The increase in the effective resistance in the primary circuit represents the dissipation of energy in the secondary circuit.

If it is assumed for the present that there are no energy losses in the transformer, then $R_{\text{p}} = R_{\text{s}} = 0$, and from eqns (3.59) and (3.60) it follows that

$$i_{\text{s}} = \frac{\text{j}\omega M v_{\text{p}}}{(\omega^2(M^2 - L_{\text{p}}L_{\text{s}}) + \text{j}\omega L_{\text{p}}Z_{\text{L}})}$$

(3.65) and

$$i_{\text{p}} = \frac{(\text{j}\omega L_{\text{s}} + Z_{\text{L}})v_{\text{p}}}{(\omega^2(M^2 - L_{\text{p}}L_{\text{s}}) + \text{j}\omega L_{\text{p}}Z_{\text{L}})}$$

Now if the secondary of the transformer is short-circuited ($Z_{\text{L}} = 0$), then, as M is increased from a small value ($M^2 \langle L_{\text{s}}L_{\text{p}}$), eqns (3.65) show that both i_{p} and i_{s} increase and would become infinite for $M^2 = L_{\text{s}}L_{\text{p}}$. Hence this condition sets an upper bound to the value of M of $\surd(L_s L_p)$. In general we write $M = k\surd(L_s L_p)$, where the **coefficient of coupling** k lies in the range $0 \leqslant k \leqslant 1$.

In an ideal transformer there would be no energy losses, no flux leakage, and perfect coupling ($k = 1$), and the inductances of the primary and the secondary windings would be so large that their reactances would swamp all other impedances in the circuit (L_p, $L_s \to \infty$ but with their ratio remaining constant). Many transformers, especially at power (or 'mains') frequencies, approach quite closely to the ideal with efficiences as high as 95 per cent, so the ideal model serves as a useful basis for further discussions of real transformers.

Equation (3.63) can be written

$$\frac{v_L}{v_p} = \frac{-j\omega\surd(L_s L_p)\cdot Z_L}{\{(R_p + j\omega L_p)(R_s + j\omega L_s + Z_L) + \omega^2 L_s L_p\}}$$

or, assuming $\omega L_p \gg R_p$, $\omega L_s \gg R_s$ and remembering that $\omega^2 = (-j\omega)^2$,

$$\frac{v_L}{v_p} \approx -\sqrt{\left(\frac{L_s}{L_p}\right)} = -\frac{n_s}{n_p}, \tag{3.66}$$

since $L_p/L_s = (n_p/n_s)^2$, where n_p, n_s are the numbers of turns on the primary and secondary windings respectively (see eqn (1.55)}. Thus for an ideal transformer the 'secondary' voltage v_B (using the labelling of Fig. 3.12) is 180° out of phase with the 'primary' voltage v_A and is 'stepped up' or 'stepped down' depending on the magnitude of the ratio n_s/n_p.

Equations (3.65) indicate that

$$\frac{i_s}{i_p} = \frac{j\omega M}{(j\omega L_s + Z_L)} \qquad \left(\begin{matrix}\text{no losses,}\\ \text{perfect coupling}\end{matrix}\right)$$

and, if $\omega L_s \gg |Z_L|$, then

$$\frac{i_s}{i_p} \approx \frac{M}{L_s} = \sqrt{\left(\frac{L_p}{L_s}\right)} = \frac{n_p}{n_s}. \tag{3.67}$$

Thus the current is stepped up if the voltage is stepped down, and vice versa. Also $v_p i_p = v_L i_s$; this is a manifestation of the assumption of no power losses, of course.

An equivalent circuit for an ideal transformer can be obtained by considering eqn (3.62) under the assumption $\omega L_p \gg R_p$:

$$Z_{in} = j\omega L_p + \frac{\omega^2 M^2}{(j\omega L_s + Z_L)}$$

or

$$\frac{1}{Z_{in}} = \frac{(j\omega L_s + Z_L)}{\{j\omega L_p(j\omega L_s + Z_L) + \omega^2 M^2\}}.$$

Thus

$$\frac{1}{Z_{in}} = \frac{L_s}{L_p Z_L} + \frac{1}{j\omega L_p}.$$

Finally,

$$\frac{1}{Z_{in}} = \frac{1}{n^2 Z_L} + \frac{1}{j\omega L_p}, \tag{3.68}$$

where $n \equiv n_p/n_s$. From eqn (3.68) an equivalent circuit for the transformer can be drawn as shown in Fig. 3.12(c); the term $n^2 Z_L$ is called the **reflected impedance** of the secondary circuit at the primary side of the transformer. If $\omega L_p \gg n^2|Z_L|$, then $Z_{in} \approx n^2 Z_L$.

From the discussion of Section 3.5 it follows that, under the reigning assumptions, the maximum transfer between the source and the load is effected if $n^2 X_L = -X_g$ and $n^2 R_L = R_g$. This shows that a transformer can be used as a matching device. Some examples of the use of transformers in this role are: matching of a low-impedance transducer (e.g. a record player pick-up of the moving coil type; a ribbon microphone) to the input stage of an amplifier and the matching of the output stage of an amplifier to a load.

Other common uses of transformers, apart from mains transformers, are in the coupling between stages in multistage amplifiers and as isolating transformers; in this latter use the primary and secondary windings have no wired connection and so provide d.c. isolation.

A **'dot' convention** is used to establish a relation between the senses of the currents and voltages in the circuit diagram of a transformer and the currents and voltages in the real transformer 'on the bench'. If a dot is placed at terminal A in the diagram of Fig. 3.12(a), then, for i_p in the sense shown, and *increasing*, the polarity at C is positive with respect to terminal B. The convention is that a dot would be placed on terminal C of the secondary winding.

In a real transformer the assignment of the dots can be made on the basis of tracing the actual senses of the primary and secondary windings or by supplying an increasing d.c. current to the primary winding and observing which one of the secondary terminals is positive with respect to the other; this is illustrated in Figs. 3.12(a) and (d) in conjunction with eqns (3.59) and (3.60).

In the case of transformers operating at audiofrequencies and radiofrequencies, their characteristics depart much further from the ideal. Referring to Fig. 3.13(a) and applying KVL to the primary and secondary circuits (and observing the dot convention), it follows (eqns (3.59) and (3.60)) that

$$v_p = i_p(R_p + j\omega L_p) - j\omega M i_s,$$
$$0 = -j\omega M i_p + i_s(R_s + j\omega L_s + Z_L).$$

So

$$Z_{in} \equiv \frac{v_p}{i_p} = (R_p + j\omega L_p) + \frac{\omega^2 M^2}{(R_s + j\omega L_s + Z_L)}.$$

Since the coupling is now assumed to be less than perfect, $M = k\sqrt{(l_p L_s)}$, and so

$$Z_{in} = (R_p + j\omega L_p) + \frac{\omega^2 k^2 L_p L_s}{(R_s + j\omega L_s + Z_L)}. \tag{3.69}$$

In order to obtain an equivalent circuit for this transformer, a useful mathematical trick, which enables eqn (3.69) to be arranged in a more helpful form, is to add and subtract a term $(j\omega k L_p)$ on the right-hand side. It then follows that

$$Z_{in} = R_p + j\omega L_p(1-k) + \frac{\omega^2 k^2 L_p L_s}{(R_s + j\omega L_s + Z_L)} + j\omega k L_p$$
$$= \{R_p + j\omega L_p(1-k)\} + \frac{j\omega k L_p\{R_s + Z_L + j\omega L_s(1-k)\}}{(R_s + j\omega L_s + Z_L)}$$
$$= \{R_p + j\omega L_p(1-k)\}$$
$$+ \left[\frac{(R_s + Z_L)}{j\omega k L_p\{R_s + Z_L + j\omega L_s(1-k)\}} + \frac{L_s}{k L_p\{R_s + Z_L + j\omega L_s(1-k)\}}\right]^{-1}.$$

Now $(1-k)$ will be considerably smaller than unity, and so, if it is assumed that $|Z_L| \gg j\omega L_s(1-k)$, and remembering that $L_p/L_s = n^2$ (where

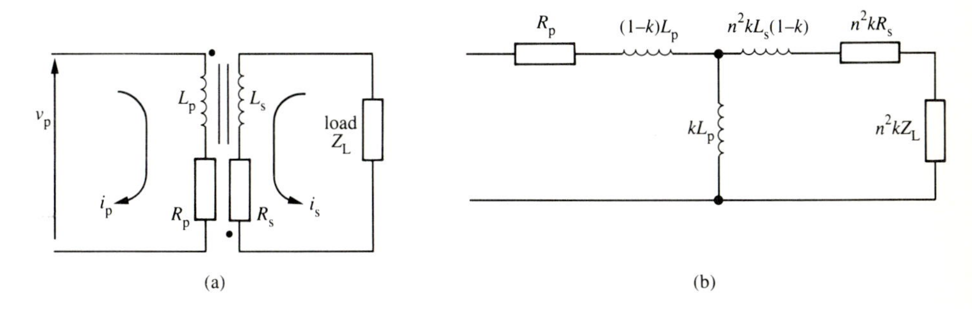

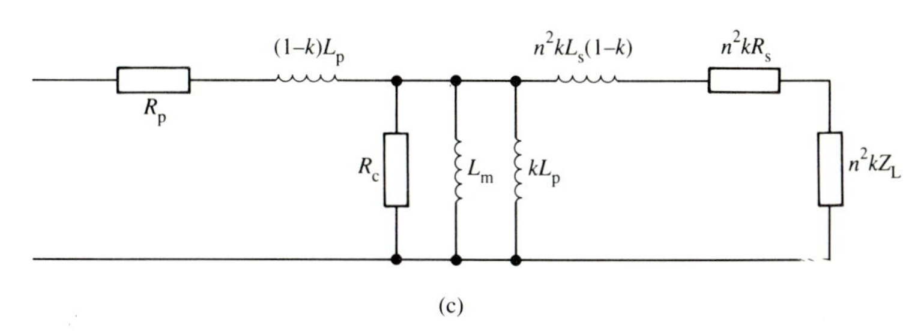

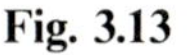
Fig. 3.13

$n \equiv n_p/n_s$), it follows that

$$Z_{in} \approx \{R_p + j\omega L_p(1-k)\} + \left[\frac{1}{j\omega kL_p} + \frac{1}{n^2k\{R_s + Z_L + j\omega L_s(1-k)\}}\right]^{-1}. \tag{3.70}$$

Note: for two impedances Z_1, Z_2 in parallel, if $\frac{1}{Z} = \frac{1}{Z_1} + \frac{1}{Z_2}$, say, then

$$Z = \frac{1}{\frac{1}{Z_1} + \frac{1}{Z_2}}.$$

By applying this result in eqn (3.70), the equivalent circuit of Fig. 3.13(b) can be drawn; the leakage flux is represented, at least in the approximation adopted, by the elements $(1-k)L_p$ and $n^2kL_s(1-k)$.

Further amendments can be made to the equivalent circuit to account for the current required to magnetize the core of the transformer and also to account for the iron losses in the core. These two effects are represented by the elements L_m and R_c, respectively, in the equivalent circuit of Fig. 3.13(c).

In an **autotransformer** one winding is common to both the primary and

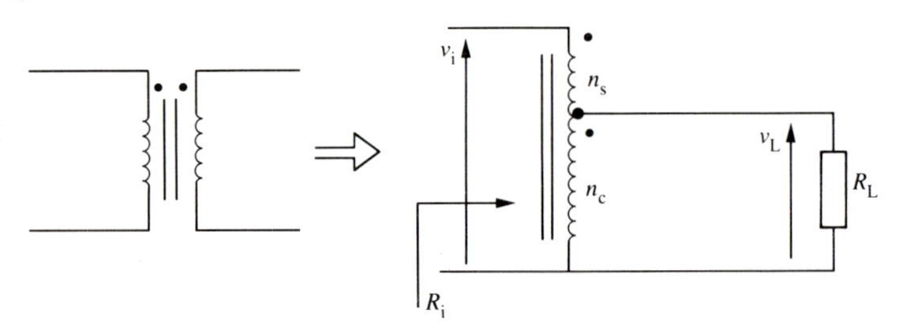

Fig. 3.14. A step-down autotransformer.

secondary circuits (see Fig. 3.14). A commonly encountered use of the autotransformer principle is in variable transformers for 'mains' supplies in which the tapping point T can move along the winding.

If n_c, n_s are the numbers of turns on the common winding and series winding respectively, then $v_L/v_i = n_c/(n_c + n_s)$. For a centre-tapped winding, $R_i = 4R_L$, which gives a 4:1 matching condition.

Example 3.6 The output stage of an audio amplifier has an output resistance of 3200 Ω and is coupled to a loudspeaker via a transformer. If the impedance of the loudspeaker is predominantly resistive with a value of 8 Ω, calculate the turns ratio for the transformer to obtain good power matching.

Assuming an ideal transformer, then we have seen that the power matching condition is that n^2R_L should be equal to the output resistance (R_g) of the 'generator', in this case the output stage of the amplifier. So we have

$$n^2 = 3200/8 \qquad \text{or} \qquad n = 20. \qquad \square$$

The efficiency of an audio transformer can be discussed on the basis of the following simplified model. The load is assumed to be purely resistive with a magnitude much less than ωL_s and it is assumed that $k \approx 1$. If it is assumed further that core losses and the magnetizing current are negligible, then the equivalent circuit of Fig. 3.15(a) follows from Fig. 3.13(c).

Since ωL_p will be ~ 10 kΩ, typically, the current in that branch will be neglected, so the efficiency, defined as the ratio of the power dissipated in the load to the power delivered to the transformer, will be given roughly by

$$\text{efficiency} \approx n^2R_L/(R_p + n^2R_s + n^2R_L).$$

Now R_p will be $\sim n^2R_s$ (since $L_p = n^2L_s$ and R_p is the resistance of the

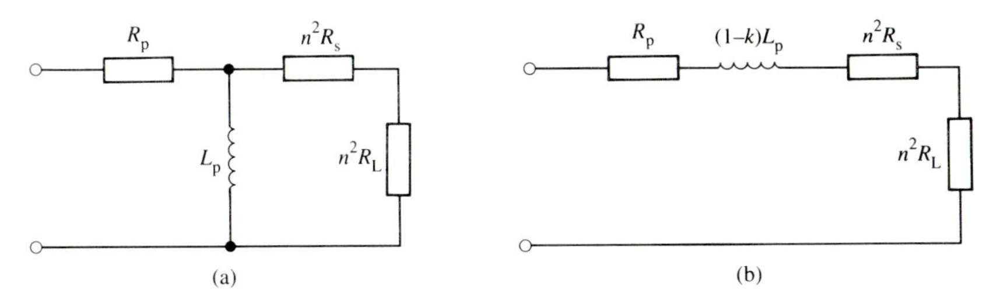

Fig. 3.15 (a) The equivalent circuit used in the definition of the efficiency of a transformer (under the assumption $k \approx 1$). (b) The high-frequency equivalent circuit.

wire constituting the primary inductance), so

$$\text{efficiency} \approx R_L/(2R_s + R_L).$$

For an efficiency of 90 per cent, say, it follows that R_L should be about $20R_s$.

It is very important to remember that the efficiency of performance of transformers is frequency-dependent, being governed by the physical properties of the core material and the inductance and self-capacitance of the windings. The low-frequency cut-off is determined by the reactance ωL_m (see Fig. 3.13), which becomes small at low frequencies and, of course, is in shunt across the input to the transformer. The increasing series reactance of the terms $\omega(1 - k)L_P$ and $\omega n^2 k(1 - k)L_s$, together with the shunting effect of the inter-winding capacitance, determines the high-frequency cut-off, see Fig. 3.15(b). In broadband radio systems that are required to cope with signals with frequency ratios of perhaps 20 000:1 (in the frequency range 10 kHz to 1 GHz, say), matching transformers can take a very different physical form, e.g. transmission line transformers.

Some aspects of the materials used in transformers and of magnetic circuits will be discussed in Section 5.3. Also there will be some discussion of the use of coupled networks in radio frequency circuits in Section 4.4.

3.8 A.c. networks

3.8.1 Examples of mesh and nodal analysis

The analysis of a.c. networks proceeds according to the basic principles described for d.c. networks in Chapter 2. However, it will be useful to consider a few examples here involving a.c. currents and voltages; many other examples will arise incidentally in later chapters.

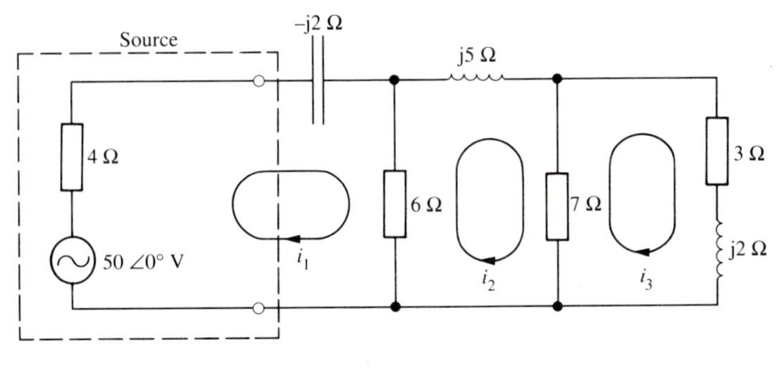

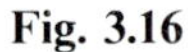
Fig. 3.16

Example 3.7 Calculate the power delivered by the source in the circuit in Fig. 3.16.

First it should be noticed that capacitances and inductances are labelled by the respective values of their reactances at the frequency of interest in the problem. For the circuit shown the mesh equations are:

$$\begin{aligned} i_1(10-\mathrm{j}2) \quad -6i_2 \quad &= 50 \\ -6i_1 + (13+\mathrm{j}5)i_2 - 7i_3 &= 0 \\ -7i_2 + (10+\mathrm{j}2)i_3 &= 0. \end{aligned}$$

Since the voltage of the source is given, it is necessary to know the modulus of i_1 and its phase angle in order to calculate the power delivered, i.e.

$$\text{power delivered} = (\tfrac{1}{2}) \times 50 \times |i_1| \cos\phi \text{ (see eqn (3.53))}.$$

Using the rules given in Section 2.4,

$$i_1 = \begin{vmatrix} 50 & -6 & 0 \\ 0 & (13+\mathrm{j}5) & -7 \\ 0 & -7 & (10+\mathrm{j}2) \end{vmatrix} \Big/ D,$$

where

$$D = \begin{vmatrix} (10-\mathrm{j}2) & -6 & 0 \\ -6 & (13+\mathrm{j}5) & -7 \\ 0 & -7 & (10+\mathrm{j}2) \end{vmatrix}$$

$$= (10-\mathrm{j}2)\{(13+\mathrm{j}5)(10+\mathrm{j}2) - 49\} + 6\{-6(10+\mathrm{j}2)\}$$

or

$$D = 502 + \mathrm{j}546.$$

Thus

$$i_1 = 50\{(13+\mathrm{j}5)(10+\mathrm{j}2) - 49\}/(502+\mathrm{j}546)$$

or

$$i_1 = \frac{50(71 + j76)}{(502 + j546)}.$$

On rationalizing,

$$i_1 = \frac{50(71 + j76)(502 - j546)}{5.50 \times 10^5}.$$

After expanding, it follows that

$$i_1 = \frac{50(7.71 \times 10^4 - j6.14 \times 10^2)}{5.50 \times 10^5}.$$

By using the procedure for calculating the modulus of a complex number as outlined in Section 3.2.2, it is found that

$$i_1 = 7.01\underline{/\tan^{-1}(-6.14/771)},$$

i.e.

$$i_1 = 7.01\underline{/-0.456°}\text{ A}.$$

Thus

$$\text{power} = (\tfrac{1}{2}) \times 50 \times 7.01 \times \cos(0.456°) = 175\text{ W}.$$ □

Example 3.8 Show that the network of Fig. 3.17 acts as a phase-shifting network, i.e. $|v_0/v_1|$ is constant and $\underline{/v_0/v_1}$ depends on the value of the variable resistance R.

The mesh equations for the network of Fig. 3.17 are:

$$2R_0 i_1 - 2R_0 i_2 = v_1$$

$$-2R_0 i_1 + \left(2R_0 + R - \frac{j}{\omega C}\right) i_2 = 0.$$

NB it has been assumed that the internal impedance of the source is negligible and also that no current is drawn by the circuit that is connected to terminals A, B of the network.

Now

$$\begin{aligned} v_0 &= (i_1 - i_2)R_0 - i_2 R \\ &= i_1 R_0 - i_2(R + R_0) \end{aligned}$$

So it is necessary to find expressions for i_1 and i_2:

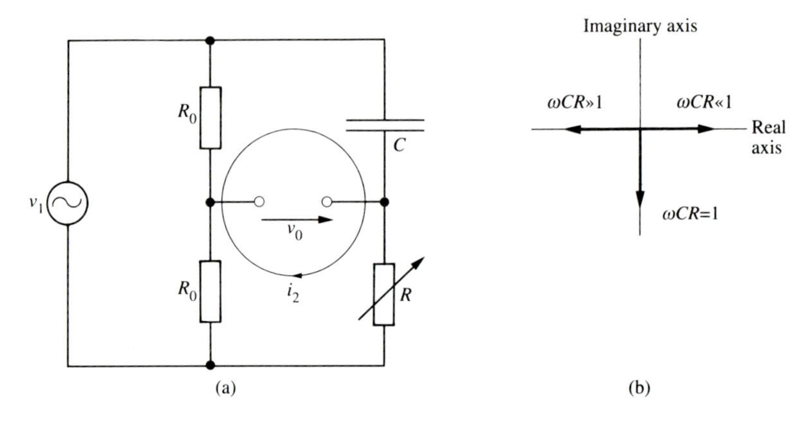

Fig. 3.17 (a) A phase-shifting network. (b) The ratio v_0/v_1 plotted in the complex plane (C kept constant, R varied).

$$i_1 = \begin{vmatrix} v_1 & -2R_0 \\ 0 & \left(2R_0 + R - \dfrac{\mathrm{j}}{\omega C}\right) \end{vmatrix} \Bigg/ \begin{vmatrix} 2R_0 & -2R_0 \\ -2R_0 & \left(2R_0 + R - \dfrac{\mathrm{j}}{\omega C}\right) \end{vmatrix}$$

$$= \frac{\left(2R_0 + R - \dfrac{\mathrm{j}}{\omega C}\right) v_1}{2R_0\left(2R_0 + R - \dfrac{\mathrm{j}}{\omega C}\right) - 4R_0^2}$$

$$= \frac{\left(2R_0 + R - \dfrac{\mathrm{j}}{\omega C}\right) v_1}{\left(2R_0 R - \dfrac{\mathrm{j}2R_0}{\omega C}\right)}.$$

Also

$$i_2 = \begin{vmatrix} 2R_0 & v_1 \\ -2R_0 & 0 \end{vmatrix} \Bigg/ \begin{vmatrix} 2R_0 & -2R_0 \\ -2R_0 & 2R_0 + \left(R - \dfrac{\mathrm{j}}{\omega C}\right) \end{vmatrix} = \frac{2R_0 v_1}{\left(2RR_0 - \dfrac{\mathrm{j}2R_0}{\omega C}\right)}.$$

Substituting for i_1, i_2 in eqn (3.70) leads to

$$\frac{v_0}{v_1} = -\frac{1}{2}\frac{\left(R + \dfrac{\mathrm{j}}{\omega C}\right)}{\left(R - \dfrac{\mathrm{j}}{\omega C}\right)}$$

and, after rationalizing,

$$\frac{v_0}{v_1} = \frac{\left(R^2 - \dfrac{1}{\omega^2 C^2}\right) + \mathrm{j}\dfrac{2R}{\omega C}}{2\left(R^2 + \dfrac{1}{\omega^2 C^2}\right)},$$

whence,

$$\left|\frac{v_0}{v_1}\right| = \frac{1}{2}$$

and the argument (ϕ) of v_0/v_1 is given by

$$\tan\phi = \left(\frac{-2R}{\omega C}\right) \Bigg/ \left(\frac{1}{\omega^2 C^2} - R^2\right)$$

or

$$\tan\phi = \frac{-2}{\left(\dfrac{1}{\omega CR} - \omega CR\right)}.$$

For $\omega CR \ll 1$, $\phi \approx 0°$ (see Fig. 3.17(b)); for $\omega CR = 1$, $\phi = -90°$; and for $\omega CR \gg 1$, $\phi \approx -180°$. Hence, for constant ω, it is possible to vary the phase of v_0 relative to v_1 through a range of 180° by varying R, without change of amplitude of v_0/v_1. □

Example 3.9 Determine the voltage v_2 in the circuit shown in Fig. 3.18. Nodal analysis will be used to solve this problem.

The voltage at node B (the voltage of interest) is v_2, and the voltage at A will be denoted by v_1.

First, the source voltage is converted into cartesian form $(x + \mathrm{j}y)$; from the specification of the problem, $\sqrt{(x^2 + y^2)} = 10$ and $y/x = \tan 60°$, which yields $x = 5.00$ V and $y = 8.65$ V. Hence the circuit of Fig. 3.18(a) can be redrawn as in Fig. 3.18(b), where the source has been converted into 'current generator' form since nodal analysis is going to be used.

The nodal equations are:

$$v_1\left(\frac{1}{(5-\mathrm{j})} + \frac{1}{10} + \frac{1}{\mathrm{j}2}\right) - \frac{v_2}{\mathrm{j}2} = \frac{(5 + \mathrm{j}8.65)}{(5-\mathrm{j})}$$

$$\frac{v_1}{\mathrm{j}2} + v_2\left(\frac{1}{\mathrm{j}2} + \frac{1}{40} + \frac{1}{(10-\mathrm{j})}\right) = 0$$

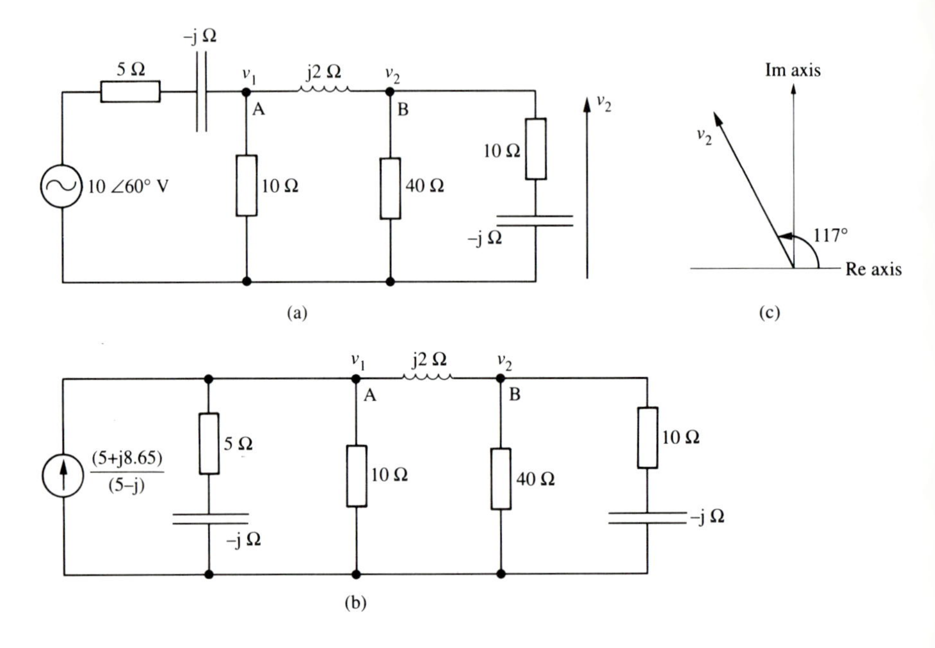

Fig. 3.18

or

$$\frac{v_1(19-\mathrm{j}30)}{65}+\mathrm{j}\frac{v_2}{2}=\frac{16.35+\mathrm{j}48.2}{26}$$

$$\mathrm{j}\frac{v_1}{2}+\frac{v_2(2-\mathrm{j}7.92)}{16.2}=0.$$

Thus

$$v_2=\begin{vmatrix}\dfrac{(19-\mathrm{j}30)}{65} & \dfrac{(16.35+\mathrm{j}48.2)}{26}\\ \mathrm{j}/2 & 0\end{vmatrix}\Bigg/\begin{vmatrix}\dfrac{(19-\mathrm{j}30)}{65} & \dfrac{\mathrm{j}}{2}\\ \dfrac{\mathrm{j}}{2} & \dfrac{(2-\mathrm{j}7.92)}{16.2}\end{vmatrix}$$

$$=\frac{(9.25-\mathrm{j}3.15)}{(1.90+\mathrm{j}2.00)}=\frac{(9.25-\mathrm{j}3.15)(1.90-\mathrm{j}2.00)}{7.61}$$

of, finally,

$$v_2=-1.48+\mathrm{j}2.85\ \mathrm{V}.$$

In polar form (see Fig. 3.18(c)),

$$v_2 = 3.21 \tan^{-1}\left(\frac{2.85}{-1.48}\right)$$

or

$$v_2 = 3.21\angle 117^\circ \text{ V}$$

□

3.8.2 Coupling between networks

This topic fits naturally into Section 7.3 dealing with two-port networks, but it is of such great practical importance that it is useful to introduce it at this earlier stage.

Consider Fig. 3.10, which depicts a source coupled to a load. Now the 'source' could be the Thévenin (or Norton) equivalent of a more complicated network (e.g. the output of the first stage of a multistage amplifier) and the 'load' could be the input impedance of a more complicated network (such as the 'next' stage of a multistage amplifier). The conditions required to maximize the coupling of power from a source to a load were discussed in Section 3.5. Now consider a multistage amplifier that amplifies a small signal from a voltage source such as a microphone (see Fig. 3.19(a)). Each stage of the amplifier is represented at its input by its effective input impedance Z_{in1}, Z_{in2} etc. At its output a stage of the amplifier is represented by its Thévenin equivalent source.

In Fig. 3.19(a) it can be seen that

$$i_1 = \frac{v_s}{(Z_s + Z_{in})}$$

and

$$v_1 = i_1 Z_{in} = \frac{Z_{in1}}{(Z_s + Z_{in})} \cdot v_s, \tag{3.71}$$

i.e. Z_{in1} and Z_s act as a 'potential-divider' (see Fig. 2.9). If no current is drawn from the output of the first stage, then $v_2/v_1 = A_v$; this is the **open-circuit gain**, or 'no-load voltage gain' of the first stage. Since the signal to be amplified is a voltage signal, the aim is to maximize the **voltage coupling** between the stages of the amplifier. To maximize the coupling between the signal source and the first stage, it is required to maximize

$$\frac{v_1}{v_s} = \frac{Z_{in1}}{(Z_s + Z_{in1})}. \tag{3.72}$$

Obviously, v_1/v_s tends towards its maximum value of unity as Z_{in_1} becomes very much longer than Z_s. This situation describes ideal voltage coupling, i.e. the impedance of the 'source' is negligible in comparison with that of the 'load'. For many practical purposes $Z_{in1}/Z_s > 20$ is good enough,

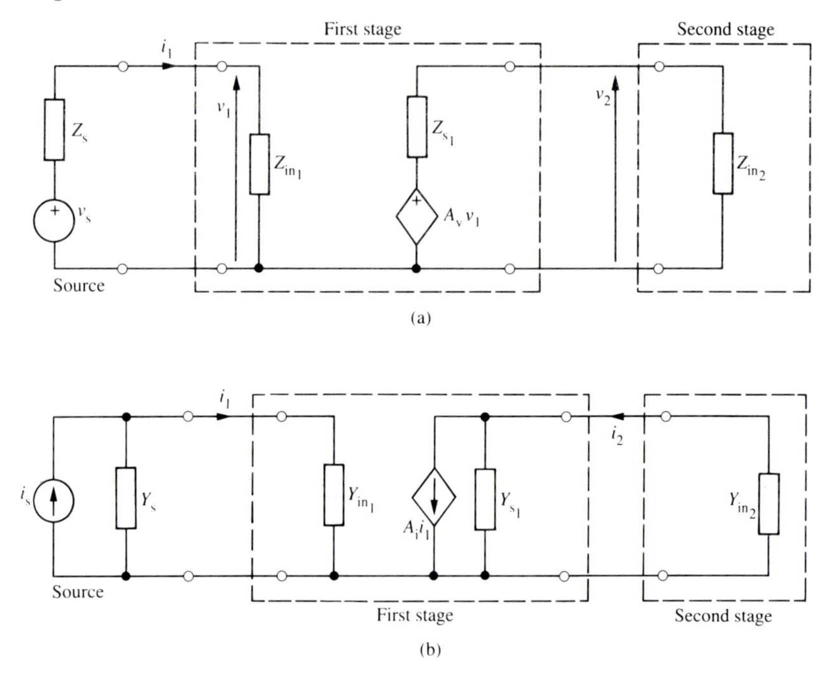

Fig. 3.19 Equivalent circuits for an amplifier represented in terms of (a) constant voltage sources and (b) constant current sources. Note that the 'circles' denote independent sources and the 'diamonds' represent dependent sources (i.e. controlled by the voltage and/or current at the input of the stage in question).

although in some specialized instruments, values of Z_{in1}/Z_s as great as 10^6 are required. It should be noted that 'Z_s' includes any series element in the interstage coupling, such as a blocking capacitor (whose reactance becomes large at low frequencies).

For the two-stage amplifier of Fig. 3.19(a) it can be seen that

$$\frac{v_2}{A_v v_1} = \frac{Z_{in2}}{(Z_{s1} + Z_{in2})}$$

and so

$$\frac{v_2}{v_s} = \frac{Z_{in1}}{(Z_s + Z_{in1})} \times A_v \times \frac{Z_{in2}}{(Z_{s1} + Z_{in2})}. \tag{3.73}$$

This argument can be extended simply for a multistage amplifier.

If the problem is to amplify a small current from a source, then good **current coupling** is required between the source and the load. Since it is current that is of interest, the source and the outputs of the stages of the amplifier are represented by their Norton equivalents; see Fig. 3.19(b). In this case

$$\frac{i_1}{i_s} = \frac{Y_{in1}}{(Y_s + Y_{in})}. \tag{3.74}$$

This equation is identical, formally, to eqn (3.72) because the circuit of Fig. 3.19(b) is the dual of that of Fig. 3.19(a): Y_s and Y_{in1} act as a **current divider network**.

If $Y_{in2} = \infty$ (short-circuit), then $i_2 = A_i i_1$, where A_i is the **short-circuit current gain** of the first stage. In general

$$\frac{i_2}{i_s} = \frac{Y_{in1}}{(Y_s + Y_{in1})} \times A_i \times \frac{Y_{in2}}{(Y_{s1} + Y_{in2})}. \tag{3.75}$$

It can be seen that for good current coupling, the admittance of the 'load' should be much greater than that of the source.

An example of a voltage-coupler situation is where a bipolar transistor is used in the 'common-emitter' configuration to amplify a voltage signal and a transistor connected in the 'common-base' configuration is a good current amplifier.

Consider the coupling situation shown in Fig. 3.20(a). It could be imagined that the capacitor couples an amplifier stage, which is represented by the source v_1 (of negligible output impedance), to a load R, which represents the input resistance of the next stage of the amplifier. It can be seen that

$$v_2 = \frac{v_1}{\left(R - \frac{j}{\omega C}\right)} \cdot R$$

or

$$\frac{v_2}{v_1} = \frac{R\left(R + \frac{j}{\omega C}\right)}{\left(R^2 + \frac{1}{\omega^2 C^2}\right)},$$

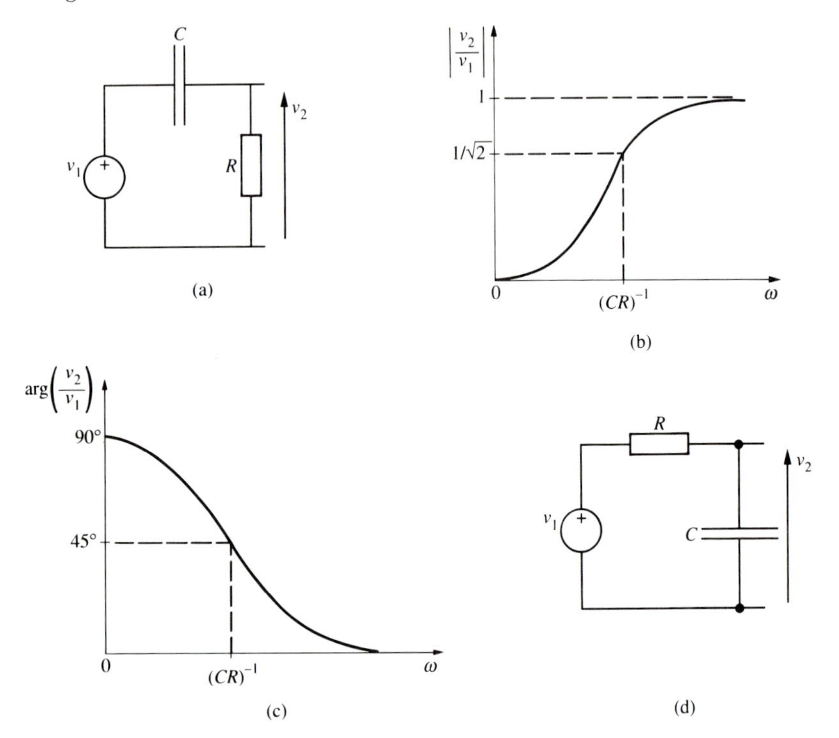

Fig. 3.20 (a) A high-pass CR network, with the variations of (b) $|v_2/v_1|$ and (c) $\arg(v_2/v_1)$ with frequency. A low-pass CR network.

so that finally

$$\left|\frac{v_2}{v_1}\right| = \frac{R\left(R^2 + \dfrac{1}{\omega^2 C^2}\right)^{1/2}}{\left(R^2 + \dfrac{1}{\omega^2 C^2}\right)}$$

$$= \frac{\omega CR}{(1 + \omega^2 C^2 R^2)^{1/2}}.$$

Now the argument of (v_2/v_1) is $\tan^{-1}(1/\omega CR)$, so

$$\frac{v_2}{v_1} = \frac{\omega CR}{(1 + \omega^2 C^2 R^2)^{1/2}} \underline{/\tan^{-1}(1/\omega CR)}.$$

The variations with frequency of $|v_2/v_1|$ and $\arg(v_2/v_1)$ are shown in Fig. 3.20(b) and (c). Note that for $\omega = (CR)^{-1}$, then $|v_2/v_1| = 1/\sqrt{2} = 0.707$

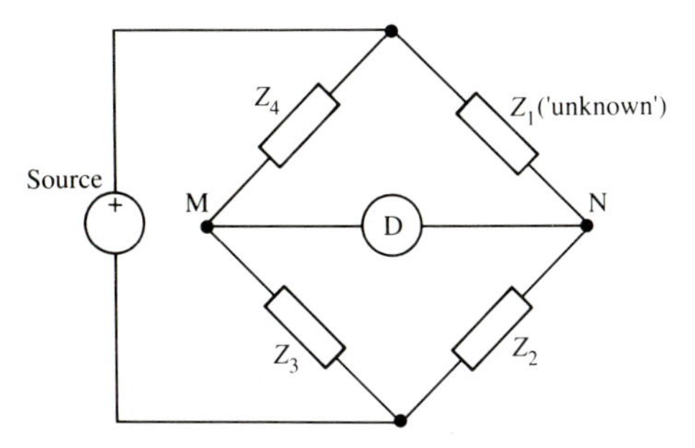

Fig. 3.21 A general form of a.c. bridge, where D is the detector.

and $\tan^{-1}(\omega CR) = 45°$, and that this network acts as a simple high-pass filter. This explains why the frequency response of a capacitor-coupled amplifier falls off at low frequencies. It is left as a problem to carry out similar calculations for the network of Fig. 3.20(d), which acts as a simple low-pass filter.

In practical coupling situations the 'source' will have a non-zero output impedance; also, there may well be significant stray capacitance in parallel with the 'load'. For example, for a typical oscilloscope with a nominal resistive input impedance of 1 MΩ there will be a shunt capacitance of ~30 pF; this significantly reduces the effective input impedance at high frequencies, of course.

3.9 A.c. bridges

The measurement of the values of inductors and capacitors is made most conveniently and accurately by means of an **a.c. bridge**, although resonance methods are useful in the case of circuit elements for use at high frequencies. For the general form of bridge circuit as shown in Fig. 3.21 the same arguments apply as used in the case of the (d.c.) Wheatstone bridge (see Section 2.2 and eqn (2.19)) and lead to the balance condition:

$$Z_1 Z_3 = Z_2 Z_4 \qquad \text{or} \qquad R_1 + \mathrm{j}X_1 = \frac{(R_2 + \mathrm{j}X_2)(R_4 + \mathrm{j}X_4)}{(R_3 + \mathrm{j}X_3)} \tag{3.76}$$

For balance, the *instantaneous* values of the voltages at M and at N must be equal, which implies equality of their amplitudes *and* of their phases. These two conditions for balance are exemplified, of course, by the equality of the real parts and of the imaginary parts of the two sides of eqn (3.76). The combinations of choices of circuit elements for R_2, R_3, R_4, X_2, X_3, and X_4 are limited, for practical purposes, by the following considerations:

(i) for convenience, only *two* elements should be adjustable in meeting the two balance conditions;

(ii) these two balancing operations should be *independent* as regards their respective influences on the approach to balance;

(iii) whilst high-accuracy pure resistances and pure capacitances are commonly available, this is not the case for inductances, so in practice the adjustable elements are resistors and capacitors.

Conditions (i) and (ii) imply that it should be arranged that in eqn (3.76) the right-hand side should be in the form $(A + jB)$, where A and B are real, and where one of the adjustable elements should appear in A but not in B and the other in B but not in A.

If the arms of the bridge are assumed to consist of inductances, capacitances and resistances in *series*, then it can be shown from eqn (3.76), by straightforward (but tedious) analysis, that the desired conditions can be met if R_2 and X_2 are the adjustable elements and the *ratio* Z_4/Z_3 is either pure real or pure imaginary (note that Z_4 and Z_2 are equivalent in this analysis and so are interchangeable).

Alternatively, the arms of the bridge may be made up of inductances, capacitances and resistances in *parallel*, in which case the balance condition, eqn (3.76), can be written in terms of admittances as:

$$\frac{1}{Y_1 Y_3} = \frac{1}{Y_2 Y_4} \quad \text{or} \quad R_1 + jX_1 = \frac{Y_3}{Y_2 Y_4}. \tag{3.77}$$

In this case $(R_1 + jX_1)$ can be written in the form $(A + jB)$ if R_3 and X_3 are the adjustable elements and the *product* $Z_2 Z_4$ is either pure real or pure imaginary.

Thus an a.c. bridge may be balanced by adjusting series elements in an arm (Z_2) adjacent to the unknown element (Z_1), the 'fixed' arms (Z_3, Z_4) appearing in the balance condition equation as a ratio (a '*ratio*' arm bridge), or by adjusting parallel elements in the arm (Z_3) opposite to the unknown element, the 'fixed' arms appearing in the balance condition equation as a product (a '*product*' arm bridge).

The usefulness of a particular bridge circuit depends critically on the response of the detector to a small deviation from the exact balance condition. By using the same method of analysis as was used for the d.c. Wheatstone bridge (see Section 2.7), the out-of-balance voltage between M and N, and the associated current in the detector branch, can be calculated. However, it will suffice here to point out that if the impedance of the detector is much greater than that of the bridge arms, then maximum sensitivity is obtained if Z_1, Z_2, Z_3, and Z_4 are of equal magnitude. If the impedance of the detector is not large compared with that of the elements in the arms, then the maximum power is available for its operation if its impedance is matched to that of the (equivalent) network that is connected to its terminals (again see Section 2.7).

One additional practical point should be mentioned here. It is highly desirable that the two balance conditions should be frequency-independent. Although drifting of the frequency of the source during the course of a measurement should not be a problem, the fact that the waveform will (inevitably) contain a small proportion of harmonics of the fundamental frequency would be troublesome if the balance condition were frequency-dependent, since the capability for detecting small out-of-balance signals would be impaired.

A.c. bridges can be used at frequencies up to ~ 100 MHz, although resonance methods are often used to measure values of inductance at frequencies from about 30 kHz upwards. Although many permutations of circuit elements may be used for $Z_1 \cdots Z_4$, only some of the more common ones will be described here. A general feature is that inductors are not used as calibrated impedances, since they have a significant (frequency-dependent) resistance as well as inductive reactance and since they are not available in variable form.

A general practical problem is the existence of stray capacitances between the detector and earth (see Fig. 3.22(a)); such stray capacitances can cause the measured value of Z_1 to be in error and/or make the balancing procedure tedious and imprecise if their magnitudes are not constant.

In a Wagner earthing system, impedances Z_5 and Z_6 are connected in series across the source and their junction E is earthed. The bridge is first balanced with the detector connected to E and then with the detector connector to M; this procedure is repeated until no further adjustment in either Z_5 or Z_6, or Z_3 or Z_4, is required in order to achieve a balanced condition. Thus M and N are also brought to earth potential and the reactances of the stray capacitances of the stray capacitances between M and E and between N and E are effectively very large, since virtually no current flows through them. It follows that C_3 and C_4 are now in shunt across Z_5 and Z_6 respectively (i.e. effectively form part of these impedances) and so no longer affect the balance condition of the bridge Z_1, Z_2, Z_3, Z_4.

For an inductor having relatively large losses (small Q-factor, $\omega L/R$), a suitable bridge for measuring its inductance and resistance is the Maxwell bridge (Fig. 3.22(b)). From eqn (3.76) the balance conditions are $L_1 = C_3 R_2 R_4$ and $R_1 = R_2 R_4/R_3$, and the Q-factor is given by $\omega L_1/R_1 = \omega R_3 C_3$. Since at balance, $R_3 = Q/(\omega C_3)$, its value may become inconveniently large if Q is large. This problem is overcome in the Hay bridge, in which C_3 and R_3 are in series and substitution in eqn (3.76) yields

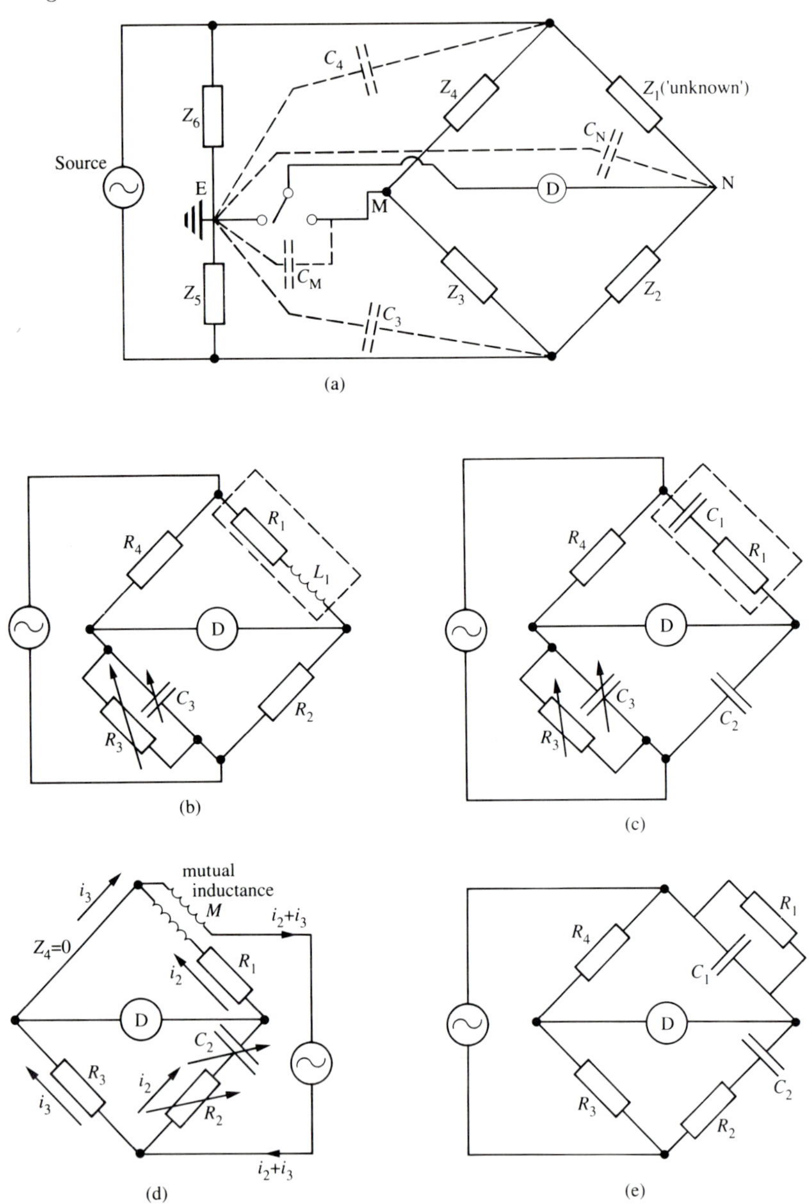

Fig. 3.22 A.c. bridges: (a) a general bridge circuit showing stray capacitances and the 'Wagner earth'; (b) the Maxwell bridge; (c) the Schering bridge; (d) the Carey–Foster bridge; (e) the Wien bridge.

$L_1 = R_2R_4R_3/(1+\omega^2R_3^2C_3^2)$. This expression for L_1 is frequency-dependent, but since ωR_3C_3 $(=R_1/\omega L_1) = 1/Q$ is much less than unity, this feature is not usually of great practical significance.

The Schering bridge (Fig. 3.22(c)) finds widespread use both for the precise measurement of capacitance and dielectric loss at low voltage amplitudes and for the study of insulating structures at high voltages:

$$C_1 = C_2R_3/R_4;\ R_1 = C_3R_4/C_2.$$

The loss angle of the dielectric (see Sections 3.6.1 and 5.1) is given by $\tan\delta = \omega C_1R_1 = \omega C_3R_3$.

A Carey–Foster bridge (Fig. 3.22(d)) is designed for the measurement of a coefficient of mutual inductance M in terms of the value of a capacitance. Assuming 'Z_4' = 0, it follows that, at balance,

$$\{i_2R_2 - \mathrm{j}i_2/(\omega C_2)\} = i_3R_3 \quad \text{and} \quad i_2(R_1+\mathrm{j}\omega L_1) - (i_2+i_3)\mathrm{j}\omega M = 0.$$

Hence it follows that $M = R_1R_3C_2$ and $L_1 = M(1+R_2/R_3) = C_2R_1(R_2+R_3)$.

For the Wien bridge (Fig. 3.22(3)) the balance condition is frequency-dependent:

$$\omega^2 = 1/R_1R_2C_1C_2 \quad \text{and} \quad C_1/C_2 = R_3/R_4 - R_2/R_1.$$

Thus the frequency of a signal can be determined in terms of the components of the bridge.

A variety of instruments may be used as detectors, e.g. earphones, high-resistance meters, CRO, or 'magic eye'. The detector must be sensitive enough for the measurement in hand, but the use of an over-sensitive detector may be not only economically unsound but also increase the difficulty, and tediousness, of the balancing adjustments.

Problems

3.1 A capacitor C is charged from a source of e.m.f. E, which is then disconnected and the capacitor is then allowed to discharge through a resistor R. Show that at a time t after the beginning of the discharge, the charge Q on the capacitor is given by

$$Q = Q_0\exp(-t/RC),$$

where Q_0 is the charge on the capacitor at time $t = 0$.

3.2 For the situation of Problem 3.1, what is the initial value of the discharge current if $E = 100$ V, $C = 1000\ \mu$F, $R = 100$ kΩ?

3.3 What is the time constant for (a) the charging (switch closed) and (b) the discharging (switch open) of the capacitor in the circuit shown in Fig. 3.23?

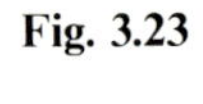

Fig. 3.23

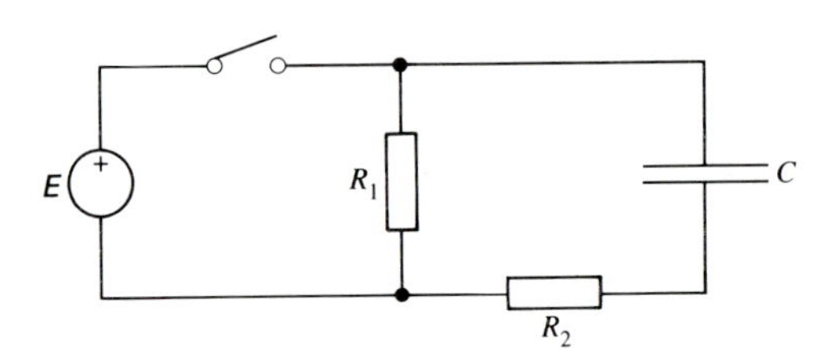

Hint: apply Kirchhoff's laws and remember that the potential difference across a capacitor C is Q/V, where Q is the charge on the capacitor plates.

3.4 Consider the circuit shown in Fig. 3.24 and then answer the following questions.

Fig. 3.24

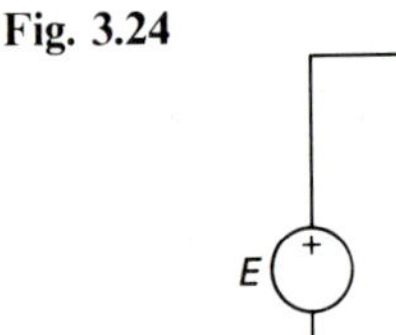

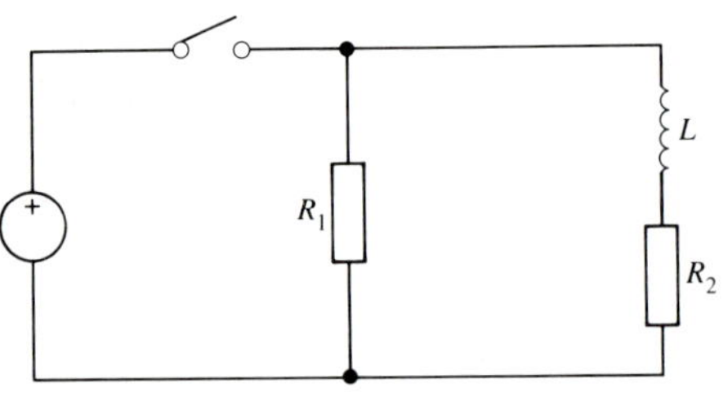

(a) What is the voltage across the inductor at the instant when the switch is closed?
(b) What is the current through the inductor after a long time has elapsed since the switch was closed?
(c) If the switch is now opened, what is the time constant of the decay of the current in the inductor?

3.5 (a) What is $(a + jb)(c - jd)$ equal to?
(b) Express $(2 + j3)(3 - j2)$ as a complex number (in Cartesian form) and find its modulus and argument.
(c) Express $(1 + j)(2 + j)/(3 + j)$ as a complex number (in Cartesian form) and find its modulus and argument.

3.6 Find the arguments of the following complex numbers:
(a) $w_1 = 2.000 - j1.155$
(b) $w_2 = -3.000 - j5.196$
(c) $w_3 = -1.500 + j0.402$.

3.7 Find the amplitude and phase of the resultant of the two phasors:

$$4\cos \omega t \qquad \text{and} \qquad 3\cos(\omega t - 3\pi/2).$$

3.8 Find the amplitude and phase of the resultant of the two phasors:

$$1.5\cos(\omega t + 30^\circ) \qquad \text{and} \qquad 2\sin(\omega t - 60^\circ).$$

3.9 What is the difference in phase between the two phasors represented by

$$\hat{I}_1 \exp\{\mathrm{j}(\omega t + \phi_1)\} \qquad \text{and} \qquad \hat{I}_2 \exp\{\mathrm{j}(\omega t - \phi_2)\}?$$

3.10 A current $i = 10^{-2}\cos \omega t\ A$, where $\omega/2\pi = 5774$ Hz, flows in the network shown in Fig. 3.25. If $R = 100\ \Omega$ and $C = 1/2\pi\ \mu$F, express the voltage v as a phasor.

Fig. 3.25

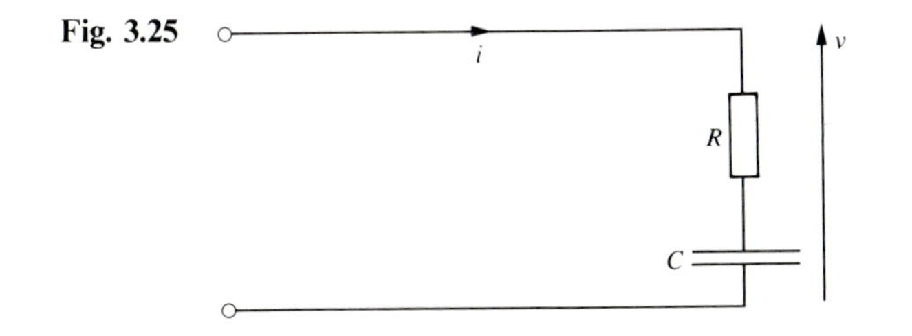

3.11 A voltage $v = \hat{V}\sin \omega t$ is applied to the network shown in Fig. 3.26. Express the current i as a phasor.

Fig. 3.26

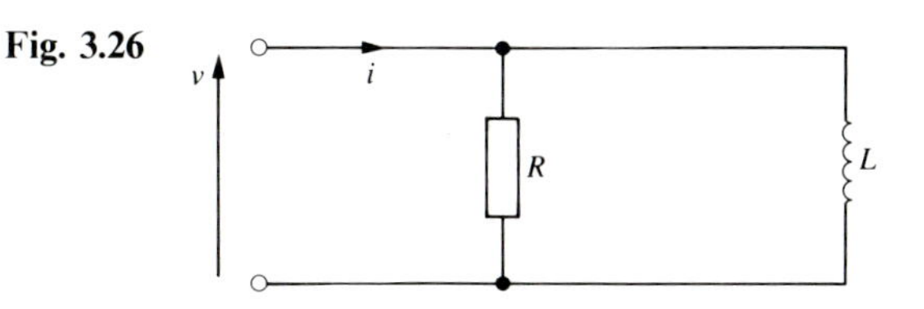

3.12 A current of amplitude 1 A at 50 Hz is flowing through a circuit element consisting of a 100 Ω resistance in series with a capacitance of 10 μF. Calculate the amplitude of the voltage across the circuit element and the phase of this voltage relative to that of the current.

3.13 A voltage of amplitude 10 V and frequency 1 kHz is applied to a circuit element consisting of a resistance of 100 Ω and an inductance of 0.1 H in series. What is the amplitude of the current flowing through the element, and what is the phase angle of the current relative to the applied voltage?

3.14 An e.m.f. of 240 V r.m.s. at 50 Hz is applied to a network containing a capacitance of 5 μF in series with a 1 kΩ resistance. Find the rate at which energy is dissipated in the network, and also find the power factor.

3.15 When an ideal capacitor and a resistance are connected in parallel to a 100 V d.c. supply, a steady current of 1 A flows. When, instead of the d.c. supply, a 100 V r.m.s., 50 Hz supply is connected, the current is 2 A r.m.s. Find (a) the value of the resistance and (b) the capacitance.

3.16 A resistor of 500 Ω and a capacitor of 1 μF are connected in series and a sinusoidal current is flowing through them. If the voltage across the resistor is $0.05 \sin(1000t - \pi/6)$ V, express the voltage across the capacitor as a phasor.

Now express in phasor form the voltage across the series combination of the resistor and the capacitor.

3.17 Obtain expressions for the conductance and the susceptance of the network shown in Fig. 3.27.

Fig. 3.27

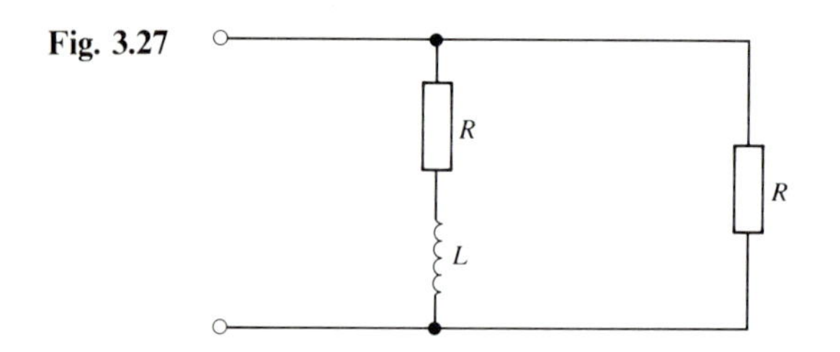

3.18 For the network of Fig. 3.28 find the amplitude of v and its phase relative to $i = \hat{I} \cos \omega t$.

Fig. 3.28

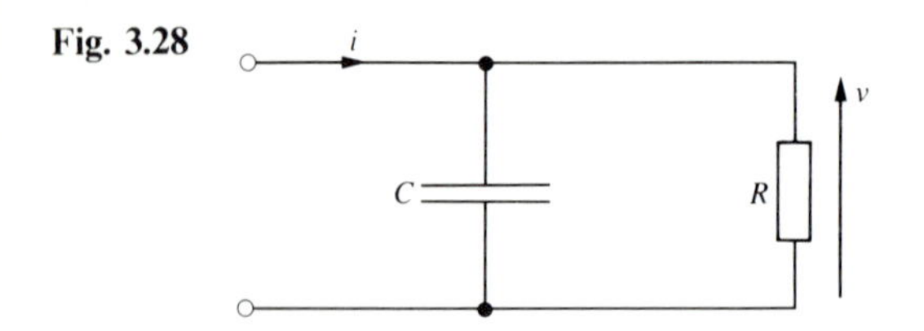

3.19 Obtain expressions for the conductance and the susceptance of the network shown in Fig. 3.29.

Fig. 3.29

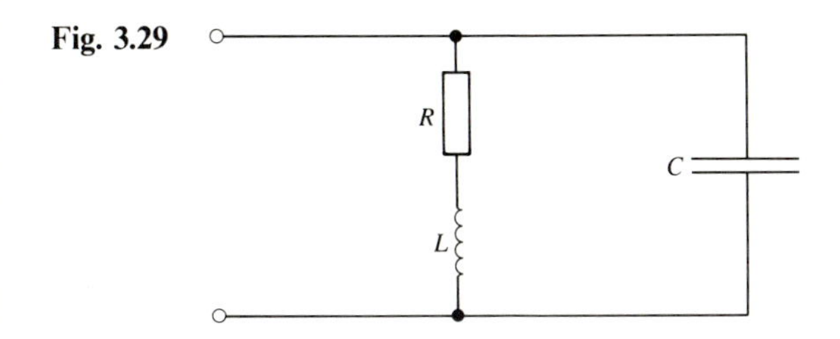

3.20 An e.m.f. of amplitude 240 V at a frequency of 50 Hz is applied to a circuit element consisting of an inductance of 0.1 H in series with a resistance of 100 Ω. Find the power factor of this circuit element and the power dissipation.

3.21 An e.m.f. of amplitude 240 V at a frequency of 50 Hz is applied to a circuit element consisting of a resistance of 1 MΩ in parallel with a capacitance of 1 μF. Find the power factor of the element and the power dissipation.

3.22 A coil has a self-inductance of 1 mH, a self-capacitance of 1 pF and a resistance of 100 Ω. Assuming that the self-capacitance is in parallel with the inductance and resistance, calculate the effective inductance and resistance of the coil at a frequency of $10/2\pi$ MHz.

3.23 Show that the capacitances and resistances of the equivalent networks of Figs. 3.11(a) and (b) are related by

$$R_s = R_p/(1 + \omega^2 C_p^2 R_p^2);\ C_s = (1 + \omega^2 C_p^2 R_p^2)/\omega^2 C_p R_p^2.$$

3.24 For the network of Fig. 3.20(d), find the modulus and argument of the ratio v_2/v_1.

3.25 Show that for the parallel equivalent circuit of a capacitor (see Fig. 3.11(b)), the loss angle δ is given by $\delta \approx \tan\delta = (\omega C_p R_p)^{-1}$.

3.26 For the network shown in Fig. 3.30 show that Y is independent of frequency if $R^2 = L/C$. What is the value of Y in these circumstances?

Fig. 3.30

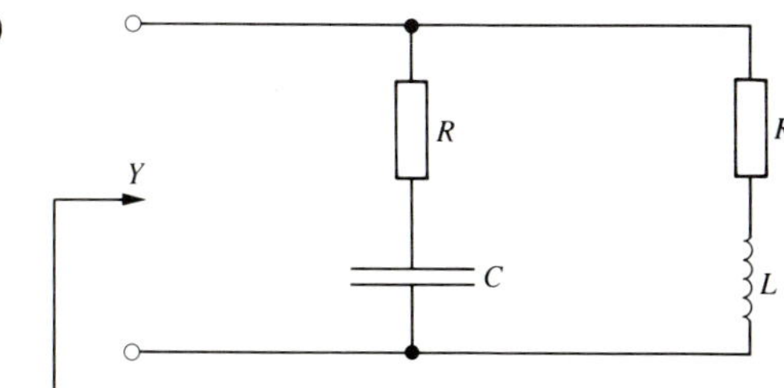

3.27 Calculate the primary and secondary currents in the transformer network of Fig. 3.31, where $L_p = 1\ H$, $L_s = 0.1\ H$, $k = 0.95$, $R_p = 10\ \Omega$, $R_L = 1\ \Omega$, and the resistance of the secondary winding can be neglected.

Fig. 3.31

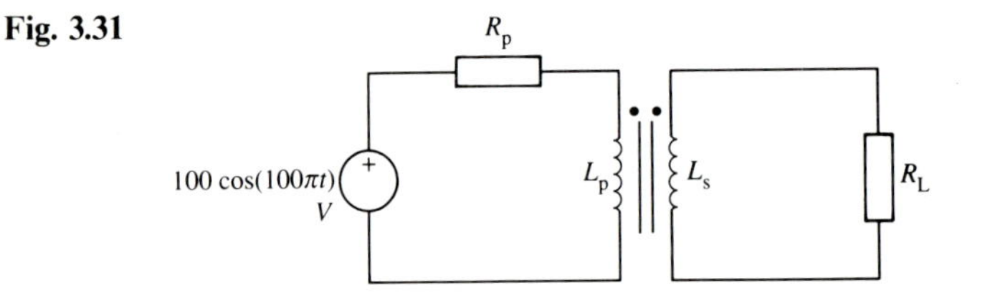

3.28 Find the voltage transfer function v_0/v_i for the network shown in Fig. 3.32.

Fig. 3.32

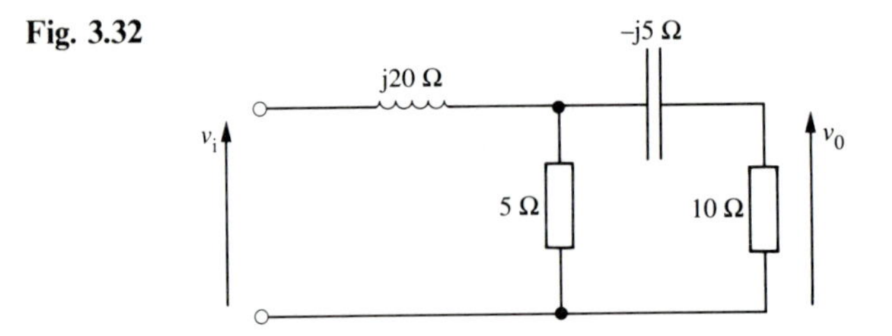

3.29 Find the Norton equivalent of the network shown in Fig. 3.33.

Fig. 3.33

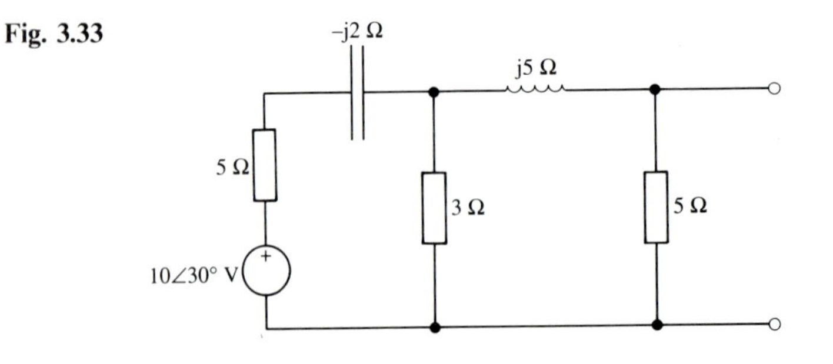

3.30 Find the Thévenin equivalent of the network shown in Fig. 3.34.

Fig. 3.34

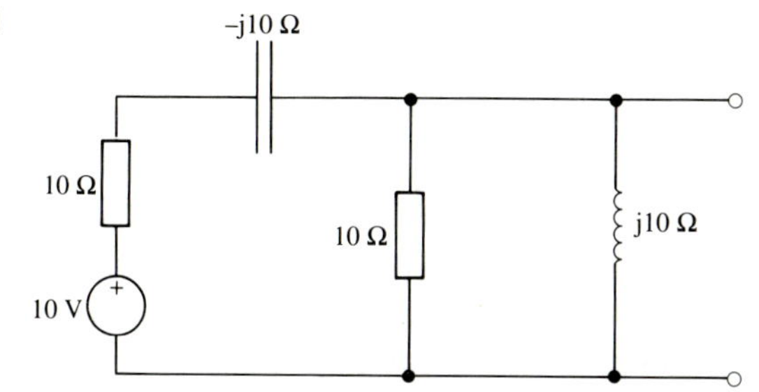

3.31 Find the current i and the power delivered by the source in the circuit of Fig. 3.35.

Fig. 3.35

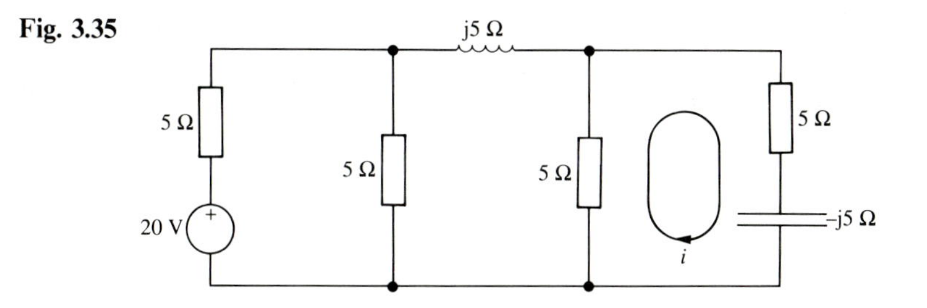

Resonance 4

4.1 Introduction: Free, damped oscillations of a system

A mass M attached to a spring, as shown schematically in Fig. 4.1(a), is a simple example of a physical system which, after being displaced from its equilibrium position and then released, will execute oscillations at a natural frequency that is characteristic of the system. Assuming that friction with the air and internal friction in the spring are negligible, then the net force acting on M is the restoring force due to the spring, which, for small-enough extensions, will be assumed to be proportional to the extension. If the force constant (force per unit extension) of the spring is denoted by K, then the extension of the spring at equilibrium (y_0) is given by

$$Ky_0 = Mg.$$

For an additional downward displacement y of M, the net downward force is

$$Mg - K(y_0 + y)$$

and, by Newton's second law of motion, this is equal to the product of the mass M and the acceleration $\mathrm{d}^2y/\mathrm{d}t^2$:

$$Mg - K(y + y_0) = M\,\mathrm{d}y/\mathrm{d}t$$

or

$$\mathrm{d}^2y/\mathrm{d}t^2 = -\frac{K}{M}\,y(t),$$

since $Ky_0 = Mg$. This is a second-order differential equation.

It can be shown by simple substitution that a function of the form

$$y(t) = A\exp(\pm \mathrm{j}\omega_0 t)$$

satisfies this differential equation, where $\omega_0^2 \equiv K/M$ is the 'natural' frequency of the (undamped) system. However, because it is a second-order differential equation, a solution must contain two constants of integration and, since it is a linear equation also, the sum of any two solutions is also a solution. Hence the general solution can be written in the form

$$y = A\exp(\mathrm{j}\omega_0 t) + B\exp(-\mathrm{j}\omega_0 t).$$

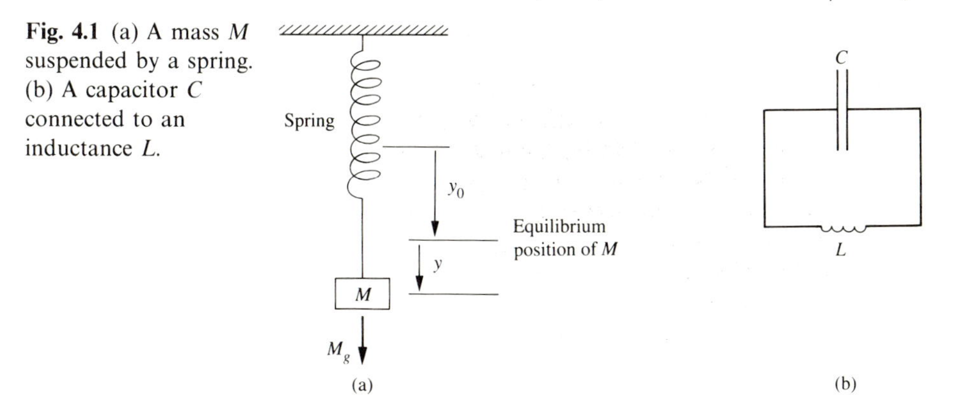

Fig. 4.1 (a) A mass M suspended by a spring. (b) A capacitor C connected to an inductance L.

If the boundary conditions are taken to be $y = \hat{y}$, $dy/dt = 0$, at time $t = 0$, then it follows that

$$y = \hat{y}\{\exp(j\omega_0 t) + \exp(-j\omega_0 t)\}/2 \quad \text{or} \quad y = \hat{y}\cos\omega_0 t.$$

Another idealized physical system that can be described by a differential equation of the same general form as above is a capacitor (C) connected to an inductance (L); see Fig. 4.1(b). On applying KVL to the potential differences across the capacitance (C) and inductance (L), it follows that

$$q/C + L\,dI/dt = 0$$

or, since $I = dq/dt$,

$$L\,d^2q/dt^2 = -q/C.$$

It can be seen that pairs of analogous quantities in these examples of mechanical and electrical systems are L and M, and C and $1/K$ (the mechanical 'compliance'). Applying the boundary conditions $q = \hat{q}$, I ($= dq/dt$) $= 0$, at $t = 0$ yields

$$q = \hat{q}\{\exp(j\omega_0 t) + \exp(-j\omega_0 t)\}/2,$$

where

$$\omega_0(= 2\pi f_0) \equiv 1/LC \tag{4.1}$$

or

$$q = \hat{q}\cos\omega_0 t = \hat{q}\,\mathrm{Re}\{\exp(j\omega_0 t)\}$$

and

$$I = dq/dt = -\omega_0\hat{q}\sin\omega_0 t = -\omega_0\hat{q}\,\mathrm{Im}\{\exp(j\omega_0 t)\} = \omega_0\hat{q}\cos(\omega_0 t + 90^0).$$

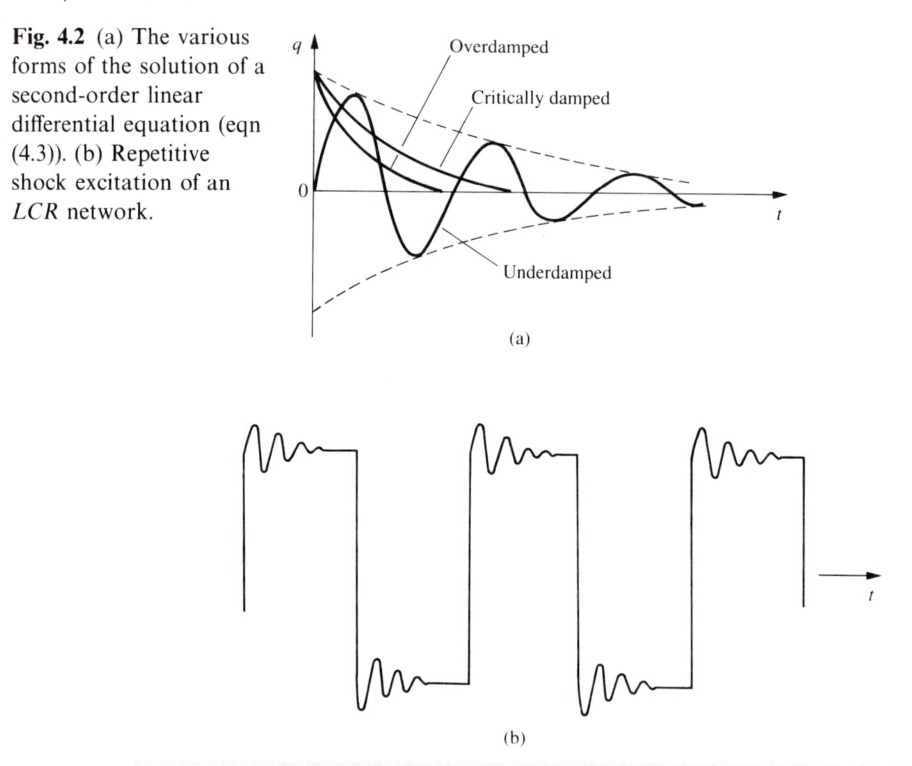

Fig. 4.2 (a) The various forms of the solution of a second-order linear differential equation (eqn (4.3)). (b) Repetitive shock excitation of an *LCR* network.

The situation where $R^2 < 4L/C$ and where the solution of the differential equation is a damped oscillation is called the *underdamped case* (see Fig. 4.2(a)).

The time-dependence of q is radically different if s is real, i.e. if $R^2 > 4L/C$; the general solution is now non-oscillatory and the circuit is said to be **overdamped** (see Fig. 4.2(a)).

The special case where $R^2 = 4L/C$ again leads to a non-oscillatory solution, the system now being described as **critically damped** (see Fig. 4.2(a)).

Note that a powerful and convenient method for obtaining the solutions of differential equations that arise in engineering contexts, namely the use of Laplace transforms, is introduced in Section 7.4.

Many mechanical and electromechanical systems are adjusted to be critically damped, since this gives the quickest approach to the equilibrium situation in response to a 'step' excitation (such as the switching-on of a voltage or a current source in a circuit). Examples are moving-coil current- and voltage-measuring instruments, pen-recorder movements and other servo-mechanical systems.

Underdamped (i.e. oscillatory) electrical circuits are frequently characterized by a **quality factor** (or Q-factor) which depends on the ratio of the stored energy to the energy losses and which is defined by:

$$Q \equiv 2\pi \times (\text{energy stored in the circuit})/(\text{energy loss per cycle}). \tag{4.7}$$

In this case, for which $R^2 < 4L/C$, the stored energy $W(t)$ in the LCR circuit at time t is given by

$$W(t) = q^2/2C + LI^2/2 \qquad \text{or} \qquad W(t) = q^2/2C + \frac{L}{2}(\mathrm{d}q/\mathrm{d}t)^2,$$

and using the solution (eqn (4.5)) for q it follows that

$$W(t) = \frac{\hat{q}^2}{2}\mathrm{e}^{-2\alpha t} \times \{(1/C + \alpha^2 L)\cos^2\omega_r t + \omega_r^2 L \sin^2\omega_r t + 2\alpha\omega_r L \sin\omega_r t \cos\omega_r t\}.$$

Since the duration of one cycle is equal to $2\pi/\omega_r$, it follows that

$$Q = 2\pi W(t) \Big/ \left\{ W(t) - W\left(t + \frac{2\pi}{\omega_r}\right)\right\} = 2\pi/\{1 - \exp(-4\pi\alpha/\omega_r)\}.$$

If $\alpha^2 \ll \omega_0^2$, then, from eqn (4.6) above, $\omega_r \approx \omega_0$ and $\alpha \ll \omega_r$ also, of course. On making use of the general approximate relationship $\mathrm{e}^{-x} \approx 1 - x$ if $x \ll 1$, it follows that, since $4\pi\alpha/\omega_r \ll 1$,

$$Q \approx \omega_0/2\alpha \qquad \text{or} \qquad Q \approx \omega_0 L/R.$$

Quality factors are of great practical interest, in that they determine the selectivity of filter networks and 'tuned' (i.e. frequency-selective) amplifiers, for instance. Q values range, typically, from ~ 10 in the audio frequency range (30 Hz to 20 kHz) to ~ 100 in the radiofrequency range (100 MHz, say) and even up to ~ 1000 in microwave circuits (1–10 GHz).

From eqn (4.6),

$$\omega_r^2 = \omega_0^2\left(1 - \frac{\alpha^2}{\omega_0^2}\right)$$

$$\approx \omega_0^2\left(1 - \frac{1}{4Q^2}\right). \tag{4.8}$$

So it can be seen that for a Q-factor even as small as 5 there is only a difference of 0.5 per cent between ω_r and ω_0, and so for $Q > 5$ it is usually safe to use

$$Q = \frac{\omega_0 L}{R}. \tag{4.9(a)}$$

By using eqn (4.1), alternative expressions for Q can be obtained:

$$Q = \frac{1}{R}\sqrt{\left(\frac{L}{C}\right)} \quad \text{and} \quad Q = \frac{1}{\omega_0 CR}. \tag{4.9(b)}$$

For systems having low Q values ($\leqslant 1$), such as many electromechanical control systems, it is usual to characterize the energy losses by a **damping factor** ζ, where $\zeta \equiv 1/2Q$.

Damped oscillations of an LCR circuit can be demonstrated by using a 'square wave' signal generator as the source, thus periodically 'shocking' the circuit into oscillation (see Fig. 4.1(b)), and displaying the voltage across R, say, on a cathode ray oscilloscope. Note that the period of the square wave must be long enough to allow sufficient cycles of oscillation to occur. The fact that a source is coupled to the LCR network means that the situation is not quite as simple as for the ideally isolated circuit that we have implicitly assumed so far; the effect of such coupling will be discussed in Section 4.3.

This phenomenon of damped oscillations following shock excitation is commonly known as **ringing**, by analogy with a bell.

4.2 Forced, damped oscillations

4.2.1 Series *LCR* circuit

Imagine that an ideal constant-voltage generator of negligible internal impedance, and angular frequency ω, is connected in series with an

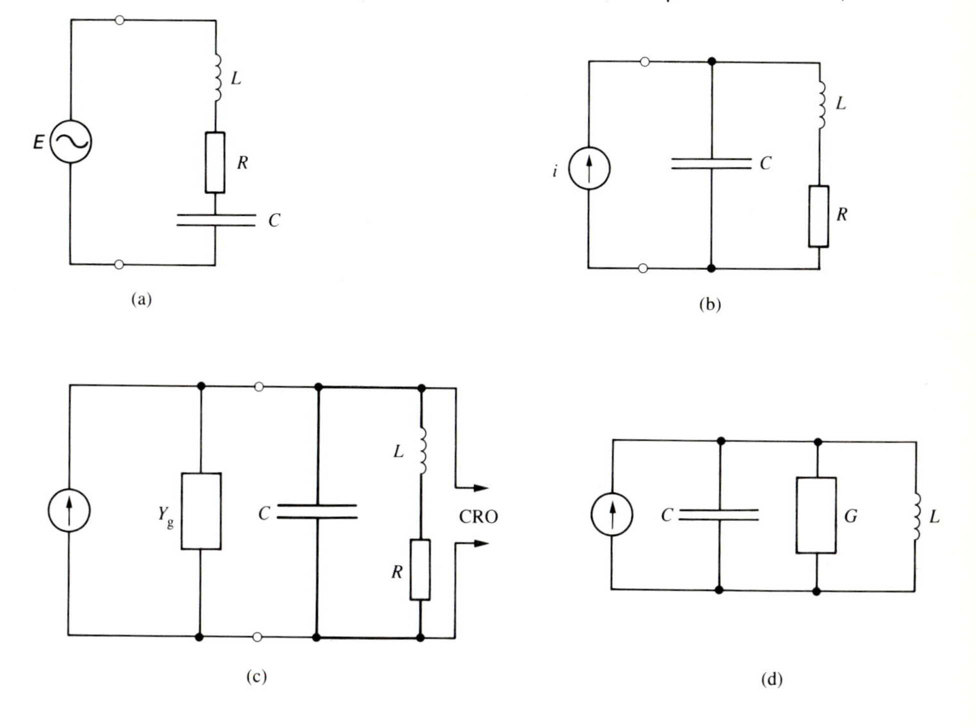

Fig. 4.3 (a) A series *LCR* network coupled to an ideal voltage source. (b) A parallel *LCR* network coupled to an ideal current source and (c) coupled to a source of admittance Y_g. (d) The dual of the series circuit of (a).

inductor L, a resistance R, and a capacitance C (see Fig. 4.3(a)). Assuming that the voltage (E) of the source is *sinusoidal*, then, using a.c. circuit theory as developed in Chapter 3, the impedance Z of the network is given by

$$Z = R + j\omega L - \frac{j}{\omega C} \tag{4.10}$$

and the current I is given by $I = E/Z$.

So

$$I = E \bigg/ \left\{ R + j\left(\omega L - \frac{1}{\omega C} \right) \right\}. \tag{4.11}$$

Remembering that $\omega_0 \equiv (LC)^{-1/2}$, then, after rationalizing this expression for I, it follows that

$$I = \frac{E\left\{R - j\omega L\left(1 - \frac{\omega_0^2}{\omega^2}\right)\right\}}{\left\{R^2 + \omega^2 L^2\left(1 - \frac{\omega_0^2}{\omega^2}\right)^2\right\}}.$$

The peak value $\hat{I}$ of the sinusoidal current is given by $|I|$ and the phase angle (ϕ) of I with respect to the voltage E is given by

$$\tan\phi = \mathrm{Re}(I)/\mathrm{Im}(I).$$

So

$$\hat{I} = \frac{\hat{E}}{\left\{R^2 + \omega^2 L^2\left(1 - \frac{\omega_0^2}{\omega^2}\right)^2\right\}^{1/2}} \tag{4.12}$$

and

$$\tan\phi = -\frac{\omega L}{R}\left(1 - \frac{\omega_0^2}{\omega^2}\right). \tag{4.13}$$

From eqn (4.12), and remembering that $\omega_0 = 1/\sqrt{(LC)}$, it can be seen that if ω or L or C is varied, then $\hat{I}$ is a maximum when the condition $\omega^2 = 1/LC$ ($\equiv \omega_0^2$) is satisfied.

This is an example, in an electrical context, of **resonance**. In general a system is said to be a resonant system if its response (electric current in the situation just discussed above) as a function of the frequency of a constant-amplitude sinusoidal excitation exhibits a maximum at a particular frequency (sometimes at more than one frequency); see Fig. 4.4(c). Resonance occurs in a wide range of types of electrical and electromechanical systems, and can be exploited usefully to give frequency-selective responses. In some contexts it is undesirable, e.g. large-amplitude mechanical vibrations in engineering structures, or unwanted resonances in loudspeakers.

In the LCR circuit of Fig. 4.3(a), resonance can be observed by monitoring the amplitude of the voltage across L or R. Actually the values of the variables ω or L or C to give maximum $\hat{V}_L$ or $\hat{V}_C$ differ from those satisfying $\omega_0 = 1/\sqrt{(LC)}$ by factors like $\{1 \pm 1/(nQ^2)\}$, where $n = 1, 2$, or 4. However, provided that $Q \geqslant 10$, the condition for resonance $\omega_r^2 = \omega_0^2 = 1/LC$ holds true to a high degree of accuracy.

The variations of $|Z|/R$, and of $\angle Z$, with ω/ω_0 for a series LCR network are shown in Fig. 4.4(b). Notice that at resonance, $\phi = 0$, i.e. the impedance of the network is purely resistive. At frequencies much less

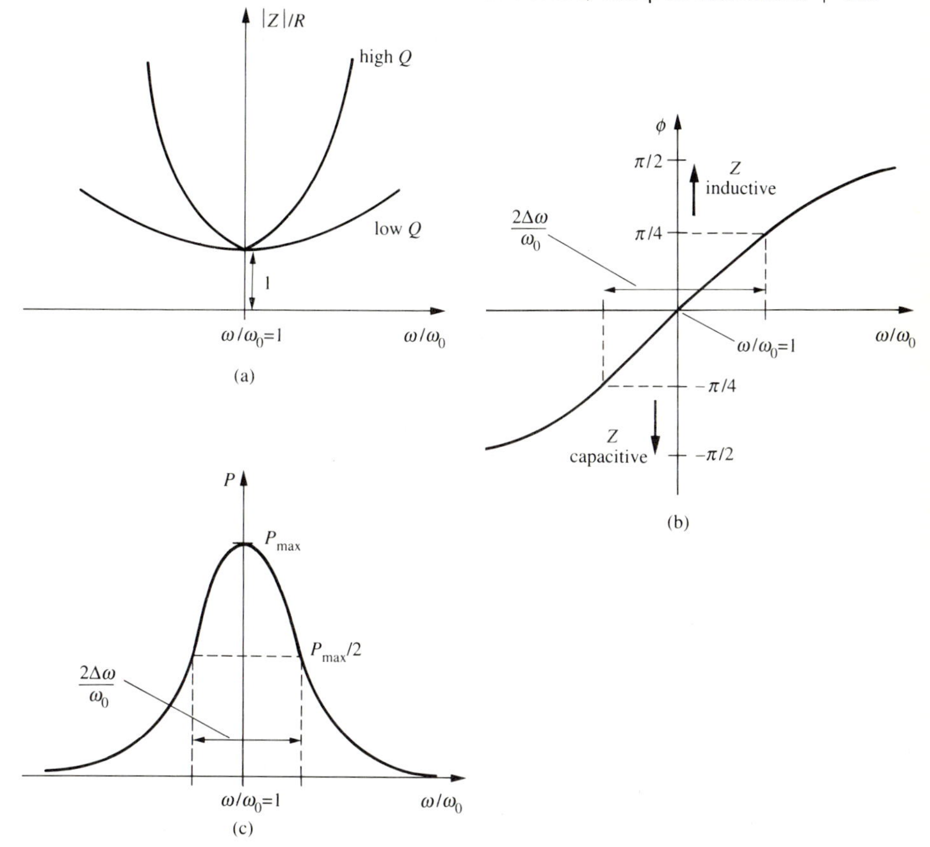

Fig. 4.4 (a) $|Z|/R$ for a series LCR network. (b) Phase angle of Z for a series LCR network. (c) Power dissipation in a resonant circuit.

than ω_0 the reactance of the capacitance becomes much greater than that of the inductance and the network behaves 'capacitively' (i.e. $\phi \to -90°$). Conversely, at frequencies much greater than ω_0 the reactance of the inductance dominates and the network behaves 'inductively' (i.e. $\phi \to +90°$). The frequency-dependence of $\angle Z$ for the parallel LCR network of Fig. 4.3(b) follows from analogous arguments.

For the series LCR network the power (P) delivered to the circuit by the source is given by $P = (\hat{E}\hat{I} \cos \phi)/2$ or, using eqns (4.12) and (4.13),

$$P = \hat{E}^2/[2R\{Q^2(\omega_0^2/\omega^2 - 1)^2\omega^2/\omega_0^2 + 1\}]. \tag{4.14}$$

Thus maximum power is delivered at resonance and is given by

$$P_{\max} = \hat{E}^2/2R. \tag{4.15}$$

The Q-factor of the network is directly related to the shape (or 'sharpness') of the power response curve (e.g. Fig. 4.4(c)), as can be seen from the following considerations. For $\tan\phi = 1$ (i.e. $\phi = 45°$), then from eqn (4.13),

$$(\omega_0^2 - \omega^2) = \frac{\omega\omega_0}{Q}$$

or

$$(\omega_0 - \omega)(\omega_0 + \omega) = \frac{\omega\omega_0}{Q}.$$

In most cases of practical interest, $Q \geqslant 10$ and the response curve falls off very quickly as ω deviates from ω_0. So it is assumed that the response is significant only for frequencies ω close to ω_0 ($\omega \approx \omega_0$), with the consequence that $(\omega_0 + \omega) \approx 2\omega_0$. Further, $\Delta\omega \equiv (\omega_0 - \omega)$, so that (see Fig. 4.4(c))

$$Q = \omega_0/2\Delta\omega. \tag{4.16}$$

It follows from eqn (4.14) that $P = P_{max}/2$ for the values of ω for which $\tan\phi = 1$ (i.e. for $\omega = \omega_0 \pm \Delta\omega$), and so the 'sharpness' of a resonance is often specified in terms of the frequency difference (or **bandwidth**) between the two frequencies at which the power absorption by the network is half the maximum (the **half-power points**).

It is useful to introduce the **decibel** notation at this juncture. In dealing with ratios of quantities whose magnitudes range over many orders of magnitude (i.e. powers of ten), it has proved convenient in many fields of engineering and science to use the logarithm of the ratio (since $\log_{10}(10^n) = n$).

As originally introduced in the field of acoustics, the ratio of two powers (P_1, P_2, say) was expressed in bels as $\log_{10}(P_2/P_1)$. However, 1 bel was found to be an inconveniently large unit and so the decibel (dB) was introduced, i.e. $10\log_{10}(P_2/P_1)$. For example, the half-power points referred to above are frequently called the **3 dB points**, since $P = P_{max}/2$ and

$$\begin{aligned} 10\log(P/P_{max}) &= 10\log(1/2) = -3.010\text{ dB} \\ &= -3\text{ dB very closely.} \end{aligned} \tag{4.17}$$

Note that logarithms to the base ten are used. Further note that the minus sign indicates that the ratio in question has a magnitude less than unity; it would be said that at the half-power points the power is 3 dB 'down' on P_{max}.

In using the decibel notation to specify the ratio of two powers, beware of situations where the associated voltages are developed across resistances of different magnitudes. For example, consider the power ratio when a voltage v_1 is developed across R_1, and v_2 is developed across R_2:

$$\text{power ratio} = v_2^2/R_2/v_1^2/R_1$$

$$\begin{aligned}\text{power ratio (dB)} &= 10\log(v_2^2/R_2) - 10\log(v_1^2/R_1) \\ &= 20\log(v_2/v_1) + 10\log(R_1/R_2).\end{aligned} \tag{4.18}$$

The voltage ratio v_2/v_1 could refer to the voltage gain of an amplifier, for instance, and this gain might be specified (in dB) as the value of 20 log (v_2/v_1). However, this dB value does not correspond to the power gain of the amplifier unless $R_1 = R_2$.

In a number of contexts in which quantities are commonly expressed in dB there are widely recognized reference levels. For example, in acoustics a commonly used reference level is the sound pressure level corresponding to the threshold of hearing. In radio receiving systems the level of an incoming signal is often expressed as a decibel ratio ('dBm') relative to a power level of 1 mW. So, for instance, a receiver input power of 1 μW is equal to

$$10\log(10^{-6}/10^{-3}) = -30\ \text{dBm}.$$

It is left as a problem to show that at resonance in a series LCR circuit the voltage across the capacitance and the voltage across the inductance are 180° out of phase with each other and that each has an amplitude Q times the voltage of the source. Thus the Q-factor can also be thought of as a 'magnification factor'.

4.2.2 Parallel *LCR* circuit

Imagine now that an ideal constant-current source is connected to a capacitor and an inductor as shown in Fig. 4.3(b); the resistance R is placed in series with L since it corresponds to the resistance of the coil of the inductance (plus any additional resistance connected in this branch of the circuit). It is convenient to analyse such a parallel circuit in terms of admittances rather than impedances, as outlined in Section 3.4:

$$Y = \frac{1}{(R + 1\mathrm{j}\omega L)} + \mathrm{j}\omega C = \frac{R}{(R^2 + \omega^2 L^2)} + \mathrm{j}\omega\left\{C - \frac{L}{(R^2 + \omega^2 L^2)}\right\} \tag{4.19}$$

The condition for **electrical resonance** in a circuit can be expressed, for convenience, as the condition that the impedance (Z) or the admittance (Y), as the case may be, is *real.*

In the situation being considered, this condition yields $R^2 + \omega_r^2 L^2 = L/C$, which leads to an expression for the resonance angular frequency (ω_r) of the form

$$\omega_r^2 = \left(\frac{1}{LC}\right)\left\{1 - \frac{1}{Q^2}\right\}, \tag{4.20}$$

where, as for the series LCR circuit, $Q = (1/R)\sqrt{(1/LC)}$. Again for $Q \geqslant 10$, $\omega_r^2 = 1/LC$ for most practical purposes.

$|Y|$ is a minimum at resonance (and hence $|Z|$ is a maximum) but, as with the series resonant circuit, slightly different conditions for resonance are obtained, depending on whether the variable is ω, L, or C. However, these differences are $\sim 1/Q^2$ and so can usually be ignored.

From eqns (4.19) and (4.20) it follows that, at resonance,

$$Y = 1/R(1 + Q^2) \approx 1/RQ^2. \tag{4.21}$$

Now, for frequencies much greater than ω_0, the reactance ($\omega_0 L$) of the inductance in the circuit of Fig. 4.3(b) is very much greater than that of the capacitance (i.e. $1/\omega_0 C$), so the network behaves capacitively. Conversely, at frequencies much less than ω_0 the network behaves inductively.

From eqn (4.19),

$$\underline{/Y} = \tan^{-1}\left\{\frac{\omega CR^2 + \omega L(\omega^2 LC - 1)}{R}\right\}$$

$$= \tan^{-1}\left[\frac{\omega}{\omega_0}\left\{\frac{1}{Q} + Q\left(\frac{\omega^2}{\omega_0^2} - 1\right)\right\}\right]. \tag{4.22}$$

It should be noted that for $\omega = \omega_0$, $\underline{/Y} = \tan^{-1}(1/Q)$. For $Q = 10$, $\underline{/Y} \approx 5°$, so in practice (i.e. $Q \geqslant 10$), $\underline{/Y}$ is very small. For the parallel LCR network of Fig. 4.3(d), which is the dual of the network of Fig. 4.3(a), the phase angle of Y is zero at resonance.

4.3 Some practical aspects of resonant circuits

First the effect on the properties of resonant circuits of using real as opposed to ideal sources will be considered. For a series circuit, if the voltage source has internal impedance $Z_g = R_g + jX_g$, then the resonance condition (imaginary part of total impedance $= 0$) becomes

$$\{\omega_r L - 1/(\omega_r C) + X_g = 0\},$$

so $|X_g|$ should be small for ω_r to be close to $\sqrt{(1/LC)}$. Also R_g should be small in order not to severely degrade the Q-factor, which will be equal to $\omega_0 L/(R + R_g)$.

For a parallel circuit, if the practical current source has an internal shunt admittance Y_g (see Fig. 4.3(c)), then the admittance Y 'seen' by the ideal source is

$$Y = Y_g + j\omega C + \frac{1}{(R + j\omega L)} = G_g + jB_g + j\omega C + \frac{(R - j\omega L)}{(R^2 + \omega^2 L^2)}. \tag{4.23}$$

Consequently Y is real if $\{B_g + \omega C - \omega L/(R^2 + \omega^2 L^2)\} = 0$, and so B_g should be relatively small in order for the resonance frequency to be close to $(LC)^{-1/2}$.

In practice the current source will often be realized by using a commercial signal generator (a voltage source of internal resistance 60 Ω typically) with a large-enough resistance R_g connected in series with its output terminals (remember that an ideal current source has an infinite output impedance; see Fig. 4.3(c)). If it can be assumed that the input impedance of the CRO (~ 1 MΩ at least, typically) is high enough so that the CRO itself does not significantly 'load' the *LCR* network,* then it can be shown (see Section 7.4.1) that the effective Q-factor (Q_L) of the 'loaded' *LCR* network (i.e. loaded by the signal generator) is given by

$$\frac{1}{Q_L} = \frac{1}{Q_u} + \frac{Q_u R}{R_g}, \tag{4.24}$$

where Q_u ($\equiv \omega_0 L/R$) is the Q-factor of the 'unloaded' (i.e. isolated) *LCR* network.

The second term in eqn (4.24) accounts for the coupling between the resonant system and the 'outside world', which in this case is represented by the signal generator. The tighter the coupling (i.e. the smaller the value of R_g in this particular situation), the smaller is the loaded Q-factor. In the limit of R_g becoming very large, then the value of Q_L tends to the unloaded Q-factor Q_u (but, of course, the larger is R_g, the smaller is the fraction of the signal generator voltage that is actually applied to the *LCR* network). So a general rule to be followed in the use of a resonant system

* From eqn (4.23) the impedance of the parallel *LCR* network at resonance is equal to $(R^2 + \omega^2 L^2)/R \approx RQ^2$ if $Q \geqslant 10$. Hence the input impedance of the CRO should be at least ten times this in order to avoid loading the *LCR* network and hence reducing the effective Q-factor; e.g. for $R = 10\ \Omega$ and $Q = 50$, then $RQ^2 = 25$ kΩ.

Fig. 4.5 An example of a pair of coupled circuits.

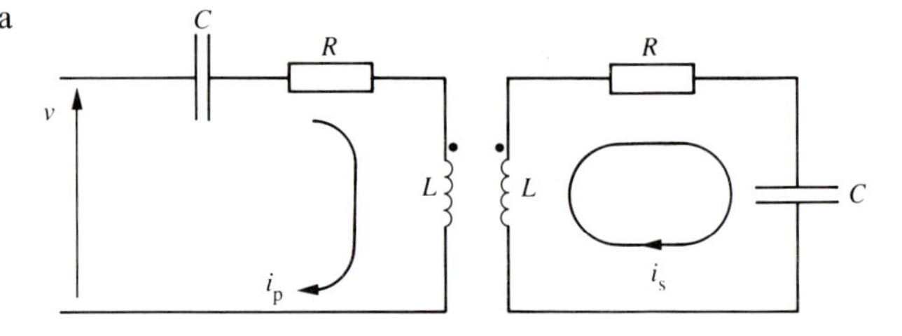

is that in order not to degrade its Q-factor, the system should be only loosely coupled to the 'outside world'.

Because of the magnification property of resonant circuits with regard to voltage and current, care must be taken to ensure that the capacitors and inductors used are rated to withstand the expected voltages and currents. For instance, in the series LCR circuit of Fig. 4.3(a) the peak voltage across the capacitor, at resonance, is $\hat{E}/\omega_r RC$ or $Q\hat{E}$.

4.4 Coupled resonant circuits

A general description of the properties of a pair of coupled resonant circuits (see Fig. 4.5) is quite complicated because of the number of variables involved. For illustrative purposes the situation will be considered where the primary and secondary circuits are identical; this will serve to indicate the most important features of coupled circuits in general.

On applying KVL to the network depicted in Fig. 4.5, it follows that

$$i_p(R + jX) - j\omega M i_s = v_p$$

$$-j\omega M i_p + i_s(R + jX) = 0,$$

where $X = \left(\omega L - \dfrac{1}{\omega C}\right)$. Thus

$$i_s = \frac{j\omega M v_p}{(R^2 - X^2 + j2XR + \omega^2 M^2)}.$$

If we consider the situation near resonance such that $\omega = \omega_0 \pm \Delta\omega$, where $\Delta\omega \ll \omega_0$, then

$$X = \left(\omega L - \frac{1}{\omega C}\right) = \frac{L(\omega^2 - \omega_0^2)}{\omega}$$

or

$$X \approx 2L\Delta\omega,$$

where it has been assumed that $(\omega + \omega_0) \approx 2\omega_0$.

Using this result,

$$\hat{I}_s = \omega_0 M v_p \left[\left\{ R^2 - 4L^2(\Delta\omega)^2 + \omega_0^2 M^2 \right\} + 16R^2L^2(\Delta\omega)^2 \right]^{-1/2}$$

In order to find the condition for $\hat{I}_s$ to be a maximum, this expression is differentiated expression with respect to $\Delta\omega$ and the result set equal to zero:

$$\Delta\omega\{4L^2(\Delta\omega)^2 + (R^2 - \omega_0^2 M^2)\} = 0.$$

Remembering that $M^2 = k^2L^2$ and that $Q \equiv \omega_0 L/R$, it is found that for $\hat{I}_s$ to be a maximum

$$\Delta\omega = 0 \qquad \text{or} \qquad \Delta\omega = \pm\frac{\omega_0}{2Q}\sqrt{\{(kQ)^2 - 1\}}.$$

Only real roots are of physical interest. If $(kQ)^2 > 1$ there are three real roots, there being a minimum of $\hat{I}_s$ for $\Delta\omega = 0$ and maxima for the other two roots. If $(kQ)^2 < 1$ there is a single real root (and a maximum) at $\Delta\omega = 0$. For $(kQ)^2 = 1$ the coupling is said to be critical and the three roots coincide.

The most important feature which emerges from this cursory discussion is that as the coupling between the two circuits is made tighter (i.e. the larger the value of k), the resonance is split into two peaks: this is a feature of coupled physical oscillators in general, ranging from the atomic level to engineering systems.

There is a wide range of variations on this theme, depending on whether both or only one of the primary and secondary circuits are resonant circuits, and depending on whether or not they are tuned to the same resonance frequency. In addition, the primary and/or secondary circuit may be a parallel circuit and not a series circuit such as has been used for illustrative purposes.

The critically coupled condition is often used in radiofrequency amplifier circuits to give a 'band pass' frequency response (see Section 7.5), since the 'roll-off' in the stop-bands is steepest for this condition.

Problems

4.1 Show that for forced resonance in a series LCR network the voltage across the inductance and the capacitance are 180° out of phase and have amplitudes Q times that of the voltage source.

4.2 Find the admittance of the parallel LCR network shown in Fig. 4.6 and hence show that the resonance frequency ω_r is given by

$$\omega_r^2 = \frac{1}{LC} - \frac{R^2}{L^2}, \qquad \text{i.e.} \quad \omega_r^2 = \omega_0^2\left(1 - \frac{1}{Q^2}\right).$$

Fig. 4.6

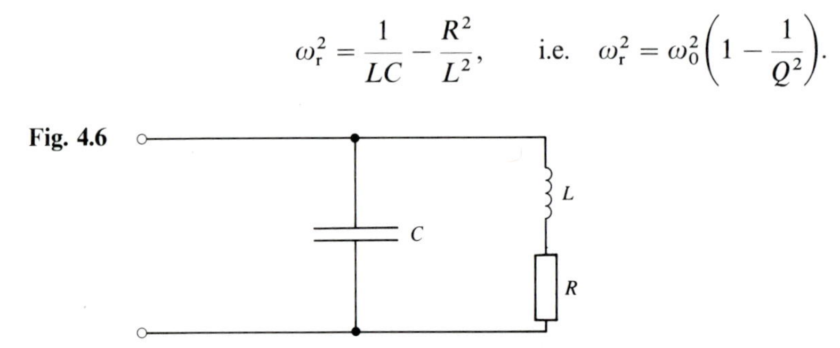

4.3 Calculate the values of the capacitances that would resonate with a 1 mH inductance at frequencies of 100 kHz and 10 MHz. What values of resistance would yield a Q-value of 100 at each of these two frequencies?

4.4 Find the reactance and the resonance frequency of the network shown in Fig. 4.7.

Fig. 4.7

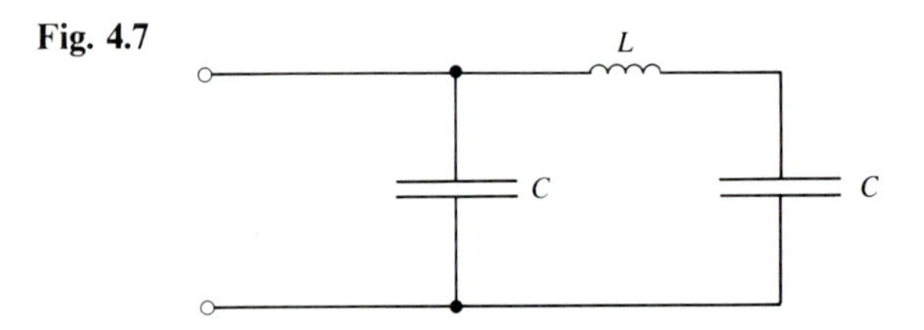

4.5 For the network shown in Fig. 4.8 calculate the values of C, R to give a resonance frequency of 200 kHz and a Q-factor equal to 20.

Fig. 4.8

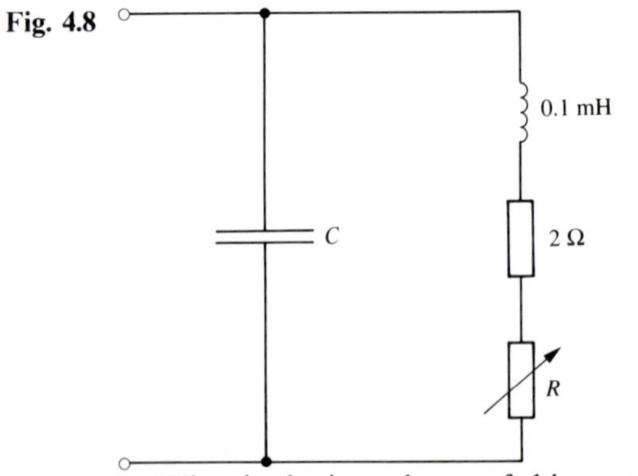

What is the impedance of this network at resonance?

Electromagnetics

5

This chapter is concerned with some of the electrical and magnetic properties of physical media in so far as they are of immediate relevance to the material components of electrical circuits. The range of physical phenomena underlying the technical features of modern circuit components is very wide, too wide to be covered in detail in a text with the aims of this one. Thus the choice of topics is highly selective and the treatments are largely phenomenological in nature. As with Chapter 1, it is more important that the basic principles should be recognized and the important results noted at a first reading; a consideration of the detailed explanations, where given, and of mathematical analyses can be deferred to a second reading.

5.1 Electric fields in conductors, semiconductors, and insulators

At 300 K the d.c. conductivities of solids range from $10^7\ \mathrm{S\,m^{-1}}$ in some metals, through values in the roughly range 10^1–$10^4\ \mathrm{S\,m^{-1}}$ for semiconductors, down to $10^{-18}\ \mathrm{S\,m^{-1}}$ in good insulators. The term **dielectric** is usually reserved for materials having conductivities less than $10^{-12}\ \mathrm{S\,m^{-1}}$, although the divisions between dielectrics, so-called semi-insulators, and semiconductors, and also between semiconductors, semimetals, and metals are very blurred indeed.

In this text the predominant interest is in passive circuit components with linear characteristics. The effects of electric fields in conducting materials are covered to sufficient depth by the definition of resistance (eqn (1.3)) and the statement of Ohm's law given in Section 1.2. Hence this section is primarily concerned with the effects of static and alternating electric fields in insulating materials, inasmuch as they are relevant to the properties of electrical circuits.

The energy losses (**dielectric losses**) that occur when alternating electric fields are applied to dielectrics are due to a number of physical mechanisms other than the ohmic losses arising in the straightforward transport of charge through the material. These physical mechanisms will not be discussed in detail, and it will be assumed that for a dielectric-filled capacitor the energy loss in the dielectric can be represented by a series resistance R_s or by a parallel resistance R_p (see Section 3.6.1).

Table 5.1. Typical electrical properties of a selection of dielectrics

Material	Static relative permittivity, ϵ_s	tan δ (at 1 kHz)	Time constant (s)
Paper	3.7	$<10^{-2}$	10^3
Mica	6	10^{-3}	5×10^2
Polyethylene	2.2	5×10^{-5}	10
Polypropylene	2.3	10^{-4}	10^4
PTFE	2.1	5×10^{-5}	10^7
Polystyrene	2.5	$<5 \times 10^{-5}$	10^7
Polycarbonate	1.5	3×10^{-3}	10^4
Corundum (Al_2O_3)	7	2×10^{-3}	10^2
Silica (SiO_2)	4	10^{-4}	10^5
Ceramics	5.5–7.5	10^{-3}	10^3

Now the relative permittivity ϵ_r of a dielectric can be defined (eqn (1.28)) through $\epsilon_r \equiv C/C_0$, where C_0 is the capacitance of a capacitor before being filled with the dielectric and C is the capacitance with the dielectric material in place. For a parallel-plate capacitor having plates of area A and separation t,

$$C = \frac{\epsilon_r \epsilon_0 A}{t} \quad \text{and} \quad R_s = \frac{t}{\sigma A};$$

see eqn (1.4); the conductivity σ *is equal to the reciprocal of the resistivity* ρ. Hence

$$CR_s = \frac{\epsilon_r \epsilon_0}{\sigma}. \tag{5.1}$$

If $\epsilon_r \approx 5$ and $\sigma < 10^{-12}\,\mathrm{S\,m^{-1}}$, then the time constant CR_s of the capacitor (the time constant of the dielectric, essentially) is greater than roughly 50 s and may be as high as 10^7 s in some materials (see Table 5.1); note that $10^6\,\mathrm{s} \approx 12$ days. In Table 5.1 the dielectric losses are represented by the loss tangent tan δ (see Section 3.6.1).

In their principal areas of use, namely for insulation and as the filling for capacitors, dielectrics also require to have the property of being able to withstand electric fields of $10^6\,\mathrm{V\,m^{-1}}$ at least. For instance, although low voltages are used in integrated circuits (a few volts), the layers of semiconductor and insulator are often very thin; a potential difference of 2 V across a film of SiO_2 insulation of 2 μm thickness gives an electric field strength of $10^6\,\mathrm{V\,m^{-1}}$. The electrical breakdown strength of thin insulating layers is one of the physical factors that limit the miniaturization of integrated circuits.

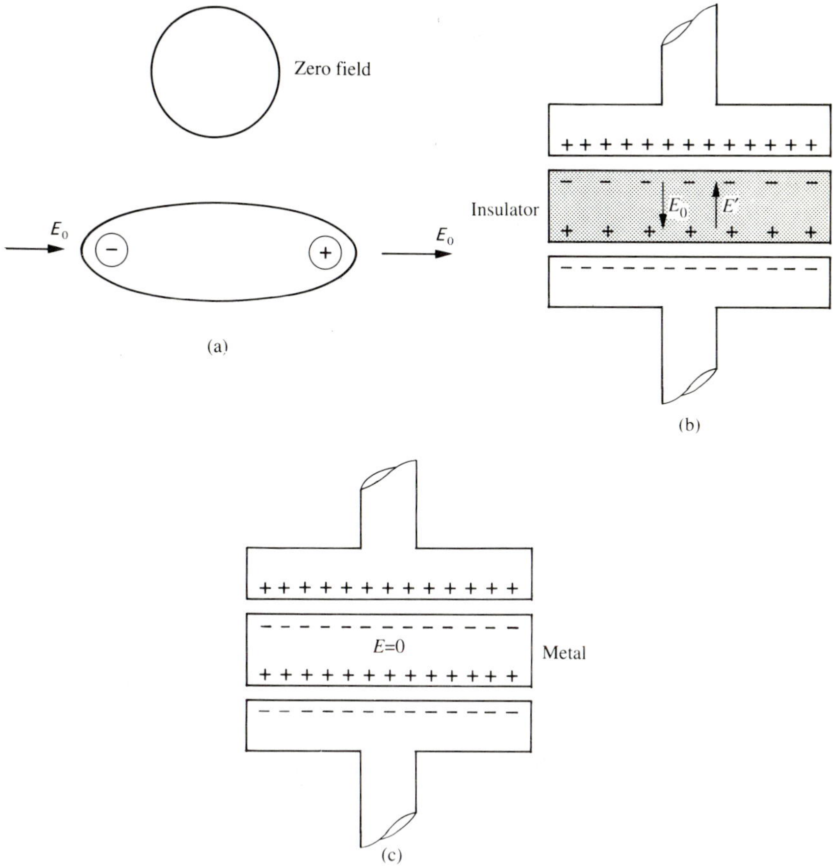

Fig. 5.1 (a) An induced **dipole** resulting from the relative displacement of positive and negative charges in a molecule, say, due to the effect of an applied electric field E_0. (b) A slab of dielectric between the metal plates of a capacitor: the field E' due to the induced charges is in the opposite direction to the applied field; hence the resultant electric field between the plates of the capacitor is less than the field E_0 that would exist if there was vacuum between the plates. (c) In the case of a slab of an ideal conductor ($\sigma = \infty$) the induced surface charges are of such a magnitude that the net electric field inside the conductor is zero.

The physical origin of the relative permittivity of a dielectric lies in the relative displacement of the negative and positive charges of the constituent atoms or molecules under the influence of an applied electric field. For an applied field E_0 the forces experienced by the negative charges (electrons, negative ions) and by the positive charges (positive ions) are in opposite directions and induced **dipoles** are created (see Fig. 5.1(a)). In the case of a rectangular slab of material between the metal plates of a parallel plate capacitor (see Fig. 5.1(b)) the end result of the creation of induced dipoles is that (polarization) charges appear on the faces of the slab of dielectric, i.e. the material has been **polarized**. Since $C \equiv Q/V$ (see eqn (1.24)), and assuming that Q is kept constant as the dielectric is introduced, then V becomes smaller (since the electric field is smaller) and it follows that the capacitance is increased by the presence of the dielectric, i.e. $\epsilon_r > 1$. The values of the relative permittivity of some commonly used dielectrics are shown in Table 5.1. Note that the values are quoted for a constant applied electric field; both ϵ_r and the loss tangent δ are frequency-dependent quantities. If a sinusoidal electric field is applied to a dielectric, the relative displacement of the positive and negative charges of the effective polarizable unit (atom, molecule, etc.) lags behind the field. Because of this phase lag, the relative permittivity can be expressed as a complex quantity (cp. complex a.c. circuit theory, Section 3.2.2):

$$\epsilon_r = \epsilon_r' - j\epsilon_r''. \tag{5.2}$$

Now the admittance Y (see Section 3.4) of a capacitance C is equal to $j\omega C$ (where $C = \epsilon_r C_0$), so for a dielectric-filled capacitor,

$$Y = j\omega C_0(\epsilon_r' - j\epsilon_r'')$$

or

$$Y = \omega\epsilon_r'' C_0 + j\omega\epsilon_r' C_0. \tag{5.3}$$

Hence, as admittances in parallel combine additively, the capacitor can be represented by a circuit element consisting of a lossless capacitance C_p $(= \epsilon_r' C_0)$ in parallel with an effective resistance R_p equal to $(\omega\epsilon_r'' C_0)^{-1}$ (see Fig. 3.11). It can easily be shown (see Problem 3.23; refer to Example 3.5) that $\tan\delta \approx (\omega C_p R_p)^{-1}$, so it follows that

$$\tan\delta \approx \frac{\epsilon_r''}{\epsilon_r'}. \tag{5.4}$$

From eqn (5.3) it can be seen that the energy losses are associated with the imaginary part of the complex relative permittivity. Values for ϵ_r' and ϵ_r'' can be determined from measurements of C_p (or C_s) and $\tan\delta$ using an a.c. bridge (see Section 3.9).

In the frequency range of interest in electrical circuits, ϵ_r' generally decreases with increase in frequency, whereas ϵ_r'' increases. For general circuit situations the smaller the loss angle the better; for example, the

Q-factor of a tuned LCR network is inversely related to the energy losses (see eqn (4.7)). On the other hand, dielectric losses are exploited in one form of radiofrequency heating with many industrial and domestic applications such as the bulk drying of powders and microwave ovens. In microwave ovens (frequency ~ 1.4 GHz), energy dissipation in the water contained in items of food is the source of the heat.

5.2 Alternating electromagnetic fields in conductors

5.2.1 The 'skin' effect

If a sinusoidal electric field is driving a current along a cylindrical conductor, say, then the alternating current is not uniformly distributed over the cross-section of the conductor, the current density being greatest at the surface. The effect becomes more pronounced as the frequency is increased, with the current being increasingly confined to the layer of the conductor nearest to the surface—hence the name **skin effect**.

Consider a straight conductor of circular cross-section of radius a, carrying a constant current I. An expression can be obtained for the magnetic field $B(r)$ at a distance r from the axis of the conductor ($r \geqslant a$) by applying Ampère's circuital law to a circular path of integration of radius r:

$$B(r) \cdot 2\pi_r = \mu_0 I \quad \text{or} \quad B(r) = \frac{\mu_0}{4\pi}\frac{2I}{r} \quad (r \geqslant a). \tag{5.5}$$

If the circular path lies inside the wire ($r < a$), then

$$B(r) \cdot 2\pi r = \mu_0 I \frac{\pi r^2}{\pi a^2} \quad \text{or} \quad B(r) = \frac{\mu_0}{4\pi}\frac{2Ir}{a^2} \quad (r < a). \tag{5.6}$$

In deriving eqn (5.6) it has been assumed that the constant current is distributed uniformly over the cross-section of the conductor; the distribution of $B(r)$ in, and around, the straight cylindrical conductor is sketched in Fig. 5.2(a).

A qualitative insight into the physical reasons for the skin effect can be gained by considering a loop of wire carrying a current, as sketched in Fig. 5.2(b). An elementary filament such as TT′ will have a net magnetic flux linkage due only to the magnetic field inside the loop of wire (i.e. external to the material of the wire itself), since the flux linkage due to the field over the cross-section of the wire averages to zero (portion PQ of the graph of $B(r)$ in Fig. 5.2(a)). Also a filament SS′ will have a flux linkage due to the magnetic field inside the loop only. The maximum flux linkage will be for filament OO′ on the axis of the wire, since there is maximum contribution from the magnetic field inside the wire itself. If the current

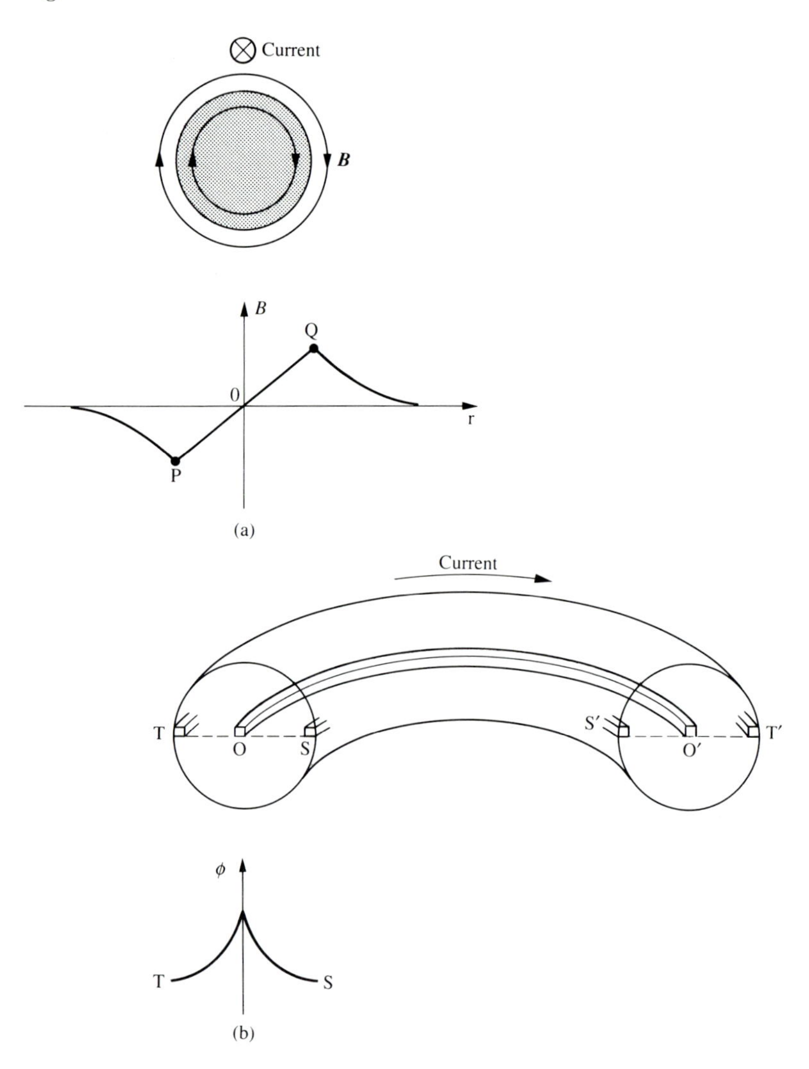

Fig. 5.2 (a) The magnetic field $B(r)$ for a long straight wire. (b) The magnetic flux linked with various filaments (TT′), (OO′), (SS′) of a loop of wire.

is sinusoidal the induced back e.m.f. will be greatest for the filament with the greatest flux linkage (i.e. OO′), so the impedance to current flow will be greatest in this filament. Hence the distribution of the alternating current over the cross-section of the wire will be non-uniform, with a tendency to be concentrated near the surface of the wire. As stated earlier, this phenomenon becomes more marked as the frequency of the alternating current is increased, since the amplitude of the back-e.m.f. is proportional to the time derivative of the alternating magnetic flux (eqn (1.49)) and hence is proportional to frequency.

Because of the non-uniform current distribution, the effective area of cross-section of a wire is less than the geometrical cross-sectional area and so the resistance $R(\omega)$ at frequency ω will be greater than the value R_0 for a constant current:

$$R(\omega) = R_0[1 + \alpha(\omega)]. \quad (5.7)$$

For a plane slab of conducting material an electromagnetic field falls off exponentially with penetration into the material, the amplitude decreasing by the factor 1/e in a distance called the **skin depth** (δ). It can be shown that

$$\delta = \sqrt{\left(\frac{2}{\mu_r \mu_0 \sigma \omega}\right)}, \quad (5.8)$$

where μ_r, σ are the relative permeability and conductivity, respectively, of the material. Equation (5.8) can be applied to cylindrical conductors, provided that δ is much less than the radius of the conductor. Some representative values of the skin depth are given in Table 5.2 (some additional values are given in Table 9.1); notice the small magnitude of the skin depth in good conductors at high frequencies.

Table 5.2. Values of the skin depth (δ) in aluminium and copper

Frequency	Aluminium	Copper
50 Hz	1 mm	1 cm
1 kHz	3 mm	2 mm
1 MHz	0.1 mm	70 μm
3 GHz	2 μm	1 μm

The results of measurements of the resistance of a small, loosely wound, air-cored coil made up of copper wire of 0.5 mm diameter are shown in Fig. 5.3. At high-enough frequencies, where $\delta \ll$ radius of the wire, the resistance is proportional to $\sqrt{}$(frequency), as expected from eqn (5.8).

5.2.2 The inductance of lines and coils

Generally speaking, if specific values of self-inductance or mutual inductance are required of coils, then use is made of tabulated data. However, it is of interest to review some particular physical situations of importance to electrical circuits in general.

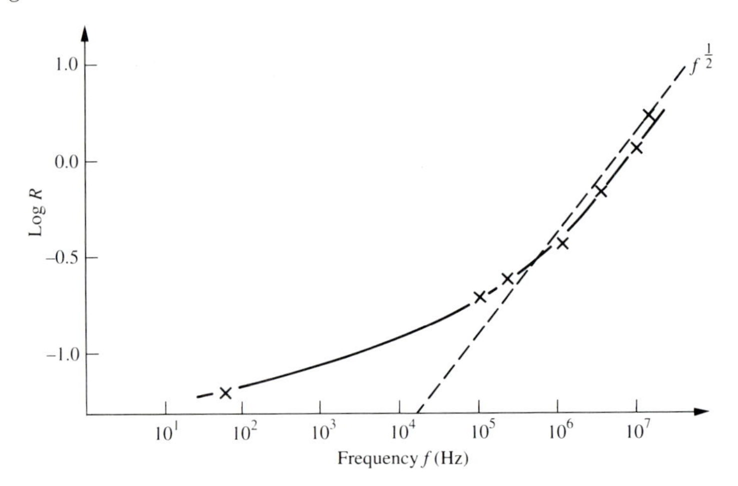

Fig. 5.3 The frequency-dependence of the resistance of a loosely wound, air-cored coil.

A long straight wire has self-inductance associated with the energy stored in the magnetic field inside the wire itself, generated by a current flowing in the wire. For a wire of radius a, eqns (5.6) and (1.56) yield for the stored energy (W) per unit length

$$W = \frac{\mu_0 I^2}{8\pi^2 a^4} \int_0^a r^2 2\pi r \, \mathrm{d}r,$$

where I is the current. Thus

$$W = \frac{\mu_0 I^2}{16\pi}.$$

Now, since $W = LI^2/2$ (see eqn (1.55)), it follows immediately that

$$L = \frac{\mu_0}{8\pi} \text{ H m}^{-1}. \tag{5.9}$$

This is usually called the **internal self-inductance** of the wire. Because of the assumption of uniform current distribution, this result applies accurately only at low frequencies, where the skin depth is large compared with the radius of the wire. For high frequencies it can be shown that $L = (4\pi^2 a\delta f)^{-1}$ H m^{-1}.

For a coaxial line there is self-inductance associated with the energy stored in the magnetic field that exists between the inner and outer conductors. It is left as an exercise to show that the self-inductance L per unit length is given by

$$L = \frac{\mu_0}{4\pi} 2 \ln(b/a) \text{ H m}^{-1}, \tag{5.10}$$

where a, b are the radii of the inner and outer conductors respectively. Note that the current in the outer conductor produces no magnetic field in the space between the conductors. If there were a field, then the symmetry of the situation would demand that the field lines were concentric with the axis. However, $\oint \boldsymbol{B} \cdot \mathrm{d}\boldsymbol{l}$ round such a path would be zero, since the path would not link the current in the outer conductor of the line, and hence B is zero. It should also be noted that in deriving eqn (5.10) it is assumed that the skin depth is very small so that the currents in the conductors can be assumed to flow on the outer surface of the inner conductor and the inner surface of the outer conductor.

Application of Ampère's circuital law to a concentric circular path outside the outer conductor yields the result that the magnetic field outside the coaxial line is zero (ideally), since the net current linkage is zero; the application of Gauss's law similarly predicts that the electric field outside the coaxial line is zero also. This feature means that the use of coaxial cable to make connections reduces electromagnetic interference (e.m.i.) due to unwanted coupling of signals between different parts of a system (see Chapter 9). Conversely, the outer conductor provides very good screening against interference emanating from outside the cable.

It is also of interest to note that the expression for L (eqn (5.10)) is the inverse of that for the capacitance per unit length of a coaxial line (eqn (1.30)), apart of course from the ways in which the fundamental electric and magnetic constants enter the expressions. This is an example of a general principle; the inductance per unit length of two parallel cylindrical conductors can be obtained from eqn (1.37) by exploiting this principle:

$$L = \frac{\mu_0}{4\pi} 4 \cosh^{-1}(b/2a) \text{ H m}^{-1}. \tag{5.11}$$

For $b/a \approx 22$, say (e.g. twin 'balanced feeder' line), then $L \approx 0.7\ \mu\text{H m}^{-1}$, whereas for a coaxial line ($b/a \approx 4$, say), eqn (5.10) predicts $L \approx 0.28\ \mu\text{H m}^{-1}$.

The expressions of eqns (5.10) and (5.11) give the 'external' self-inductance of a coaxial line and a twin line respectively. If the skin depth is not small enough, then to those should be added the internal self-inductance of the conductors involved, but usually they will be small in comparison.

If the space between the conductors of a coil or line is filled with a material of relative permeability μ_r, then to a good approximation the self-inductance is increased by a factor equal to the value of μ_r. A discussion of the materials useful as core materials is given in Section 5.4. For the present it should be noted that because of increasing energy losses due to eddy currents, the use of ferromagnetic materials is confined to frequencies below about 20 kHz, and that ferrite cores can be used up to about 50 MHz.

As hinted above, the design of coils to obtain maximum inductance, minimum resistance, and minimum distributed capacitance of the windings is, to a significant extent, an empirical procedure. Apart from tabulated data there are formulae that apply with reasonable accuracy to common shapes of coils. For example, for a single-layer solenoid with n turns as shown in Fig. 5.4(a), the inductance L is given by

$$L = 3.95a^2n^2K/l\ \ \mu\text{H}, \tag{5.12}$$

where l, a are measured in metres. The dependence on l and a of the numerical factor K is shown in Fig. 5.4(c). In the case of a multilayer coil as sketched in Fig. 5.4(b),

$$L \approx \frac{32a^2n^2}{(6a + 9l + 10t)}\ \mu\text{H}. \tag{5.13}$$

In the audiofrequency range, coils with laminated iron cores have Q-factors that can be as high as 100. At higher frequencies, where ferrite-cored or 'air-cored' coils are used, the inter-turn capacitance becomes a significant factor as well as the additional losses associated with the increasing radiofrequency resistance due to the skin effect. The latter effect is compounded with the so-called proximity effect; the current distribution in a turn of a coil is affected not only by the magnetic flux

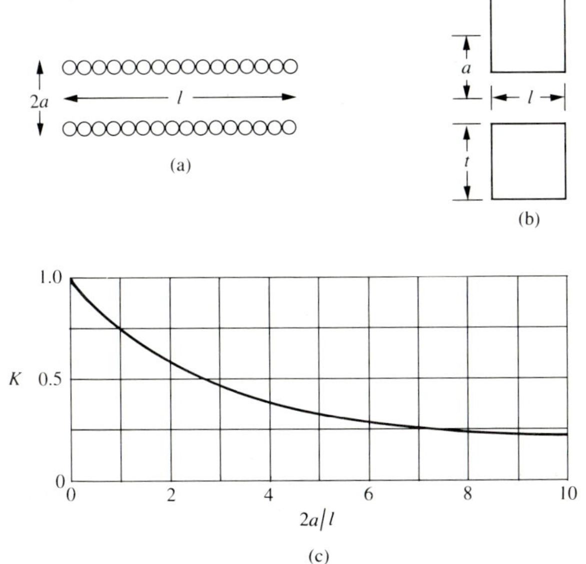

Fig. 5.4 (a) and (b) show the shapes of coils for which the self-inductances are given by eqns (5.12) and (5.13) respectively. The dependence on l and a of the geometrical factor K for a coil in the shape of (a) is shown in (c).

due to the current flowing in that turn but also by the current flowing in neighbouring turns (see the discussion of the proximity effect in Section 1.5). In single-layer coils, a gauge of wire should be chosen so that the coil is loosely wound (diameter of wire one-half to three-quarters of the spacing of the turns, say; the decrease in losses due to the reduction of the proximity effect will more than offset the increase in losses via the skin effect due to the smaller area of cross-section of the wire used. The losses due to the skin effect can be reduced by using multi-stranded insulated wire (Litz wire); the conductor is compounded of a number of strands of fine wire connected in parallel and interwoven. Thus the current divides amongst the strands and, provided that the diameter of a strand is less than or of the order of the skin depth, then the current divides fairly uniformly over the cross-section of the compound conductor; the a.c. resistance can be made to approach the d.c. value at moderate radiofrequencies. So, in well-designed coils, Q-factors of 100 to 300 can be obtained in the frequency range up to about 1 MHz; above this frequency the capacitance between the strands is a major factor in reducing the effectiveness of Litz wire.

The self-capacitance (or distributed capacitance) of a coil should be made as small as possible, for two reasons in particular. First, the value of the self-capacitance determines the highest frequency at which the coil can be made to resonate. Second, the dielectric losses in the material of the coil (coil former and insulating materials), which are determined by the electric fields existing in the capacitance between the windings, can give rise to a significant additional effective resistance and hence a reduction in the Q-factor of the coil. The general rule to be followed in order to minimize self-capacitance is to wind the coil so that turns that are close to each other are not at widely differing potentials; e.g. use a 'bank' winding.

5.3 Magnetic fields in solids

The principal magnetic property of solids that is of interest in electrical circuits is the relative permeability μ_r. As was mentioned in Section 1.7, the relative permeability of a material can be defined through $\mu_r \equiv L/L_0$, where L is the self-inductance of a toroidal coil (see Fig. 1.24) when it is filled with the material in question, and L_0 is the self-inductance with the material removed. For most materials, $\mu_r = 1$ to within a fraction of one per cent, but for the ferromagnetic solids, $\mu_r \approx 10^3$ commonly, and can be $\approx 10^4$ in some cases. There is also a class of electrically insulating solids called ferrites, for which the values of μ_r range from a few hundred up to $\sim 10^3$. The usefulness of materials having large values of μ_r is that, since $L = \mu_r L_0$, coils can be made much more compact for a given value of L; alternatively, large values of L can be obtained for a given size of coil. In the case of transformers, better flux linkage between the coils can be obtained. An advantage of physically compact coils is that the inter-winding capacitance is smaller, as it is fairly directly related to the number of turns on a coil.

The relative permittivity ϵ_r of most of the dielectrics used in circuits is independent of the strength of the applied electric field, that is, the materials are said to be 'linear' in this respect. The non-linear properties of some materials form the basis of some electro-optic devices, but this topic lies outside the scope of this text. In contrast, the relative permeability of ferromagnetic and ferrimagnetic materials does depend on the strength of the applied magnetic field, and this causes some complications in the design of components that exploit the large values of μ_r.

The origin of the magnetic properties of a material lies in the internal magnetic fields generated by the motion of electrons bound to the atoms and molecules of the material. For instance, an electron orbiting an atomic nucleus is equivalent to a small current loop, and there is an associated magnetic field (see Fig. 5.5(b)). At distances much greater than the radius of the loop, the magnetic field of a current loop has the same spatial distribution as the electric field due to an **electric dipole** (see Fig. 5.5(a)), and so a 'small' current loop is designated a **magnetic dipole**, with magnetic dipole moment $m \equiv Ia$ A m^2, where a is the area of the loop. There is also a magnetic dipole moment associated with the spinning motion of an electron; this is an important contribution to the magnetic properties of ferromagnetic and ferrimagnetic materials.

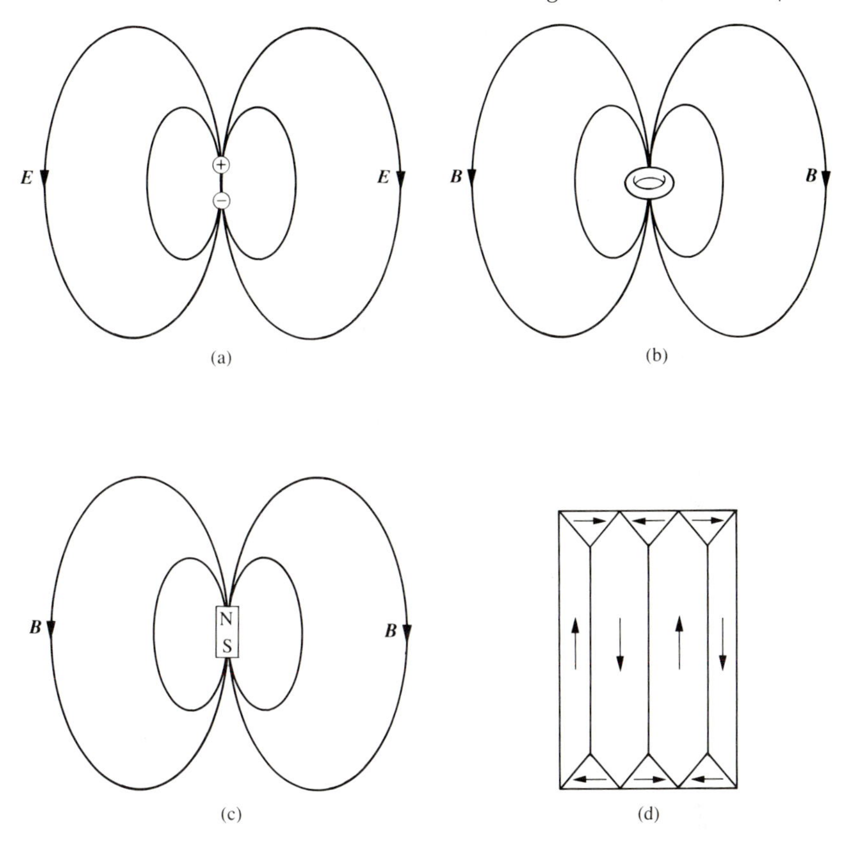

Fig. 5.5 (a) and (b) The external fields of an electric and a magnetic dipole are of the same form. (c) For a uniformly magnetized sample (e.g. a bar magnet) there is an external field B, which is shown schematically. (d) The formation of magnetic domains is shown schematically; the overall magnetization of the sample is small and so, correspondingly, is the external magnetic field.

In the so-called ferromagnetic and ferrimagnetic materials there is a strong, cooperative, interaction between the constituent atoms, with the consequence that within a region called a **domain** all the atomic dipole moments align parallel to each other in a particular direction. In other words, the material in a domain becomes **spontaneously magnetized** in that particular direction; the magnetization vector is usually denoted by the symbol $\boldsymbol{M}$. Domains are of macroscopic size with dimensions which can be in the region of 1 mm and even larger. In a sample of material the number of domains and their directions of magnetization are determined by the condition that the free energy should be a minimum. Bearing in mind that energy $\mu_0 B^2/2$ J m^{-3} is associated with a magnetic field, another way of viewing the energy constraint is to say that the domains arrange themselves so that the external magnetic field B is as small as possible. This condition is illustrated in an idealistic and schematic way in Fig. 5.5(d); the overall, net, magnetization of the sample is very small (zero ideally).

If a magnetizing field of gradually increasing strength is applied to the sample, by placing the sample inside a coil and increasing the current through the coil, for instance, then the sample acquires a net magnetization that increases with the strength of the magnetizing field, up to a limit value called the **saturation magnetization**. The net magnetization arises from a combination of processes, including changes of domain size and rotation of the direction of domain magnetization towards that of the magnetizing field. Saturation corresponds to the maximum degree of parallel orientation of the domain magnetization vectors (see Fig. 5.5(d)).

The magnetizing field is specified by an auxiliary field quantity usually denoted by the symbol $\boldsymbol{H}$. The utility of $\boldsymbol{H}$ is illustrated by the feature that in cases where the magnetizing field is generated by a current flowing in a coil, the line integral of $\boldsymbol{H}$ round a closed path is equal to the total current linked with the path:

$$\oint \boldsymbol{H} \cdot \mathrm{d}\boldsymbol{l} = NI. \tag{5.14}$$

For instance, in a 'vacuum-filled' toroidal coil of radius R with N turns and current I (see Fig. 1.24), $\oint \boldsymbol{H} \cdot \mathrm{d}\boldsymbol{l} = H \cdot 2\pi R$, and so $H = NI/2\pi R$ ampère-turns per metre. Ampère's circuital law has already been quoted in the form (eqn (1.52))

$$\oint \boldsymbol{B} \cdot \mathrm{d}\boldsymbol{l} = \mu_0 I,$$

and so the implication is that

$$\boldsymbol{B} = \mu_0 \boldsymbol{H} \tag{5.15}$$

(*in free space or, in practice, in media for which* $\mu_r = 1$, *very closely*).

If the toroid is wound on a material medium of relative permeability μ_r, then

$$\mu_r \equiv \frac{L}{L_0}, \tag{5.16}$$

where L_0, L are the self-inductances of the toroidal coil with and without the medium, respectively.

The self-inductance of the toroidal coil is proportional to the magnetic flux linked with the coil (eqn (1.51)), and for given values of I, N, and A (the area of cross-section of the coil) this means that

$$\mu_r = \frac{B}{B_0}, \tag{5.17}$$

where B, B_0 are the magnetic fluxes in the coil in the respective situations. Hence from eqns (5.15) and (5.17) it follows that

$$B = \mu_r \mu_0 H. \tag{5.18}$$

It should be appreciated that this relation is really a definition of a variable quantity $\mu_r(B)$ and this is not an indication of a general linear relation between B and H.

The concept of a **magnetic circuit** is very useful with regard to the design of transformers and also to many other devices used in electrical engineering, such as electromagnets, electromagnetic actuators and motors and generators. Consider a ring made of four different ferromagnetic materials, where each material has a different cross-sectional area and length (see Fig. 5.6); the application of Ampère's circuital law (eqn (5.14)) to the indicated closed path yields

$$H_1 l_1 + H_2 l_2 + H_3 l_3 + H_4 l_4 = NI.$$

If the relative permeabilities of the four materials are denoted by $\mu_1, \mu_2, \mu_3, \mu_4$ respectively, then, using eqn (5.18),

$$H_1 l_1 = \frac{B_1}{\mu_1 \mu_0} l_1$$

or, using eqn (1.48a),

$$H_1 l_1 = \frac{\phi_1 l_1}{\mu_1 \mu_0 A_1},$$

where A_1 is the cross-sectional area and ϕ_1 is the magnetic flux linked with the material (it being assumed that the lines of B are confined to the material, which will be a reasonable assumption for a material having a relative permeability greater than 10^3). Since lines of B are continuous,

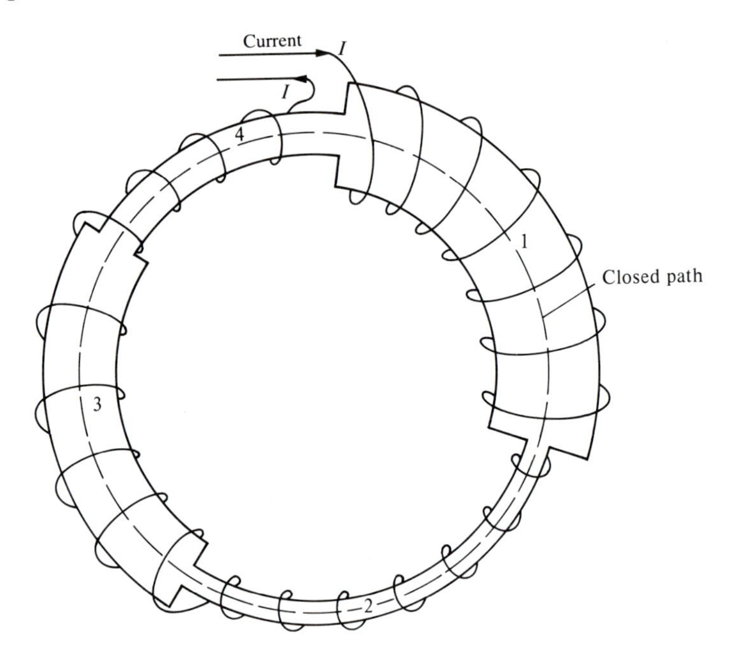

Fig. 5.6 A closed ring composed of four different ferromagnetic materials and wound with a coil of N turns.

the magnetic flux will have the same value (simply denoted by ϕ) in each segment of the ring, so it follows that

$$\frac{\phi}{\mu_0}\left\{\frac{l_1}{\mu_1 A_1}+\frac{l_2}{\mu_2 A_2}+\frac{l_3}{\mu_3 A_3}+\frac{l_4}{\mu_4 A_4}\right\}=NI.$$

This equation can be written,

$$\phi\mathscr{R}=\mathscr{F}_{\mathrm{m}} \tag{5.19}$$

and, by analogy with $IR = E$ in electrical circuits, $\mathscr{F}_{\mathrm{m}}$ ($\equiv NI$) is called the **magnetomotive force** (m.m.f.) and $\mathscr{R}_{\mathrm{m}}$ the **reluctance** of the magnetic circuit. In

$$\mathscr{R}_{\mathrm{m}}=\sum_i \frac{l_i}{\mu_i\mu_0 A_i}, \tag{5.20}$$

where the summation is over the elements of the circuit.

It is of interest to examine more closely the analogy between magnetic and electrical circuits: for a conductor of uniform cross-sectional area A, length l and conductivity σ, the electrical resistance R is

$$R = \frac{l}{\sigma A},$$

and for an element of a magnetic circuit of the same length and cross-sectional area but having relative permeability μ_r the reluctance is

$$\mathcal{R}_m = \frac{l}{\mu_r \mu_0 A}.$$

In the usual electric circuit situation, the conductivity of the material of the circuit element is probably 10^{20} times that of the surrounding medium (air or insulation). However, for most magnetic circuits the relative permeability will be no more than 10^3 to 10^4 times that of the surrounding medium. Also, the ratio of length to transverse dimension is usually much smaller for magnetic than for electrical circuits. The practical significance of these features is that whereas in an electrical circuit the current, for practical purposes, flows through the circuit elements only and not at all through the surrounding medium, in magnetic circuits there is significant magnetic flux outside the circuit elements. Such **flux leakage** is difficult to calculate in practice, but use can be made of tabulated data in many design situations.

Example 5.1 A simple example can be used to give some idea of the order of magnitude of the physical quantities involved. Consider a coil wound on a core of the form shown in Fig. 5.7(a), where $l = 5$ cm, $A = 2$ cm^2, $N = 10^2$, $I = 0.1$ A, $\mu_r = 10^3$.

Using eqns (5.19) and (5.20),

$$\phi = \frac{\mu_r \mu_0 ANI}{4l}, \tag{5.21}$$

where it has been assumed that there is negligible flux leakage and an obvious approximation has been made in taking the effective path length for the flux to be $4l$.

On substituting the given values, then $\phi = 4\pi \times 10^{-6}$ Wb and the magnetic field B in the core ($= \phi/A$) is $2\pi \times 10^{-2}$ T. The self-inductance L of the coil is equal to $N\, d\phi/dI$ (see Section 1.7), i.e.

$$L = \mu_r \mu_0 AN^2/4l \qquad \text{or} \qquad L \approx 25 \text{ mH}. \qquad \square$$

If in the core specified in Example 5.1 there is a small air gap of width

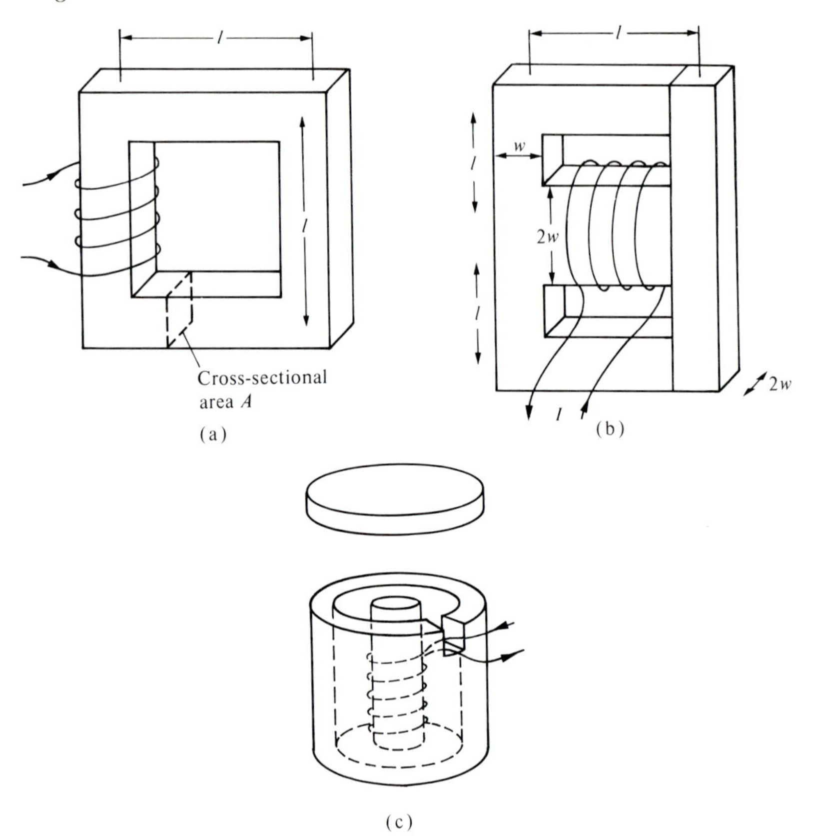

Fig. 5.7 (a) A simple form of core. (b) A core with branches in parallel. (c) A ferrite 'pot' core.

l_g ($l_g \ll l$), then the flux ϕ' is given by

$$\phi' = \frac{\mu_0 ANI}{\left[\dfrac{4l - l_g}{\mu_r} + l_g\right]}$$

or

$$\phi' \approx \frac{\mu_r \mu_0 ANI}{4l\left[1 + \dfrac{\mu_r l_g}{4l}\right]} \tag{5.22}$$

(*if* l_g *is neglected compared with* $4l$).

For $l_g = 4l/10^3$, $\phi' = \phi/2$, where ϕ is given by eqn (5.21). This rough calculation illustrates the fact that the flux in a magnetic circuit is reduced dramatically by the existence of even a small air gap and so also is the self-inductance of the coil. Another way of viewing this situation is to define an effective relative permeability μ_{eff} for the core such that

$$\phi' = \frac{\mu_{\text{eff}}\mu_0 ANI}{4l}$$

where

$$\mu_{\text{eff}} = \frac{\mu_r}{1 + \dfrac{\mu_r l_g}{4l}}.$$

Since the self-inductance of the coil is proportional to μ_{eff}, adjustment of the value of L can be made by altering the magnitude of l_g.

Magnetic circuits of a parallel nature can be analysed by direct analogy with parallel electrical circuits.

Example 5.2 Obtain an expression for the magnetic flux ϕ in the central limb of the core shown in Fig. 5.7(b).

The magnetic circuit can be drawn as follows:

$$\mathcal{R}_{m_1} \approx l/4\mu_r\mu_0 w^2$$

$$\mathcal{R}_{m_2} \approx l/2\mu_r\mu_0 w^2$$

Thus $\phi = NI/\mathcal{R}_m$ where

$$\mathcal{R}_m = \mathcal{R}_{m_1} + \frac{3\mathcal{R}_{m_2}}{2}$$

Hence $\phi \approx 4\mu_r\mu_0 w^2 NI/4l$. □

The advantage of the form of core shown in Fig. 5.7(b) over that shown in (a) is that there is smaller flux leakage.

The relation between the magnetic field B and the magnetizing field H in a ferromagnetic or a ferrimagnetic material is of the general form shown in Fig. 5.8(a). An applied magnetizing field of increasing magnitude produces an increasing net magnetization of the sample in the direction of the applied field, owing to changes in the domain structure and the directions of magnetization of the individual domains. The saturated condition is reached when the magnetization vectors of all the domains in a sample are aligned in the same direction. If H is increased to the saturation level (segment OD in Fig. 5.8(b)) and then gradually decreased to zero, the value of B does not retrace the path OD. If H is reversed and

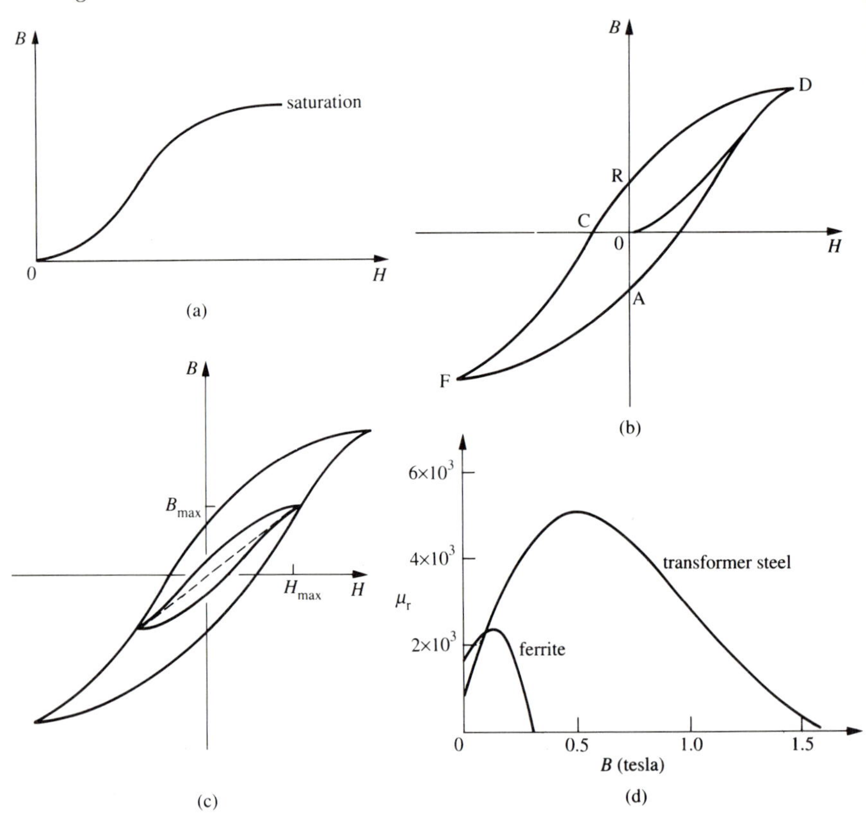

Fig. 5.8 (a) The general form of the relation between B and H for a magnetizable material. (b) A hysteresis loop. (c) The effective relative permeability is taken as $B_{max}/\mu_0 H_{max}$. (d) Relative permeability as a function of magnetic field B for a typical transformer steel and for a typical ferrite.

then taken through a complete cycle, the loop FADRCF is followed. It can be shown quite easily (see below) that the net work done per unit volume on the sample in taking it round the cycle of magnetization is equal to $\int H\,dB$, where the integral is taken round the loop; this is the area of the loop. This work denotes that energy is required to magnetize the sample and that this energy is not totally recovered when the sample is demagnetized. The loop is called a **hysteresis loop**.

To obtain the expression for the work done, consider a toroidal coil of radius R and area of cross-section A with N turns wrapped around a core of magnetizable material (see Fig. 1.24).

The flux ϕ in the core is related to the magnetic field B through $\phi = BA$. A source of e.m.f. $E(t)$ is required to drive a time-dependent current $I(t)$ through the coil because of the induced back-e.m.f. and $E(t) = N\,\mathrm{d}\phi/\mathrm{d}t$, i.e. $E(t) = AN\,\mathrm{d}B/\mathrm{d}t$. Now the rate of working (power) of the source is equal to $E(t)I(t)$, so the work done $\mathrm{d}W$ over an element of time $\mathrm{d}t$ is

$$\mathrm{d}W = E(t)I(t)\,\mathrm{d}t \qquad \text{or} \qquad \mathrm{d}W = NAI(t)\,\mathrm{d}B.$$

Hence

$$W = NA\int I(t)\,\mathrm{d}B.$$

It has already been seen (eqn (5.14) *et seq.*) that the magnetizing field H in the core material is given by $H = NI/2\pi R$ and also the volume $\mathscr{V}$ of the material is equal to $2\pi RA$. Thus

$$W = \mathscr{V}\int H\,\mathrm{d}B. \qquad (5.23)$$

Bearing in mind that if the magnetizing field is generated by an alternating current flowing in a coil wrapped around all or part of a core, the energy dissipated in the core material in taking it repeatedly round its hysteresis loop is one of the energy loss mechanisms that determine the Q-factor of the coil or, if the coil is part of a transformer, reduces the efficiency of the transformer. Hence 'narrow loop' materials are used where it is required to minimize energy losses.

A core may be taken through repeated cycles of magnetization and demagnetization in which the saturated condition is not attained (see Fig. 5.8(c)); the effective relative permeability μ_{eff} is taken as $B_{\mathrm{max}}/\mu_0 H_{\mathrm{max}}$ and the general form of the dependence of μ_{eff} in B is sketched in Fig. 5.8(d). The effective relative permeability is

$1/\mu_0 \times$ (slope of the straight line joining the tips of the hysteresis loop).

Alternating magnetic fields induce alternating currents (eddy currents) in the core material of a transformer, say, as well as in the windings, and hence there are associated energy losses. In the description of the circuit properties of transformers, hysteresis and eddy current losses are often lumped together under the heading of 'iron losses'. So the electrical resistance of the core material should be as high as possible to reduce eddy current losses; this is an advantage of ferrites, which have electrical conductivities in the range $1\ \mathrm{S\,m^{-1}}$ to $10^{-3}\ \mathrm{S\,m^{-1}}$. However, ferromagnetic materials have better magnetic properties than ferrites, and their effective resistance can be increased, so reducing the amplitude of the eddy currents, by making cores from thin laminations orientated so that the direction of flow of the eddy currents is generally at right angles to the plane of the laminations. For sinusoidal magnetic fields, the eddy current amplitude is proportional to the frequency and hence the power loss is proportional to $(\text{frequency})^2$ (see Section 1.7); this sets an upper frequency limit to the use of ferromagnetic cores made from iron, or some steel alloys, of not much more than 20 kHz.

5.4 Maxwell's equations and electromagnetic waves

Maxwell's four equations encapsulate in a very compact form the general relation between the components of the electromagnetic field, namely the electric field E and the magnetic fields B and H. Only the free-space context will be considered here, so that B is directly related to H: $B = \mu_0 H$ (eqn (5.15)). Historically, probably the most important consequence of the formulation of Maxwell's equations in the nineteenth century was the prediction that a disturbance in an electromagnetic field propagated at a speed that was equal, it turned out, to the known measured value for the speed of light waves. This was the first indication that visible light propagated as electromagnetic waves; it is well known that visible light contitutes a (small) part of the electromagnetic spectrum, which in principle stretches from zero frequency to infinite frequency. The practical limits in circuits and in telecommunication and radar systems are, currently, from a fraction of one hertz to the visible ('optical') region of the spectrum ($\sim 10^{16}$ Hz).

Since the electromagnetic field quantities are vectors, it is impossible to discuss Maxwell's equations further without recourse to some vector analysis; the necessary concepts and definitions are introduced briefly in the next section.

5.4.1 Differential vector operators

$\boldsymbol{E}$, $\boldsymbol{B}$, and $\boldsymbol{H}$ are all vector quantities and so each can be expressed in terms of three components; a common choice is for the three components to be directed along the mutually perpendicular axes of a Cartesian frame of coordinates (see Fig. 5.9(a)). Further, it transpires that Maxwell's equations involve spatial and time derivatives of the field quantities $\boldsymbol{E}$, $\boldsymbol{B}$, and $\boldsymbol{H}$. The differential vector operators defined below allow the relationships in Maxwell's equations to be expressed very compactly.

The **gradient** operator ('**grad**')

It has been seen earlier (eqn (1.15)) that in a one-dimensional situation the electric field E is equal to the gradient of the electric potential V: $E = -\mathrm{d}V/\mathrm{d}x$. In a general three-dimensional situation the (scalar) potential will be a function of x, y, and z, so the (vector) $\boldsymbol{E}$ will have components in the x-, y-, and z-directions:

$$E_x = -\frac{\partial V(x, y, z)}{\partial x}; \qquad E_y = -\frac{\partial V(x, y, z)}{\partial y}; \qquad E_z = -\frac{\partial V(x, y, z)}{\partial z}. \tag{5.24}$$

Here $\partial/\partial x$ denotes the partial derivative of $V(x, y, x)$ with respect to x; that is, the function $V(x, y, z)$ is differentiated with respect to x, with y and z treated as constants, and analogously for the partial derivatives with respect to y and z.

The vector $\boldsymbol{E}$ is written in vectorial form as

$$\boldsymbol{E} = \hat{\boldsymbol{i}}E_x + \hat{\boldsymbol{j}}E_y + \hat{\boldsymbol{k}}E_z, \tag{5.25}$$

where $\hat{\boldsymbol{i}}, \hat{\boldsymbol{j}}, \hat{\boldsymbol{k}}$ are unit vectors in the x-, y-, and z-directions respectively, i.e. vectors defined to have unit magnitude (see Fig. 5.10). Hence, on substituting for E_x, E_y, E_z from eqn (5.24),

$$\boldsymbol{E} = -\hat{\boldsymbol{i}}\frac{\partial V}{\partial x} - \hat{\boldsymbol{j}}\frac{\partial V}{\partial y} - \hat{\boldsymbol{k}}\frac{\partial V}{\partial z} \tag{5.26}$$

or

$$\boldsymbol{E} = -\nabla V, \tag{5.27}$$

where the gradient operator ∇ (grad) is defined by

$$\nabla \equiv \hat{\boldsymbol{i}}\frac{\partial}{\partial x} + \hat{\boldsymbol{j}}\frac{\partial}{\partial y} + \hat{\boldsymbol{k}}\frac{\partial}{\partial z} \tag{5.28}$$

(*NB the symbol* ∇ *is sometimes called 'del' or 'nabla'*).

Note that even at this stage eqn (5.27) is more compact than (5.26).

The **divergence** operator ('**div**')

A general theorem in vector analysis called the divergence theorem is expressed as

$$\int_S \boldsymbol{E}\cdot \mathrm{d}\boldsymbol{S} = \int_{\mathscr{V}} \nabla\cdot\boldsymbol{E}\,\mathrm{d}\mathscr{V}, \tag{5.29}$$

where $\mathscr{V}$ is the volume enclosed by the closed surface S.

Using the definition of ∇ and eqn (5.25) for $\boldsymbol{E}$,

$$\nabla\cdot\boldsymbol{E} = \left(\hat{\boldsymbol{i}}\frac{\partial}{\partial x} + \hat{\boldsymbol{j}}\frac{\partial}{\partial y} + \hat{\boldsymbol{k}}\frac{\partial}{\partial z}\right)\cdot(\hat{\boldsymbol{i}}E_x + \hat{\boldsymbol{i}}E_y + \hat{\boldsymbol{k}}E_z)$$

and, on multiplying out,

$$\nabla \cdot \boldsymbol{E} = \frac{\partial E_x}{\partial x} + \frac{\partial E_y}{\partial y} + \frac{\partial E_z}{\partial z}, \tag{5.30}$$

since the scalar product (see eqns (1.16) and (1.17)) of two parallel unit vectors (e.g. $\hat{\boldsymbol{i}} \cdot \hat{\boldsymbol{i}}$) is equal to unity and the scalar product of two perpendicular vectors (e.g. $\hat{\boldsymbol{j}} \cdot \hat{\boldsymbol{k}}$) is zero.

The quantity $\nabla \cdot \mathbf{E}$ is called the 'divergence' of the vector $\boldsymbol{E}$ because it expresses the flux of the vector quantity $\boldsymbol{E}$ out of a volume $\mathscr{V}$ bounded by a closed surface S; see eqn (5.29). Note that since the volume $\mathscr{V}$ can be shrunk indefinitely, from a mathematical point of view, $\nabla \cdot \boldsymbol{E}(x, y, z)$ is a function that is defined at a point.

Example 5.3 An electric field is specified by the potential function

$$V(x, y, z) = \frac{A}{(x^2 + y^2 + z^2)^{1/2}}.$$

Obtain the electric field $\boldsymbol{E}$ and show that in this case $\nabla \cdot \mathbf{E} = 0$.

The x-component of $\boldsymbol{E}$ is given by

$$E_x = -\frac{\partial V}{\partial x}.$$

Using the rule for the differentiation of a quotient,

$$\frac{\partial V}{\partial x} = \frac{-A \cdot \frac{1}{2}(x^2 + y^2 + z^2)^{-1/2} \cdot 2x}{(x^2 + y^2 + z^2)} = \frac{-Ax}{(x^2 + y^2 + z^2)^{3/2}},$$

and similarly for $\partial V/\partial y$ and $\partial V/\partial z$. Hence

$$\boldsymbol{E} = \frac{-A}{(x^2 + y^2 + z^2)^{3/2}} (\hat{\boldsymbol{i}}x + \hat{\boldsymbol{j}}y + \hat{\boldsymbol{k}}z).$$

To obtain the divergence of $\boldsymbol{E}$ it is necessary first to obtain $\partial E_x/\partial x$, and again using the rule for the differentiation of a quotient,

$$\frac{\partial E_x}{\partial x} = \frac{-(x^2 + y^2 + z^2)^{3/2} + Ax \cdot \frac{3}{2}(x^2 + y^2 + z^2)^{1/2} \cdot 2x}{(x^2 + y^2 + z^2)^3}$$

$$= \frac{-(x^2 + y^2 + z^2) \cdot A + 3Ax^2}{(x^2 + y^2 + z^2)^{5/2}}.$$

After carrying out the equivalent derivations for $\partial E_y/\partial y$ and $\partial E_z/\partial z$, then the result is that $\nabla \cdot \boldsymbol{E} = 0$. □

The operator **curl**

For the electric field $\boldsymbol{E}$, curl $\boldsymbol{E}$ (which is written as $\nabla \times \boldsymbol{E}$ in vector notation) is given in Cartesian cordinates by

$$\nabla \times \mathbf{E} = \hat{\boldsymbol{i}}\left(\frac{\partial E_z}{\partial y} - \frac{\partial E_y}{\partial z}\right) + \hat{\boldsymbol{j}}\left(\frac{\partial E_x}{\partial z} - \frac{\partial E_z}{\partial x}\right) + \hat{\boldsymbol{k}}\left(\frac{\partial E_y}{\partial x} - \frac{\partial E_x}{\partial y}\right). \tag{5.31}$$

Stokes's theorem states that for a vector $\boldsymbol{G}$,

$$\oint_l \boldsymbol{G} \cdot \mathrm{d}\boldsymbol{l} = \int_S (\nabla \times \mathbf{G}) \cdot \mathrm{d}\boldsymbol{S}; \tag{5.32}$$

i.e. if the closed path denoted by l bounds a surface S, then the integral of curl $\boldsymbol{G}$ over S is equal to the line integral of $\boldsymbol{G}$ round the path l. This indicates that curl $\boldsymbol{G}$ is related to the 'circulation' of $\boldsymbol{G}$; it is rather difficult to give a pictorial illustration of this property of a vector.

Example 5.4 For the vector fields

$$\boldsymbol{E}_1 = -\hat{\boldsymbol{i}}y + \hat{\boldsymbol{j}}x + \hat{\boldsymbol{k}}z$$

and

$$\boldsymbol{E}_2 = \hat{\boldsymbol{j}}2x,$$

show that curl $\boldsymbol{E}_1$ = curl $\boldsymbol{E}_2$.

For $\boldsymbol{E}_1$: $\partial E_z/\partial y = \partial E_y/\partial z = \partial E_x/\partial z = \partial E_z/\partial x = 0$; $\partial E_y/\partial x = 1$; $\partial E_x/\partial y = -1$.
So $\nabla \times \boldsymbol{E}_1 = \hat{\boldsymbol{k}}2$.
For $\boldsymbol{E}_2$: $\partial E_z/\partial y = \partial E_y/\partial z = \partial E_x/\partial z = \partial E_z/\partial x = \partial E_x/\partial y = 0$; $\partial E_y/\partial x = 2$.
Hence $\nabla \times \boldsymbol{E}_2 = \hat{\boldsymbol{k}}2$ also.
QED. □

5.4.2 Maxwell's equations

Gauss's law in electrostatics (eqn (1.20)) can be rewritten

$$\int_S \boldsymbol{E} \cdot \mathrm{d}\boldsymbol{S} = \frac{Q}{\epsilon_0} \quad \text{or} \quad \int_S \boldsymbol{E} \cdot d\boldsymbol{S} = \frac{1}{\epsilon_0}\int \rho(x, y, z)\, \mathrm{d}\mathscr{V} \tag{5.33}$$

where $\rho(x, y, z)$ is the **charge density function**, which describes the distribution of charge through the volume $\mathscr{V}$, i.e.

$$Q = \int_{\mathscr{V}} \rho \, \mathrm{d}\mathscr{V}.$$

Now applying the divergence theorem to the left-hand side of eqn (5.33) yields

$$\int_{\mathscr{V}} \nabla \cdot \boldsymbol{E} \, \mathrm{d}\mathscr{V} = \frac{1}{\epsilon_0} \int_{\mathscr{V}} \rho(x, y, z) \, \mathrm{d}\mathscr{V}.$$

Since this equation is true for any volume $\mathscr{V}$ it must therefore be true, in the limit, at a point, i.e.

$$\nabla \cdot \boldsymbol{E} = \frac{\rho}{\epsilon_0}. \tag{5.34}$$

This is the first of Maxwell's equations; eqn (5.33) (Gauss's law) is the integral form of the equation.

For a magnetostatic field the net flux $\int \boldsymbol{B} \cdot \mathrm{d}\boldsymbol{S}$ over a closed surface is zero. A pictorial justification of this statement is that, since lines of $\boldsymbol{B}$ are continuous, a line that at one point enters a volume of space bounded by a closed surface must leave the volume by crossing the enclosing surface at another point; hence the net flux across a closed surface is zero:

$$\int_S \boldsymbol{B} \cdot \mathrm{d}\boldsymbol{S} = 0. \tag{5.35}$$

On applying the divergence theorem,

$$\int_S \boldsymbol{B} \cdot \mathrm{d}\boldsymbol{S} = \int_{\mathscr{V}} \nabla \cdot \boldsymbol{B} \, \mathrm{d}\mathscr{V} \qquad (= 0).$$

Since this equation is true for any volume $\mathscr{V}$ bounded by a closed surface S, it is also true at a point, and so

$$\nabla \cdot \boldsymbol{B} = 0. \tag{5.36}$$

This is the second of Maxwell's equations; eqn (5.35) is the integral form.

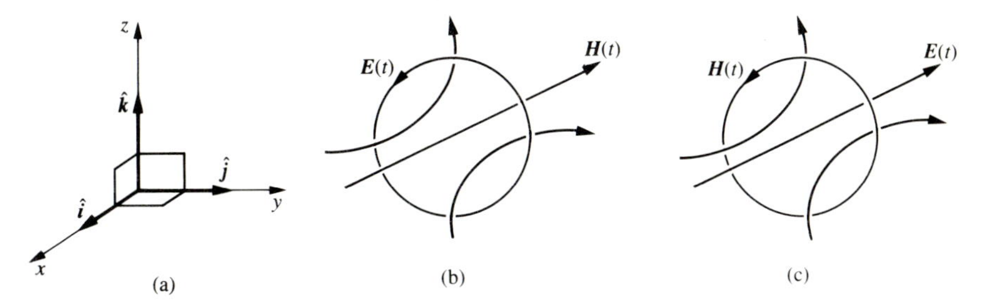

Fig. 5.9 (a) The unit vectors $\hat{i}, \hat{j}, \hat{k}$ in a Cartesian frame of coordinates. The relationships of eqns (5.38) and (5.41) are illustrated schematically in (b) and (c) respectively (remember that $B = \mu_0 H$ in free space).

From eqns (1.49(c)) and (1.50), and with a change of notation for the area) from A to S, it follows that if the magnetic flux linked with a loop of wire bounding an area S is changing with time, then the induced e.m.f. $\mathcal{E}$ in the loop is given by

$$\mathcal{E} = \frac{\partial}{\partial t}\int_S \boldsymbol{B}\cdot \mathrm{d}\boldsymbol{S}. \tag{5.37}$$

It is very important to realise that the e.m.f. $\mathcal{E}$ exists in the path even if it is just a mathematical closed path in space and does not coincide with a physical conductor (see Fig. 5.9(b)). Of course, only if there is a conductor coinciding with the path does the induced e.m.f. cause a current to flow.

By adapting the concept of electromotive force as introduced in eqn (1.19), the e.m.f. $\mathcal{E}$ can be written as the line integral of electric field $\mathbf{E}$ round the loop:

$$\mathcal{E} = -\oint_l \boldsymbol{E}\cdot \mathrm{d}\boldsymbol{l}$$

and

$$\oint_l \boldsymbol{E}\cdot \mathrm{d}\boldsymbol{l} = -\frac{\partial}{\partial t}\int_S \boldsymbol{B}\cdot \mathrm{d}\boldsymbol{S}. \tag{5.38}$$

On applying Stokes's theorem to the expression $\oint \boldsymbol{E}\cdot \mathrm{d}\boldsymbol{l}$,

$$\mathcal{E} = -\int_S (\nabla \times \boldsymbol{E})\cdot \mathrm{d}\boldsymbol{S}.$$

So, on substituting in eqn (5.38),

$$-\int_S (\nabla \times \boldsymbol{E})\cdot \mathrm{d}\boldsymbol{S} = \frac{\partial}{\partial t}\int_S \boldsymbol{B}\cdot \mathrm{d}\boldsymbol{S}$$

and so

$$\nabla \times \mathbf{E} = -\frac{\partial \boldsymbol{B}}{\partial t}. \tag{5.39}$$

This is the third of Maxwell's equations and is the differential form of the equations for electromagnetic induction. Equation (5.38) is the integral form and embodies Faraday's, Lenz's and Neumann's laws; see Section 1.7.

Imagine now a current I_c flowing along a wire into a capacitor and flowing out from the other plate of the capacitor along another wire; for simplicity assume that it is a parallel-plate capacitor, with the area of the plates equal to A. The magnetic field around the wire is given by Ampère's circuital law (eqn (5.14)). Historically, a problem arose in trying to account for the magnetic field on a path around the capacitor that did not encircle the 'conduction' current I_c flowing in the wire; there is no conduction current flowing across the gap between the plates of the capacitor. Now I_c is equal to the rate of change of charge on the capacitor plates, and so if the charge density on the plates is denoted by $\sigma(t)$ and the area of the plates is A, then

$$I_c = A\frac{d\sigma(t)}{dt}$$

The electric field E between the plates is given by $E = \sigma(t)/\epsilon_0$ (see eqns (1.22) and (1.26)), and so

$$A\frac{d\sigma(t)}{dt} = A\epsilon_0\frac{dE}{dt}.$$

The quantity $\epsilon_0(dE/dt)$ is called the **displacement current density** in the space between the capacitor plates. In free space Ampère's circuital law takes the general form

$$\oint_l \boldsymbol{H}\cdot d\boldsymbol{l} = \oint_S \frac{\partial(\epsilon_0 \boldsymbol{E})}{\partial t}\cdot d\boldsymbol{S}. \tag{5.40}$$

where S is any surface bounded by the closed path l. On applying Stokes's theorem to the left-hand side of eqn (5.40),

$$\nabla \times \boldsymbol{H} = \epsilon_0\frac{\partial \boldsymbol{E}}{dt}. \tag{5.41}$$

This is the fourth of Maxwell's equations and expresses the phenomenon of **magnetoelectric induction**: eqn (5.40) is the integral form of the equation.

Electromagnetic induction and magnetoelectric induction are illustrated schematically in Figs. 5.9(b) and (c).

It must be stressed that the forms of Maxwell's equations as given above are relevant only to unbounded free space (strictly speaking *in vacuo*). However, for most purposes they apply accurately enough in air at large enough distances from material objects (particularly metallic structures).

In material media, Maxwell's equations are modified to account for conduction currents and for polarization and magnetization, which manifest themselves through the relative permittivity ϵ_r and the relative permeability μ_r, respectively. The displacement current density is then equal to $\epsilon_r \epsilon_0\, \partial \boldsymbol{E}/\partial t$ and is denoted by the vector $\boldsymbol{D}$.

It is useful to remember (see Section 1.5 and Figs. 1.12 and 1.13, for example) that the electric field at the surface of a good conductor (strictly an ideal conductor) is **normal** to the surface and that the magnetic field at the surface is **parallel** to the surface.

5.4.3 Electromagnetic waves in free space

It can be deduced from Maxwell's equations that, in free space, disturbances in the electromagnetic field propagate at a constant speed equal to $(\mu_0 \epsilon_0)^{-1/2}$; this is the speed of light (2.998×10^8 m s^{-1}). In unbounded free space a sinusoidal disturbance propagates as a transverse electromagnetic wave (t.e.m. wave). This means that the electric field vector and the magnetic field vector in the wave are perpendicular to each other and lie in a plane perpendicular to the direction of propagation of the wave: a 'plane' wave (see Fig. 5.10). If the sinusoidal electric field vector is in the x–z plane, and the wave is propagating in the $+z$-direction, then it can be expressed as

$$E_x(z, t) = \hat{E}_x\, e^{[j(\omega t - kz)]}, \tag{5.42(a)}$$

where $\hat{E}_x$ is the amplitude, ω is the angular frequency, and k ($\equiv \omega/c$) is the so-called 'wave vector'; c is the speed of light *in vacuo*. Since for a wave, speed of propagation = (frequency, f) $\times$ (wavelength, λ), it follows that $k = 2\pi/\lambda$.

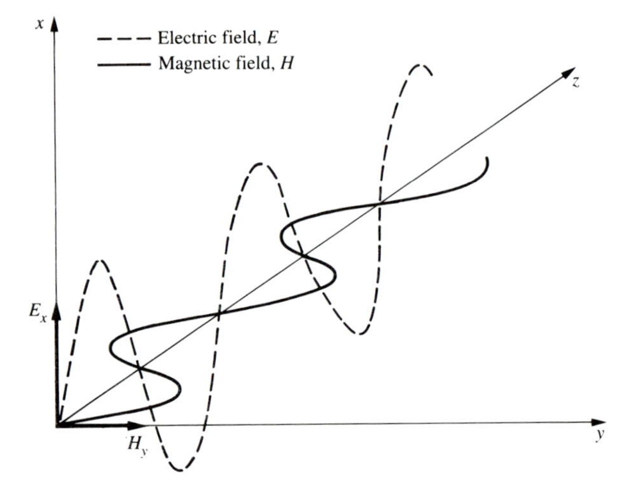

Fig. 5.10 The relation between the electric field (E_x), the magnetic field (H_y), and the direction of propagation (z-axis) for a plane electromagnetic wave.

The concomitant magnetic field vector is given by

$$H_y(z, t) = \hat{H}_y \, \mathrm{e}^{[\mathrm{j}(\omega t - kz)]}. \tag{5.42(b)}$$

The real or the imaginary part of the above wave expressions may be chosen for convenience, e.g.

$$E_x = \hat{E}_x \, {}^{\sin}_{\cos} \, (\omega t - kz). \tag{5.42(c)}$$

A travelling electromagnetic wave (e.m. wave) transports energy through space, since there is stored energy associated with the electric and magnetic field components of the wave (see eqns (1.31) and (1.55)). This flow of energy is represented by the **Poynting vector** $\boldsymbol{S}$, which is a vector in the direction of propagation of the wave; symbolically,

$$\boldsymbol{S} = \boldsymbol{E} \times \boldsymbol{H} \ \mathrm{W\,m^{-2}}. \tag{5.43}$$

$\boldsymbol{E} \times \boldsymbol{H}$ is the so-called vector product of $\boldsymbol{E}$ and $\boldsymbol{H}$, i.e. a vector of magnitude $EH \sin\theta$ pointing in a direction at right angles to both $\boldsymbol{E}$ and $\boldsymbol{H}$ (θ is the angle between $\boldsymbol{E}$ and $\boldsymbol{H}$); see Fig. 5.10.

It can be shown that the energy density associated with the magnetic component of a t.e.m. wave in free space is equal to that associated with the electric component. Hence, using eqn (1.31), an expression for the average total power density P_{av} (average power per unit area over a plane perpendicular to the direction of propagation) in a wave can be obtained, as follows.

Consider a cylinder, with area of cross-section A, in the space through which the wave is travelling. All the energy in a cylinder of length equal to c will cross a transverse section in one second (and power equals energy per unit time, remember). The instantaneous value of the total energy W in a volume cA is given by

$$W = \left(\frac{\epsilon_0 E^2}{2} + \frac{\mu_0 H^2}{2}\right) cA$$

or, from above

$$W = \epsilon_0 E^2 cA.$$

Hence the average power density P_{av} is given by

$$P_{av} = c\epsilon_0 \overline{E^2}$$

or

$$P_{av} = \frac{c\epsilon_0 \hat{E}^2}{2} \ W\,m^{-2}. \tag{5.44}$$

For a plane wave close to a transmitting antenna, the power density could be ~ 1 kW m^{-2}, which gives $\hat{E} \approx 1$ kV m^{-1}, whereas for an incoming radio wave at a receiving antenna, $\hat{E}$ may be of the order of 1 μV m^{-1} or less.

For plane e.m. waves in free space, the ratio E/H is equal to $\sqrt{(\mu_0/\epsilon_0)}$, which has the value 120π ohms (377 Ω, very closely); this is known as the (wave) **impedance of free space**. If a wave is incident at an interface with a medium in which the ratio $E/H \neq 120\pi$ Ω, then the wave will be partially reflected and partially transmitted on into the medium; the degree of reflection/transmission depends on the degree of 'mismatch' between free space and the medium in question. The same general argument applies to an interface between any two media that have different values for their respective wave impedances. For example, radio wave propagation 'over the horizon' at frequencies ~ 10 MHz is greatly enhanced by reflections from ionized layers in the ionosphere that have different values of wave impedance to the un-ionized space below them.

T.e.m. waves can propagate in some bounded systems such as coaxial cable and other types of transmission line (see Chapter 8), but this is not the case in general. In commonly used coaxial cable, for example, at frequencies below a critical frequency (which is ~ 100 GHz) the mode of propagation is a t.e.m. mode; i.e. the electric field and the magnetic field in the insulating medium (dielectric) filling the space between the conductors are everywhere at right angles (see Fig. 1.13). A consequence is that the speed of propagation along the cable is equal to the speed of light in the unbounded dielectric, namely $c/(\mu_r\epsilon_r)^{1/2}$ (and $\mu_r = 1$, for practical purposes, for the dielectrics used, e.g. polyethylene).

For non-t.e.m. waves there is a component of $\boldsymbol{E}$ or $\boldsymbol{H}$ in the direction of propagation, and the speed of propagation is not equal to the speed of light in the medium in question.

Although the propagation of signals/energy in circuits has been treated in the earlier parts of the text in terms of voltages and currents, it will now be appreciated that there are always accompanying electric and magnetic fields in and around the circuit elements and connecting wires, cables and buses. In 'lumped element' circuits (that is, at low frequencies), voltages and currents (and phase differences) can be conveniently measured, but in the frequency ranges where the 'distributed' circuit approach is appropriate, a wave-like description of the transmission of signals has to be used (see Section 8.1).

In a uniform transmission line the 'characteristic impedance' mirrors the wave impedance, and the proportion of reflection and onward transmission at a discontinuity in a transmission line system is governed by the value of the characteristic impedances on either side of the interface (see Section 8.3.3).

A wave such as that sketched in Fig. 5.10 is called a **plane-polarized wave**; the plane of polarization is the x–z plane, i.e. the plane in which the E-vector lies. A well-known example of plane-polarized waves is in television transmissions; 'vertically' and 'horizontally' polarized receiving antennas are easily distinguishable.

Other forms of polarization find many uses in telecommunication and radar systems; **circular polarization** is probably the most commonly used. Imagine looking along the z-axis in Fig. 5.10, and further imagine that there are two plane-polarized e.m. waves propagating in that direction, one with its E-vector in the x-direction and the other with its E-vector in the y-direction and lagging in phase by 90°. For a transverse plane located at any value of z, say $z = 0$ for convenience, the time variation of the electric field $E(t)$ is obtained from eqn (5.42(c)):

$$\boldsymbol{E} = \underbrace{\hat{\boldsymbol{\imath}}\hat{E}\cos\omega t}_{`E_x\text{'}} + \underbrace{\hat{\boldsymbol{\jmath}}\hat{E}\cos(\omega t - 90^\circ)}_{`E_y\text{'} = \hat{E}\sin\omega t} \tag{5.45}$$

and $|E| = (E_x^2 + E_y^2)^{1/2} = \hat{E}$, since $\cos^2\omega t + \sin^2\omega t = 1$.

At $t = 0$, $\boldsymbol{E} = \hat{\boldsymbol{\imath}}\hat{E}$, and as t increases, E_x and E_y develop in accordance with eqn (5.45); see Fig. 5.11(a). So the tip of the vector $\boldsymbol{E}$ traces out a circle of radius $\hat{E}$ in the clockwise sense, the period of a complete rotation being equal $2\pi/\omega$: a right-hand circularly polarized wave (r.h.c.p.).

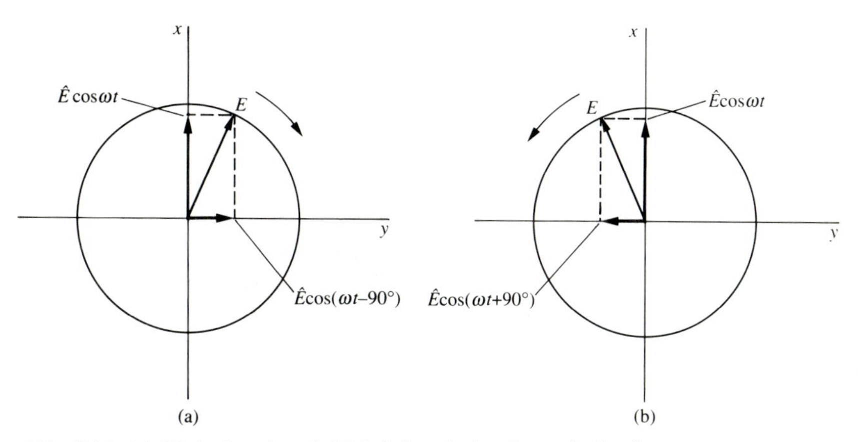

Fig. 5.11 (a) Right-hand and (b) left-hand circular polarization.

If E_y leads E_x by 90°, then, since $\cos(\omega t + 90°) = -\sin \omega t$,

$$\boldsymbol{E} = \hat{\boldsymbol{\imath}}\hat{E} \cos \omega t - \hat{\boldsymbol{\jmath}}\hat{E} \sin \omega t. \tag{5.46}$$

In this case the tip of the vector $\boldsymbol{E}$ traces out a circle in the anticlockwise sense (see Fig. 5.11(b)): a left-hand circularly polarized wave (l.h.c.p.).

Notice that if eqns (5.45) and (5.46) are added together, then

$$2\boldsymbol{E} = \hat{\boldsymbol{\imath}}\, 2\hat{E} \cos \omega t. \tag{5.47}$$

Hence it can be seen that a plane-polarized wave (eqn (5.47)) can be resolved into two oppositely rotating circularly polarized waves (eqns (5.45) and (5.46)). This feature can be exploited to generate a circularly polarized wave; a polarizing filter can be used to reject the unwanted circularly polarized component.

Problems

5.1 Show that for the equivalent series network for a lossy capacitor,

$$R_s = \frac{\epsilon_r''}{\omega C_0(\epsilon_r''^2 + \epsilon_r'^2)} \qquad \text{and} \qquad \tan\delta \approx \frac{\epsilon_r''}{\epsilon_r'}$$

(see Section 3.6 and Problems 3.23 and 3.25).

5.2 Calculate the skin depth for alternating electromagnetic fields in copper (electrical conductivity $\sigma = 5.8 \times 10^7\ \mathrm{S\,m^{-1}}$, relative permeability $\mu_r = 1$) at frequencies of 50 Hz, 5 kHz, 500 kHz, 50 MHz, 5 GHz.

5.3 Show that the distributed self-inductance L of a coaxial line is given by

$$L = \left(\frac{\mu_0}{4\pi}\right) 2\ln(b/a)\ \mathrm{H\,m^{-1}},$$

where a, b are the radii of the inner and outer conductors respectively.

5.4 For the transformer core shown in Fig. 5.7(b), show that the magnetic flux ϕ in the central limb is given by

$$\phi = 2\mu_r\mu_0 w^2 NI/7l.$$

(Make the usual approximations for the effective path length for the magnetic flux.)

5.5 In a region of space the electrostatic potential V is given by

$$V = \frac{ax}{y} - bxz.$$

Obtain the expression for the electric field E.

5.6 Obtain expressions for $\nabla \cdot \boldsymbol{E}$ and $\nabla \times \boldsymbol{E}$ if the electrostatic potential is given by

$$V = x^2y^2z^2 + Ax + By + Cz,$$

where A, B, C are constants.

5.7 The power received by a radio antenna of effective area $1\ \mathrm{m^2}$ is -80 dBm. Calculate the electric field strength in the incoming radio signal.

Pulses and transients

6.1 Introduction

So far the discussions of time-dependent voltages and currents have been concerned almost entirely with the circuit theorems applicable to sinusoidal voltages and currents of a single frequency and having time-independent amplitudes, i.e. so-called steady-state sinusoids. Such mathematically pure sinusoids cannot be obtained in practice for two reasons: (1) real circuit elements are non-linear to some degree (see Section 2.6) and this causes the generation of harmonics at integer multiples of the frequency of the fundamental sinusoid; (2) a voltage or a current source has to be switched on (and off also, ultimately) and, as will be emphasized in Section 6.6, a sinusoid of limited duration is not a mathematically pure sinusoid. In any case it is very important to note that a signal (or 'information') cannot be transmitted by a constant voltage or current; in order to transmit information it is necessary to modulate the voltage/current (see Fig. 6.1). Perhaps the simplest illustration of this principle is the 'on/off' modulation exemplified by the combinations of dots and dashes of Morse code. Another example is the transmission of information in amplitude-modulated radio broadcasts; here the amplitude of a high frequency sinusoidal carrier wave is modulated by the (relatively) low-frequency 'baseband signal' constituting the information, e.g. speech.

The signal waveforms depicted in Fig. 6.1 are all represented as functions of time (t); that is, they are represented in the **time domain**. In this chapter it will be seen that a signal can also be represented by its **spectrum** in the **frequency domain** (see Fig. 6.5 for example). The choice of which representation to use will be governed by the context, and it is vital to be able to transform from one representation to the other.

Periodic waveforms can be analysed into their respective Fourier series to yield their spectra in the frequency domain; the underlying theory and Examples are given in Section 6.2. The use of Fourier transforms to give frequency-domain descriptions of transient waveforms is described in Section 6.6; this finds extensive application in the analysis and processing of signals in a wide variety of engineering contexts.

6.2 Periodic waveforms

According to **Fourier's theorem** a periodic mathematical function (i.e. a function $v(t)$ such that $v(t + T) = v(t)$, where T is a period; see Fig. 6.2(a))

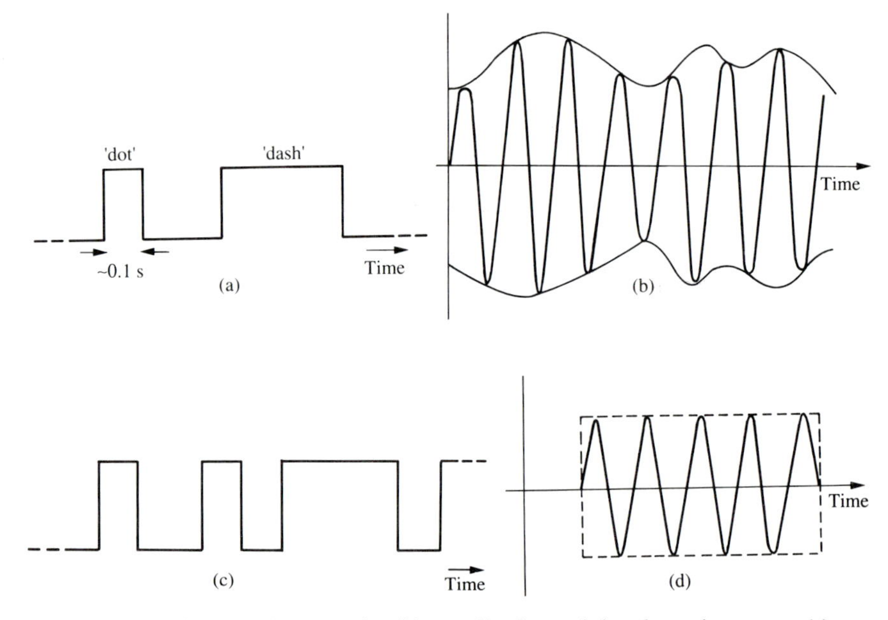

Fig. 6.1 Signals: (a) Morse code; (b) amplitude-modulated carrier wave; (c) a digital signal; (d) a radiofrequency (r.f.) pulse.

can be represented by a series of sinusoidal Fourier components, viz.

$$v(t) = \frac{a_0}{2} + \sum_{n=1}^{\infty} \left(a_n \cos \frac{2\pi nt}{T} + b_n \sin \frac{2\pi nt}{T} \right). \tag{6.1}$$

Equation (6.1) displays the so-called 'trigonometric form' of the series (the 'exponential' form will be introduced in Section 6.3). If a fundamental frequency ω is defined for the periodic function through $\omega \equiv 2\pi/T$, then it can be seen that the terms in the series in eqn (6.1) have frequencies of value $n\omega$; i.e. the Fourier components of the periodic waveform have frequencies that are **harmonics** of the fundamental frequency. The formulae giving the coefficients a_0, a_n, b_n are given below (eqns (6.5), (6.6), (6.7)).

The practical relevance of Fourier's theorem relates to the fact that all circuits, whether passive or active, have a limited 'pass band'; i.e. the magnitude of the transfer function $T(\omega)$ (where $T(\omega) \equiv$ output/input) decreases inevitably to zero at high-enough frequencies and, in many cases, at low-enough frequencies also. The high-frequency 'roll-off' stems

---

Fig. 6.2 Some examples of periodic, non-sinusoidal waveforms; (a) a general form; (b), (f), half-wave rectified sine waves; (c), (d), (e) 'square' waves; (g) a sawtooth waveform.

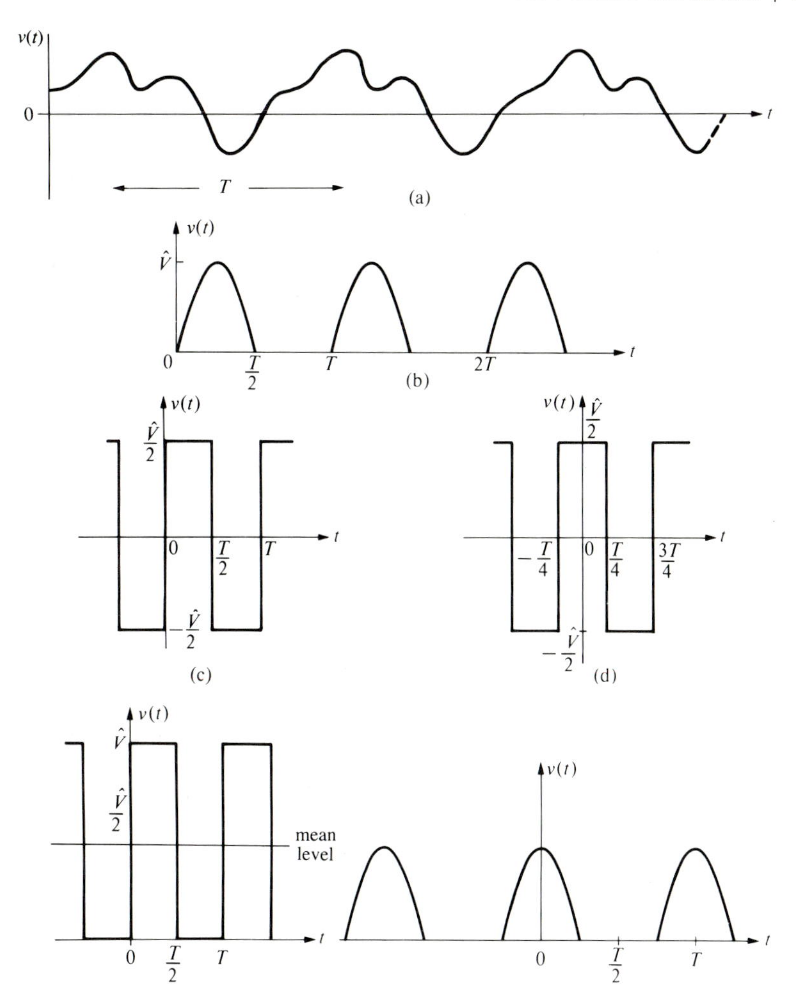
v(t)
0
t
T
(a)
v(t)
V̂
0
T/2
T
2T
t
(b)
v(t)
V̂/2
0
T/2
T
−V̂/2
t
(c)
v(t)
V̂/2
−T/4
0
T/4
3T/4
−V̂/2
t
(d)
v(t)
V̂
V̂/2
mean level
0
T/2
T
t
(e)
v(t)
0
T/2
T
t
(f)

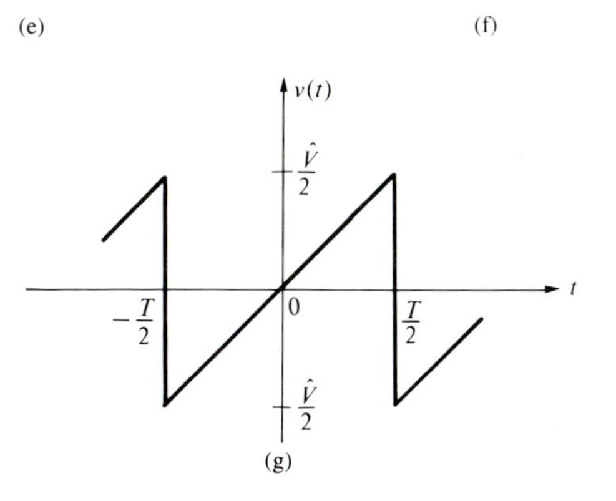
v(t)
V̂/2
−T/2
0
T/2
V̂/2
t
(g)

from inescapable stray shunt capacitances, which, however, small they are in magnitude, have a shorting effect at high-enough frequencies. Of course, in the case of filters, $T(\omega)$ is deliberately designed to have a well-defined pass band (see Section 7.5). So for a real circuit for which the input has a periodic waveform, the output will not contain all the Fourier components up to infinite frequency; this means that the output will always be a distorted version of the input. The engineering problem in a particular situation is to assess how much distortion is acceptable; this demands knowledge of the bandwidth of the circuit or system in question as well as of the spectrum of Fourier components of the input waveform. Figure 6.5(d) illustrates this situation, albeit rather crudely. If the input to a circuit has a sawtooth waveform of fundamental frequency ω and the pass band of the circuit cuts off beyond 3ω, then only the first three Fourier components of the input will be transmitted. The output will have the form shown by the dotted line in Fig. 6.5(d); this may be an acceptable degree of distortion. If it is unacceptable, then the bandwidth of the circuit will have to be increased to encompass further Fourier components. Fortunately the amplitudes of the Fourier components decrease with increasing harmonic number n, apart from some exceptions for small values of n (see eqns (6.2), (6.3), (6.4)), and so the omission of high-order components has a relatively small effect on the shape of the output waveform.

Some examples of periodic waveforms are shown in Fig. 6.2; their Fourier series are as follows.

Half-wave rectified sinusoid (see Example 6.1):

$$v(t) = \hat{V} \sin \omega t \qquad (0 \leqslant t \leqslant T/2)$$

$$v(t) = \frac{\hat{V}}{\pi} + \frac{\hat{V}}{2} \sin \omega t - \frac{2\hat{V}}{3\pi} \cos 2\omega t - \frac{2\hat{V}}{15\pi} \cos 4\omega t - \frac{2\hat{V}}{35\pi} \cos 6\omega t \ldots \tag{6.2}$$

Sawtooth (see Problem 6.1(b)):

$$v(t) = \frac{\hat{V}}{\pi} \sin \omega t - \frac{\hat{V}}{2\pi} \sin 2\omega t + \frac{\hat{V}}{3\pi} \sin 3\omega t - \frac{\hat{V}}{4\pi} \sin 4\omega t \ldots \tag{6.3}$$

Square wave (see Problem 6.1(a)):

$$v(t) = \frac{2\hat{V}}{\pi}\cos\omega t - \frac{2\hat{V}}{3\pi}\cos 3\omega t + \frac{2\hat{V}}{5\pi}\cos 5\omega t \ldots \tag{6.4}$$

Leaving aside for the moment the reasons for the appearance of sine and cosine terms, and plus and minus signs, the above expressions emphasize that the terms in the series are harmonics of the fundamental (but note that not all harmonics necessarily occur) and that there is a general decrease of amplitude with harmonic number. The sharper the 'corners' of a waveform, the greater is the number of Fourier components that have a significant amplitude; a comparison of eqns (6.2) and (6.4) gives a rough indication of this phenomenon.

The formulae for a_0, a_n, b_n are:

$$a_0 = \frac{2}{T}\int_0^T v(t)\,\mathrm{d}t \tag{6.5}$$

$$a_n = \frac{2}{T}\int_0^T v(t)\cos\frac{2\pi nt}{T}\,\mathrm{d}t \tag{6.6}$$

$$b_n = \frac{2}{T}\int_0^T v(t)\sin\frac{2\pi nt}{T}\,\mathrm{d}t. \tag{6.7}$$

Note that the integrals must be taken over a complete period of the waveform; the particular interval 0 to T has been chosen for convenience.

Example 6.1 Obtain expressions for the amplitudes a_0, a_n, b_n of the Fourier components of a half-wave rectified sine wave (see Fig. 6.2(b)).

The procedure is quite straightforward but it can be rather tedious and so care must be exercised. Using the formula for a_0,

$$a_0 = \frac{2}{T}\int_0^T \hat{V}\sin\omega t\,\mathrm{d}t$$

or

$$a_0 = \frac{2\hat{V}}{T}\int_0^{T/2} \sin\omega t\,\mathrm{d}t,$$

since $v(t)$ is zero for $T/2 \leqslant t \leqslant T$. Hence, on changing the variable to ωt,

$$a_0 = \frac{2\hat{V}}{\omega T}\int_{\omega t=0}^{\pi} \sin \omega t \, \mathrm{d}(\omega t)$$

$$= \frac{2\hat{V}}{\omega T}[-\cos \omega t]_{\omega t=0}^{\pi}$$

or $$a_0 = \frac{2\hat{V}}{\pi} \quad \text{and} \quad a_0/2 = \frac{\hat{V}}{\pi}.$$

For $n = 1$,

$$a_1 = \frac{2}{T}\int_0^{T/2} \hat{V} \sin \omega t \cos \omega t \, \mathrm{d}t$$

$$= \frac{\hat{V}}{\pi}\int_{\omega t=0}^{\pi} \sin \omega t \cos \omega t \; \mathrm{d}(\omega t)$$

$$= \frac{\hat{V}}{4\pi}\int_{2\omega t=0}^{2\pi} \sin 2\omega t \, \mathrm{d}(2\omega t)$$

$$= \frac{\hat{V}}{4\pi}[-\cos 2\omega t]_{2\omega t=0}^{2\pi}.$$

Hence $a_1 = 0$.

For $n > 1$ (notice the term $(n - 1)$ appearing in the *denominator* in the following analysis),

$$a_n = \frac{2\hat{V}}{T}\int_0^{T/2} \sin \omega t \cos n\omega t \, \mathrm{d}(t)$$

$$= \frac{\hat{V}}{\pi}\int_{\omega t=0}^{\pi} \sin \omega t \cos n\omega t \, \mathrm{d}(\omega t)$$

$$= \frac{\hat{V}}{\pi}\int_0^{\pi} \{\sin(n+1)\omega t - \sin(n-1)\omega t\} \, \mathrm{d}(\omega t)$$

$$= \frac{\hat{V}}{\pi}\left[-\frac{\cos(n+1)\omega t}{(n+1)} + \frac{\cos(n-1)\omega t}{(n-1)}\right]_0^{\pi}.$$

On evaluating this expression between the limits,

$$a_n = \frac{-2\hat{V}}{\pi(n^2 - 1)} \qquad (n \text{ even})$$

$$a_n = 0 \qquad (n \text{ odd}).$$

Now consider the coefficients b_n. For $n = 1$

$$b_1 = \frac{\hat{V}}{\pi}\int_0^{\pi} \sin^2 \omega t \, \mathrm{d}(\omega t)$$

$$= \frac{\hat{V}}{\pi}[\omega t - \sin \omega t \cos \omega t]_0^{\pi},$$

whence $b_1 = \hat{V}/2$.

For $n > 1$,

$$b_n = \frac{\hat{V}}{\pi}\int_{\omega t = 0}^{\pi} \sin \omega t \sin n\omega t \, \mathrm{d}(\omega t)$$

$$= \frac{\hat{V}}{\pi}\int_0^{\pi} \{\cos(n-1)\omega t - \cos(n+1)\omega t\} \, \mathrm{d}(\omega t)$$

$$= \frac{\hat{V}}{\pi}\left[\frac{\sin(n-1)\omega t}{(n-1)} - \frac{\sin(n+1)\omega t}{(n+1)}\right]_0^{\pi},$$

whence $b_n = 0$ for n odd or even.

So, finally, the Fourier series for $v(t)$ is

$$v(t) = \frac{\hat{V}}{\pi}\left[1 + \frac{\pi}{2}\sin \omega t - \frac{2}{3}\cos 2\omega t - \frac{2}{15}\cos 4\omega t - \frac{2}{35}\cos 6\omega t - \cdots\right],$$

which is eqn (6.2). □

It is useful to note that the term $a_0/2$ in a Fourier series is equal to the mean value of $v(t)$, that is, the average of $v(t)$ over a complete period.

Example 6.2 As another example consider the square wave shown in Fig. 6.2(c). Since, in this case, $v(t)$ is an **odd function** of t (i.e. $v(-t) = -v(t)$), it follows that, apart from a_0, each term of the Fourier series must be an odd function of t also. Hence the series must consist of sine terms only, since $\sin \phi$ is an odd function of ϕ whereas $\cos \phi$ is an even function of ϕ (i.e. $\cos(-\phi) = \cos \phi$). This means, of course, that only the b_n need to be evaluated.

Note that in the determination of whether a function is odd or even, the 'baseline' should be taken at the 'mean value' of the function. This is illustrated in Fig. 6.2(e), which is an odd function (compare with (c) and (d)).

In this example,

$$a_0 = \frac{2}{T}\int_0^{T/2} (\hat{V}/2)\,\mathrm{d}(t) - \frac{2}{T}\int_{T/2}^{T} (\hat{V}/2)\,\mathrm{d}(t)$$

and so $a_0 = 0$.

Also,

$$\begin{aligned} b_n &= \frac{2}{T}\int_0^{T/2} \frac{\hat{V}}{2} \sin n\omega t\,\mathrm{d}t - \frac{2}{T}\int_{T/2}^{T} \frac{\hat{V}}{2} \sin n\omega t\,\mathrm{d}t \\ &= \frac{\hat{V}}{2\pi}\int_0^{\pi} \sin n\omega t\,\mathrm{d}(\omega t) - \frac{\hat{V}}{2\pi}\int_{\pi}^{2\pi} \sin n\omega t\,\mathrm{d}(\omega t) \\ &= \frac{\hat{V}}{2\pi}\left[\frac{-\cos n\omega t}{n}\right]_0^{\pi} - \frac{\hat{V}}{2\pi}\left[-\frac{\cos n\omega t}{n}\right]_{\pi}^{2\pi}. \end{aligned}$$

It follows that $b_n = 0$ (n even) and $b_n = 2\hat{V}/n\pi$ (n odd), and the Fourier series is

$$v(t) = \frac{2\hat{V}}{\pi}\left\{\sin \omega t + \frac{\sin 3\omega t}{3} + \frac{\sin 5\omega t}{5} + \frac{\sin 7\omega t}{7} + \cdots\right\} \quad (6.8)$$

□

For the square waves shown in Fig. 6.2(d) and (e), and for the half-wave rectified sine wave shown in Fig. 6.2(f), it is left for the reader to show (see Problems 6.1 and 6.3) that the Fourier series, are, respectively,

$$v(t) = \frac{2\hat{V}}{\pi}\left\{\cos \omega t - \frac{\cos 3\omega t}{3} + \frac{\cos 5\omega t}{5} - \frac{\cos 7\omega t}{7} + \cdots\right\} \tag{6.9}$$

$$v(t) = \frac{\hat{V}}{2} + \frac{2\hat{V}}{\pi}\left\{\sin \omega t + \frac{\sin 3\omega t}{3} + \frac{\sin 5\omega t}{5} + \frac{\sin 7\omega t}{7} + \cdots\right\} \tag{6.10}$$

$$v(t) = \frac{\hat{V}}{\pi} + \frac{\hat{V}}{2}\cos \omega t + \frac{2\hat{V}}{\pi}\left\{\frac{\cos 2\omega t}{3} - \frac{\cos 4\omega t}{15} + \frac{\cos 6\omega t}{35} - \cdots\right\} \tag{6.11}$$

It can be seen by inspecting the examples met so far that the amplitudes and frequencies of the components in a Fourier series corresponding to a particular waveform are independent of the choice of origin for the time axis and of baseline for the v axis; the appearance of sine/cosine terms, and differences in sign, merely represent differences in phase.

If a periodic function $v(t)$ has the property $v(t + T/2) = -v(t)$, then the function is said to exhibit **half-wave symmetry** (see Fig. 6.3); in this event the Fourier series contains odd harmonics only (a square wave is a particular example: see Example 6.2).

The analysis of a periodic waveform can be rather tedious, but from the foregoing discussions, and the Examples, it should have become apparent that short cuts can sometimes be made simply from an inspection of the waveform:

Average value of the waveform $\rightarrow a_0$

Is the function even or odd? even (a_n only), odd (b_n *only*)

Half-wave symmetry? yes: n odd only

Consider the following examples:

The square wave of Fig. 6.2(e). As was argued above, this waveform has odd symmetry (about a baseline drawn at the mean value level). So the $a_n = 0$ for all n (apart from $a_0 = \hat{V}/2$, the mean value). Further, the waveform also exhibits half-wave symmetry, and so $b_n = 0$ for n even.

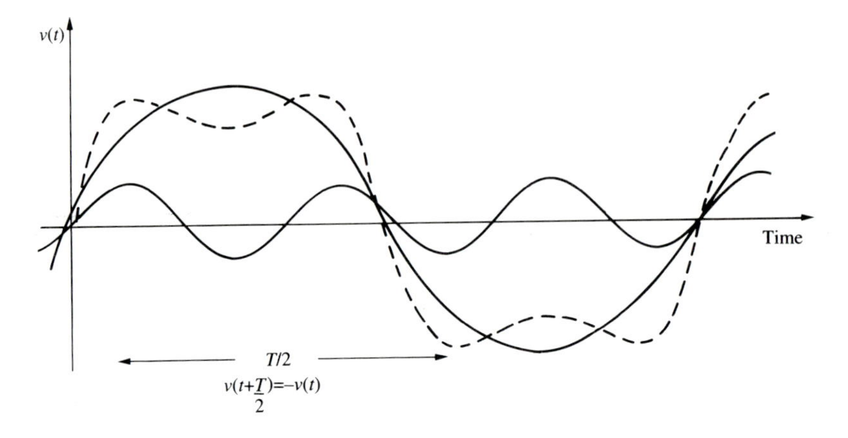

Fig. 6.3 A waveform (dashed line) exhibiting half-wave symmetry; notice that it can be analysed into odd harmonic components. (NB the 'baseline' should be taken at the mean value of the waveform.)

The sawtooth wave of Fig. 6.4(a). Here $v(t)$ is an odd function and so the $a_n = 0$ for all n (apart from $a_0 = \hat{V}/2$, the mean value).

The full-wave rectified wave of Fig. 6.4(b). Here $v(t)$ is an even function and so $b_n = 0$ for all n.

The triangular wave of Fig. 6.4(c). Here $v(t)$ is an even function and so $b_n = 0$ for all n. The mean value = 0 and so $a_0 = 0$. Further, $v(t)$ exhibits half-wave symmetry and so $a_n = 0$ for n even.

6.3 The frequency domain; amplitude and phase spectra

If eqn (6.1) is modified by defining

$$c_n \equiv (a_n - \mathrm{j}b_n)/2; \qquad c_{-n} \equiv (a_n + \mathrm{j}b_n)/2; \qquad c_0 \equiv a_0 \tag{6.12}$$

and use is made of the relations

$$\begin{aligned} \cos \omega t &= (\mathrm{e}^{\mathrm{j}\omega t} + \mathrm{e}^{-\mathrm{j}\omega t})/2 \\ \sin \omega t &= (\mathrm{e}^{\mathrm{j}\omega t} - \mathrm{e}^{-\mathrm{j}\omega t})/\mathrm{j}2, \end{aligned} \tag{6.13}$$

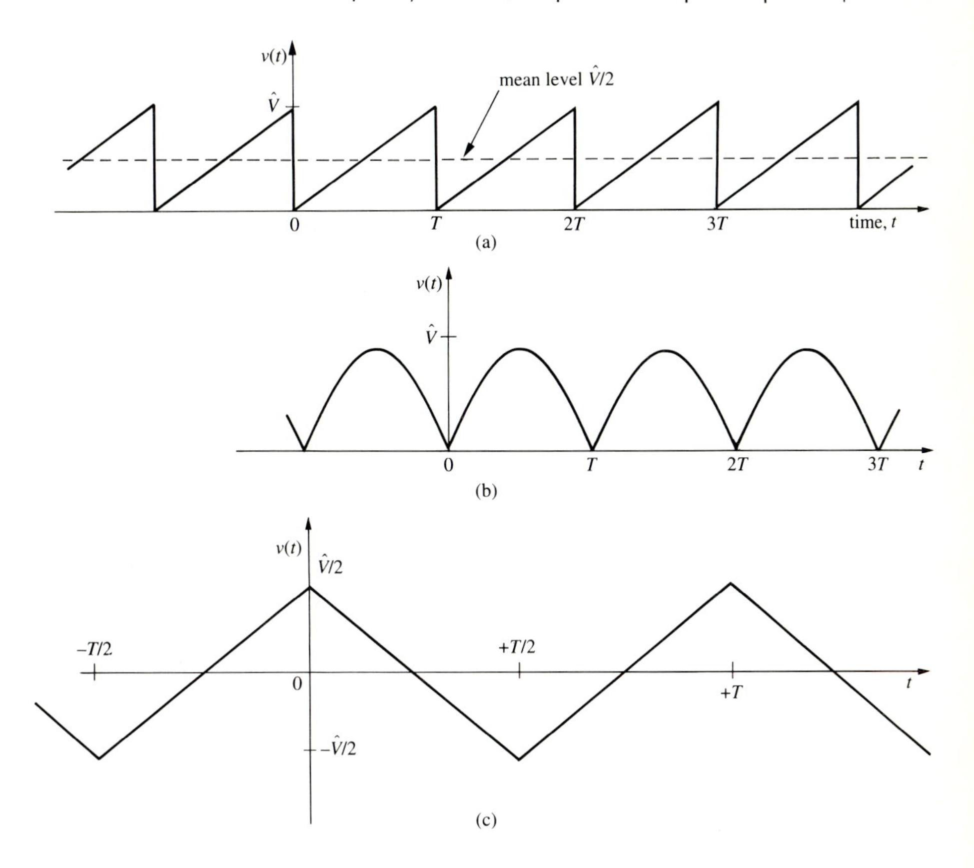

Fig. 6.4 (a) Sawtooth wave; $v(t)$ is an odd function: $v(t + T/2) \neq -v(t)$. (b) Full-wave rectified sine wave: $v(t + T/2) \neq -v(t)$. (c) Triangular wave; $v(t)$ is an even function: $v(t + T/2) = v(t)$.

then

$$v(t) = \sum_{n=-\infty}^{\infty} c_n \, e^{jn\omega t}, \tag{6.14}$$

where

$$c_n = \frac{1}{T} \int_{\text{period}} v(t) \, e^{-jn\omega t} \, dt. \tag{6.15}$$

Eqn (6.14) expresses what is often called the 'exponential form', or 'complex form', of the Fourier series.

Now c_n is complex, in general, with

$$\text{(real part)} \quad \text{Re}\, c_n = \frac{1}{T}\int v(t)\cos n\omega t\, \mathrm{d}t$$

$$\text{(imaginary part)} \quad \text{Im}\, c_n = -\frac{1}{T}\int v(t)\sin n\omega t\, \mathrm{d}t.$$

The amplitude spectrum of $v(t)$ is defined by

$$|c_n| = \{[\text{Re}(c_n)]^2 + [\text{Im}(c_n)]^2\}^{1/2} \tag{6.16}$$

and the phase spectrum by

$$\phi_n = \tan^{-1}\{\text{Im}(c_n)/\text{Re}(c_n)\}. \tag{6.17}$$

Consider again the square wave illustrated in Fig. 6.2(c), discussed earlier, for which it was found that $a_n = 0$ for all n, $b_n = 0$ for n even, and $b_n = 2\hat{V}/n\pi$ for n odd. In this case using eqns (6.12),

$$\text{Re}(c_n) = 0, \qquad \text{Im}(c_n) = -\frac{\hat{V}}{n\pi} \quad (n \text{ odd}), \text{ giving } |c_n| = \hat{V}/n\pi \ (n \text{ odd})$$

$$\phi_n = \begin{cases} \tan^{-1}(-\infty) & (n \text{ positive}) = -\pi/2 \\ \tan^{-1}(+\infty) & (n \text{ negative}) = +\pi/2. \end{cases}$$

These amplitude and phase spectra are illustrated in Fig. 6.5(b) and (c).

Example 6.3 As a further example of the kind of analysis involved in these situations, consider the derivation of the amplitude and phase spectra of the sawtooth waveform shown in Fig. 6.5(d).

Over the range of time $-T/2$ to $+T/2$, $v(t) = \dfrac{\hat{V}}{T}\cdot t$, and so

$$c_n = \frac{\hat{V}}{T^2}\int_{-T/2}^{T/2} t\, \mathrm{e}^{-jn\omega t}\, \mathrm{d}t$$

or

$$c_n = \frac{\hat{V}}{4\pi^2}\int_{\omega t = -\pi}^{+\pi} \omega t\, \mathrm{e}^{-jn\omega t}\, \mathrm{d}(\omega t).$$

It can be shown quite easily, by integrating by parts, that

$$\int x\, \mathrm{e}^{-jnx}\, \mathrm{d}x = \left(\frac{1}{n} + jx\right)\frac{\mathrm{e}^{-jnx}}{n}.$$

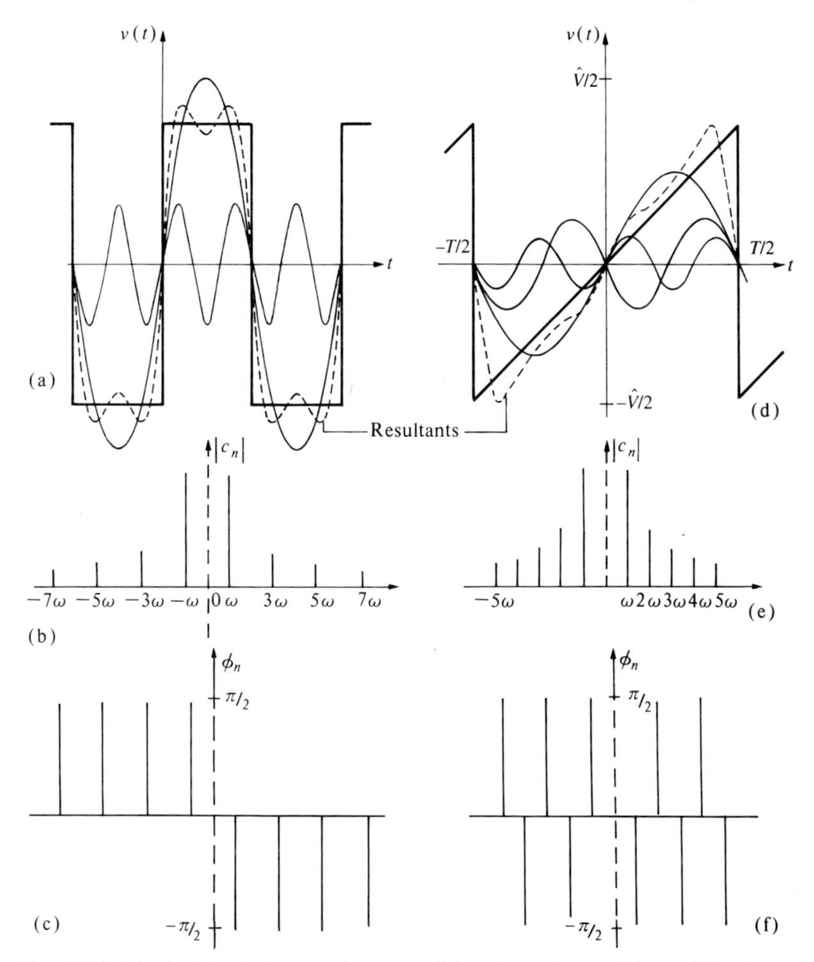

Fig. 6.5 (a) A sketch of the waveform resulting from the addition of the first two Fourier components of a square wave. (b), (c) The amplitude and phase spectra, respectively, of a square wave. (d) The waveform resulting from the addition of the first three Fourier components of a sawtooth wave. (e), (f) The amplitude and phase spectra, respectively, of a sawtooth wave.

Hence

$$c_n = \frac{\hat{V}}{4\pi^2}\left\{\left(\frac{1}{n} + \mathrm{j}\omega t\right)\frac{\mathrm{e}^{-\mathrm{j}n\omega t}}{n}\right\}_{-\pi}^{+\pi}$$

and it follows that

$$c_n = \frac{\mathrm{j}\hat{V}}{2\pi n} \quad (n \text{ even}) \qquad c_n = -\frac{\mathrm{j}\hat{V}}{2\pi n} \quad (n \text{ odd})$$

$$c_0 = a_0 = 0 \text{ (by inspection).}$$

The phase spectrum is given by

$$\phi_n = \begin{matrix} +\pi/2 & (n \text{ even}) \\ -\pi/2 & (\text{n odd}) \end{matrix} \qquad \phi_{-n} = \begin{matrix} -\pi/2 & (n \text{ even}) \\ +\pi/2 & (n \text{ odd}). \end{matrix}$$ □

Displays of amplitude spectra, such as are shown in Fig. 6.5, are useful in that they give useful guidance as to where an (infinite) Fourier series can be truncated without causing serious distortion to the waveform; e.g. a good idea can be obtained of the frequency bandwidth required of an amplifier if it is not to seriously distort the form of the input wave.

It is very useful to acquire the facility of transforming waveforms from the time domain to the frequency domain and vice versa.

The idea of a negative frequency that has been introduced may cause some disquiet, but it can be explained quite simply through the representation of $\mathrm{e}^{\mathrm{j}\omega t}$ on an Argand diagram. Consider $e^{\mathrm{j}\omega t}$ $(= \cos \omega t + \mathrm{j} \sin \omega t)$, which is represented in Fig. 6.6 as a vector in the complex plane having unit magnitude and with its phase angle increasing in the positive sense of rotation at rate ω. Similarly $\mathrm{e}^{-\mathrm{j}\omega t}$ (note the 'negative' frequency) is represented as a vector rotating in the negative sense.

To recover a 'real', physical cosine or sine variation, $\mathrm{e}^{\mathrm{j}\omega t}$ and $\mathrm{e}^{-\mathrm{j}\omega t}$ are combined:

$$\cos \omega t = (\mathrm{e}^{\mathrm{j}\omega t} + \mathrm{e}^{-\mathrm{j}\omega t})/2$$

$$\sin \omega t = (\mathrm{e}^{\mathrm{j}\omega t} - \mathrm{e}^{-\mathrm{j}\omega t})/\mathrm{j}2.$$

Example 6.4 Derive the exponential form of the Fourier series for a half-wave rectified sine wave and obtain the amplitude and phase spectra.

By using eqn (6.15) and integrating over the range $t = 0$ to T (see Fig. 6.2(b)), and remembering that $\omega \equiv 2\pi/T$,

$$c_n = \frac{\hat{V}}{2\pi} \int_{\omega t = 0}^{\pi} \sin \omega t \cdot \mathrm{e}^{-\mathrm{j}n\omega t} \, \mathrm{d}(\omega t).$$

Defining $I \equiv \int_0^\pi \mathrm{e}^{-\mathrm{j}nx} \sin x \, \mathrm{d}x$ and integrating by parts:

$$I = -\mathrm{e}^{-\mathrm{j}nx} \cos x - \int \cos x \cdot \mathrm{j}n \cdot \mathrm{e}^{-\mathrm{j}nx} \, \mathrm{d}x \Big|_0^\pi$$

$$= -\cos x \, \mathrm{e}^{-\mathrm{j}nx} - \mathrm{j}n \cdot \Big[\mathrm{e}^{-\mathrm{j}nx} \cdot \sin x + \mathrm{j}n \cdot \underbrace{\int \sin x \cdot \mathrm{e}^{-\mathrm{j}nx} \, \mathrm{d}x}_{`I\text{'}} \Big] \Big|_0^\pi$$

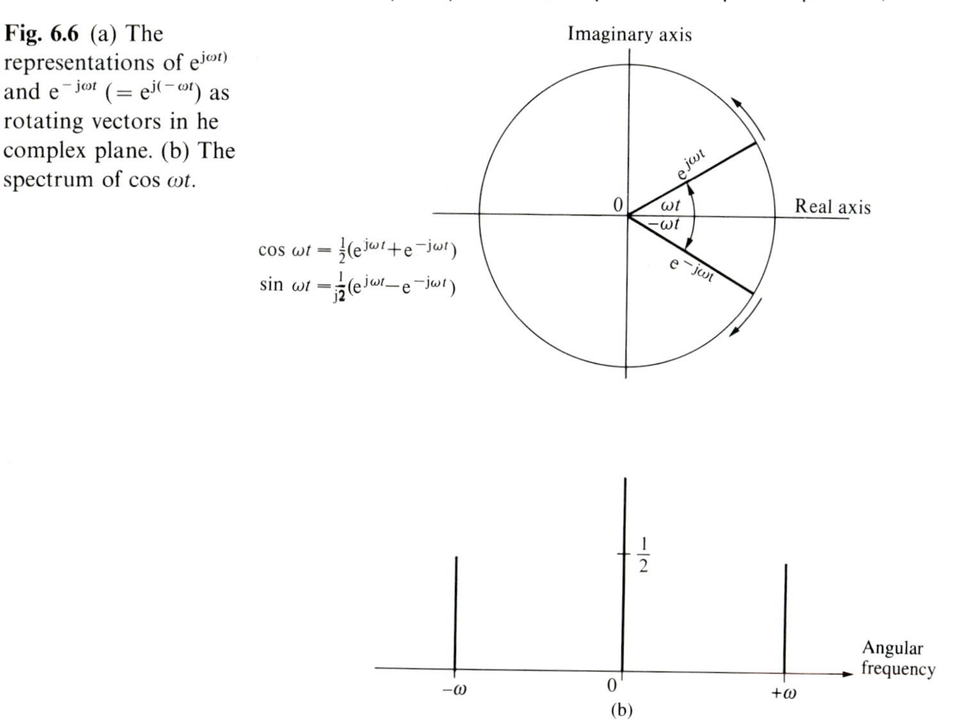

Fig. 6.6 (a) The representations of $e^{j\omega t)}$ and $e^{-j\omega t}$ $(= e^{j(-\omega t)})$ as rotating vectors in he complex plane. (b) The spectrum of cos ωt.

Therefore

$$I = \left. \frac{e^{-jnx}(\cos x + jn \sin x)}{(n^2 - 1)} \right|_0^{\pi}$$

(NB this expression cannot be used when $|n| = 1$.)

$n \geqslant 2$: $\quad \operatorname{Re} I = \dfrac{-2}{(n^2 - 1)}$ (n even); $\operatorname{Re} I = 0$ (n odd)

$\operatorname{Im} I = 0$ (n even or odd)

$n \leqslant -2$: $\quad \operatorname{Re} I = \dfrac{-2}{(n^2 - 1)}$ ($|n|$ even); $\operatorname{Re} I = 0$ ($|n|$ odd)

$\operatorname{Im} I = 0$ ($|n|$ even or odd)

Therefore $c_{-n} = c_n$ for these cases.

$n = \pm 1$:
$$I = \int_0^{\pi} e^{\pm jx} \sin x \, dx$$
$$= \int_0^{\pi} (\cos x \pm j \sin x) \sin x \, dx$$
$$= \frac{1}{2} \int \sin 2x \, dx \pm \frac{j}{2} \int (1 - \cos 2x) \, dx \Big|_0^{\pi}$$
$$= \frac{1}{4} \int \sin 2x \, d(2x) \mp \frac{j}{4} \int \cos 2x \, dx \pm \frac{j}{2} \int dx \Big|_0^{\pi}$$

Therefore
$$I = -j\pi/2 \qquad n = +1$$
$$I = +j\pi/2 \qquad n = -1.$$

So, finally, $c_1 = -j\hat{V}/4$; $c_{-1} = j\hat{V}/4$. Also $c_0 = \hat{V}/\pi$.

$|n| \geqslant 2$:
$$c_n = c_{-n} = -\frac{\hat{V}}{\pi(n^2 - 1)}.$$
□

6.4 Pulse trains

A situation that is often approached fairly closely in practice is that of an infinite train of narrow pulses. It is left as a problem (Problem 6.6) to show that if the pulses are 'rectangular', having duration λ, period T, and amplitude $\hat{V}$ (see Fig. 6.7(a)), then

$$a_0 = \hat{V}\lambda/T; \qquad b_n = 0 \quad \text{(even function)}$$
$$a_n = (2\hat{V}/n\pi) \sin(n\pi\lambda/T). \tag{6.18}$$

The function a_n of eqn (6.18) is sketched in Fig. 6.7(b). For $(n\pi\lambda/T = p\pi)$, where p is an integer, then a_n is zero. Hence if T/λ is an integer, the harmonics given by $n = pT/\lambda$ are missing from the amplitude spectrum. For example, if $T/\lambda = 2$ (a square wave), then the even harmonics are missing, as has been seen earlier.

If the 'mark-to-space' ratio T/λ is very large (i.e. the 'duty cycle' λ/T is very small), as is the case in many practical situations, then although the amplitudes of the Fourier components are much less than $\hat{V}$, they do not decrease very rapidly with increasing n. For instance, for $\lambda/T = 10^{-3}$ the amplitude of the 3500th harmonic (F in Fig. 6.7(b)) is 3/7 of the amplitude of the 1500th harmonic (B in Fig. 6.7(b)). Using eqn (6.18), the amplitude a_1 of the first harmonic, or 'fundamental' is given by $a_1 \approx 2\hat{V}\lambda/T$ (using

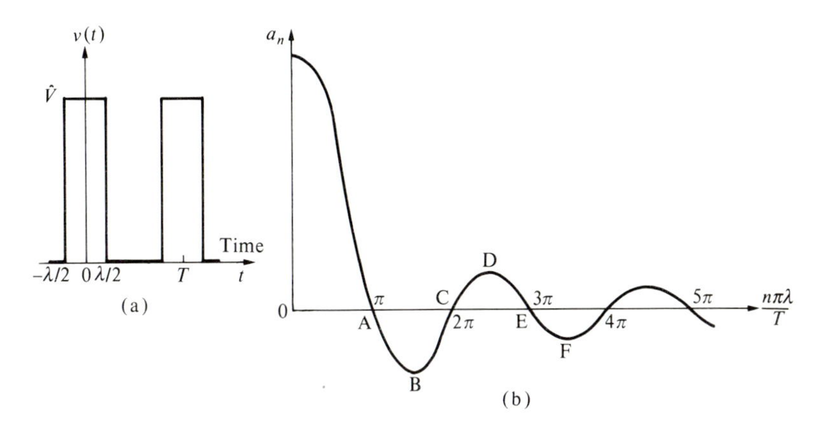

Fig. 6.7 (a) Rectangular pulses; the mark-to-space ratio is λ/T. (b) The amplitudes a_n of the Fourier components.

the approximation $\sin\phi \approx \phi$ for ϕ small), and so $a_1 \approx 2\hat{V} \times 10^{-3}$. Hence, for $a_n/a_1 \leqslant 10^{-1}$, say, it follows that $n \geqslant 3 \times 10^3$ approximately. Thus a rough, but useful, indication of the frequency bandwidth required of a circuit, or system, in order to give reasonably distortionless transmission of a train of 'narrow' pulses ($\lambda/T < 10^{-2}$, say) is 3000 times the pulse repetition frequency (p.r.f.) $1/T$; e.g. for $\lambda/T = 10^{-3}$ and p.r.f. $= 10^3\ \mathrm{s}^{-1}$, the required bandwidth is 3 MHz.

6.5 Power spectra

It was seen in Section 3.5 that the average power in a sinusoidal voltage or current is related directly to the integral over a complete cycle of the square of the voltage/current.

Before an expression for the power associated with a non-sinusoidal, periodic voltage or current waveform can be obtained, it is necessary to understand what is meant by the term **orthogonal functions**. Two functions $v_1(t)$, $v_2(t)$ are said to be orthogonal in an interval T if

$$\int_t^{t+T} v_1(t) \cdot v_2(t)\,\mathrm{d}t = 0.$$

For example, it is easy to show that $\sin\omega t$ and $\cos\omega t$ are orthogonal in an interval $T = 2\pi/\omega$ (and also over intervals that are integer multiples of T). Incidentally, 'orthogonal' means 'at right angles' and, of course, the phasors $\sin\omega t$ and $\cos\omega t$ are at right angles on a phasor diagram (see Figs. 3.4 and 3.5).

If two non-sinusoidal periodic waveforms $v_1(t)$, $v_2(t)$ are represented as trigonometrical Fourier series, then the average value of $v_1(t) \cdot v_2(t)$ is given by

$$\overline{v_1(t) \cdot v_2(t)}$$

$$= \frac{1}{T} \int_t^{t+T} \left[\frac{a_{0_1}}{2} + (a_{1_1} \cos \omega t + b_{1_1} \sin \omega t + a_{2_1} \cos 2\omega t + b_{2_1} \sin 2\omega t \ldots) \right]$$

$$\times \left[\frac{a_{0_2}}{2} + (a_{1_2} \cos \omega t + b_{1_2} \sin \omega t + a_{2_2} \cos 2\omega t + b_{2_2} \sin 2\omega t \ldots) \right] dt$$

$$= \frac{a_{0_1} a_{0_2}}{4} + \frac{a_{1_1} a_{1_1}}{2} + \frac{b_{1_1} b_{1_2}}{2} + \frac{a_{2_1} a_{2_2}}{2} + \frac{b_{2_1} b_{2_2}}{2} + \frac{a_{3_1} a_{3_2}}{2} \cdots,$$

since

$$\frac{1}{T} \int_0^T \begin{Bmatrix} \sin^2 n\omega t \\ \cos^2 n\omega t \end{Bmatrix} dt = \frac{1}{2}.$$

Notice that the integrals of the 'cross' terms (such as ($\cos \omega t \cdot \sin \omega t$), ($\sin \omega t \cdot \sin 2\omega t$), ($\sin 2\omega t \cdot \cos 3\omega t$), etc.) are zero, since the respective pairs of functions are orthogonal.

If $v_1(t) = v_2(t) \equiv v(t)$, then the above expression yields the average value $\overline{v^2(t)}$:

$$\overline{v^2(t)} = \frac{a_0^2}{4} + \frac{1}{2} \sum_{n=1}^{\infty} (a_n^2 + b_n^2)$$

(6.19) or

$$\overline{v^2(t)} = \frac{a_0^2}{4} + \frac{1}{2} \sum_{n=1}^{\infty} |c_n^2|.$$

This indicates that the power in the waveform is equal to the sum of the powers in the individual Fourier components. It also means, in practical terms, that a filter centred on the frequency ω_n can be used to filter out power at that frequency (provided that the bandwidth of the filter is narrow enough to exclude the adjacent Fourier components).

6.6 Fourier integrals and transforms

So far in this chapter, only waveforms that are periodic in the time domain have been considered, and it has been seen that their spectra in the frequency domain are 'discrete', or 'line' spectra. A non-periodic, or transient, waveform can be thought of as being periodic but with an infinitely long period. Hence the separation of the Fourier components, which is equal to $2\pi/T$, will tend to zero and the amplitude spectrum will become **continuous**.

From eqns (6.14) and (6.15) (with $f(t)$ replacing $v(t)$, for generality, and ω_0 replacing ω, for convenience) it follows that

$$f(t) = \frac{1}{T} \sum_{n=-\infty}^{\infty} \left\{ \int_{-T/2}^{T/2} f(t)\, \mathrm{e}^{-\mathrm{j}n\omega_0 t}\, \mathrm{d}t \right\} \mathrm{e}^{\mathrm{j}n\omega_0 t}.$$

Since $T = 2\pi/\omega_0$, and defining $\omega \equiv n\omega_0$,

$$f(t) = \frac{1}{2\pi} \sum_{n=-\infty}^{\infty} F(\omega)\, \mathrm{e}^{\mathrm{j}\omega t} \cdot \frac{\omega}{n},$$

where

(6.20)
$$F(\omega) \equiv \int_{T/2}^{T/2} f(t)\, \mathrm{e}^{-\mathrm{j}\omega t}\, \mathrm{d}t.$$

Letting $\Delta\omega$ ($\equiv \omega_0 = \omega/n$) denote the spacing between the Fourier components, then, as T becomes larger, $\Delta\omega$ becomes smaller and, in the limit of $T \to \infty$, $\Delta\omega$ can be replaced by the infinitesimally small element $\mathrm{d}\omega$. In these circumstances the summation sign in the expression for $f(t)$ is replaced by an integral sign and

(6.21)
$$f(t) = \frac{1}{2\pi} \int_{-\infty}^{\infty} F(\omega)\, \mathrm{e}^{\mathrm{j}\omega t}\, \mathrm{d}\omega.$$

The pair of functions $f(t)$, $F(\omega)$ are called a **Fourier transform pair**, and eqns (6.20), (6.21) define the Fourier transform.

$F(\omega)$ is the **amplitude density function** of $f(t)$; synonyms are **spectral density** and **frequency distribution function**.

A restrictive condition on the type of function $f(t)$ for which a Fourier transform exists is that $\int_{-\infty}^{\infty} |f(t)|^2\, \mathrm{d}t$ must be finite. This integral is essentially the sum of the squares of the amplitudes of the Fourier components, which in turn is directly related to the total energy in the real signal that is being represented by $f(t)$; this energy must be finite, of course. A mathematical 'step' function (see Fig. 6.11) does not meet the above condition, since it does not return to zero within a finite time duration. Nevertheless, practical approximations to step functions are very useful as excitations to test the responses of circuits and systems (see Section 7.4).

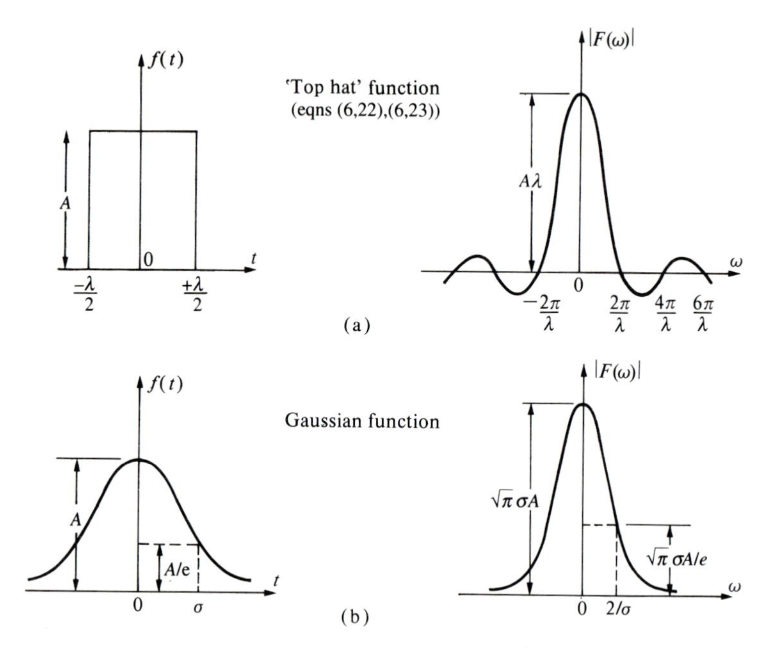

Fig. 6.8 Two functions and their Fourier transforms: (a) the 'top hat' function; (b) the 'Gaussian' function.

Some transient functions for which Fourier transforms do exist are shown in Fig. 6.8. Consider the 'top hat' function (see Fig. 6.8(a)):

$$F(\omega) = \int_{-\infty}^{\infty} f(t)\,\mathrm{e}^{-\mathrm{j}\omega t}\,\mathrm{d}t$$

$$= A\int_{t=-\lambda/2}^{+\lambda/2} \mathrm{e}^{-\mathrm{j}\omega t}\,\mathrm{d}t \rightarrow -\frac{A}{\mathrm{j}\omega}\int_{-\lambda/2}^{+\lambda/2} \mathrm{e}^{-\mathrm{j}\omega t}\,\mathrm{d}(-\mathrm{j}\omega t)$$

$$= \frac{\mathrm{j}A}{\omega}\{\mathrm{e}^{-\mathrm{j}\omega t}\}_{t=-\lambda/2}^{+\lambda/2}$$

$$= \frac{A\lambda\sin(\omega\lambda/2)}{\omega\lambda/2}$$

or

$$F(\omega) = A\lambda\,\mathrm{sinc}(\omega\lambda/2), \tag{6.22}$$

where

$$\operatorname{sinc} \phi \equiv \frac{\sin \phi}{\phi} \tag{6.23}$$

To find the value of $F(\omega)$ for $\omega = 0$, use the series expansion for $\sin \phi$, which gives

$$\operatorname{sinc} \phi \approx 1 - \frac{\phi^2}{3!} + \frac{\phi^4}{5!} - \cdots \tag{6.24}$$

So $\operatorname{sinc} \phi \to 1$ for $\phi \to 0$. Therefore

$$F(0) = A\lambda.$$

Also, $\sin \omega\lambda/2 = 0$ for $\omega = n(2\pi/\lambda)$, which defines the zeros of $F(\omega)$.

The form of $F(\omega)$ is shown in Fig. 6.8(a). Notice that as λ decreases, the spectrum spreads out and the peak value $(A\lambda)$ decreases. This illustrates a general feature of transient waveforms: the shorter the duration of the 'pulse', the wider its spectrum in the frequency domain. Very often this relationship is referred to as **reciprocal spreading**. Consider, for example, the top hat function. Most of the energy is contained in (angular) frequencies less than $2\pi/\lambda$, the frequency at which the first zero of $F(\omega)$ occurs. Hence the frequency bandwidth Δf of the pulse can be considered to be $\sim 1/\lambda$ Hz. Since the time duration Δt of the pulse is λ, it follows that $\Delta t \Delta f = 1$. Bearing in mind the degree of arbitrariness in the above definition of bandwidth, it is usual to write

$$\Delta t \Delta f \approx 1 \quad \text{or} \quad \Delta t \Delta \omega = 2\pi \tag{6.25}$$

This approximate relationship applies to transient waveforms of any shape, e.g. the 'Gaussian' waveform shown in Fig. 6.8(b).

The so-called 'unit impulse' function, $\delta(t)$, (see Fig. 6.9) is defined, in mathematical terms, as a top hat function of infinitely short duration (but having unit area, note) whose value is zero for all values of t except $t = 0$.

Using eqn (6.22) for $F(\omega)$, but with $A\lambda = 1$, and noting from eqn (6.24) that $\operatorname{sinc}(\omega\lambda/2) \to 1$ for $\lambda \to 0$, it can be seen that $F(\omega) = 1$ for a unit impulse; i.e. the frequency spectrum is 'flat' from $\omega = -\infty$ to $\omega = +\infty$.

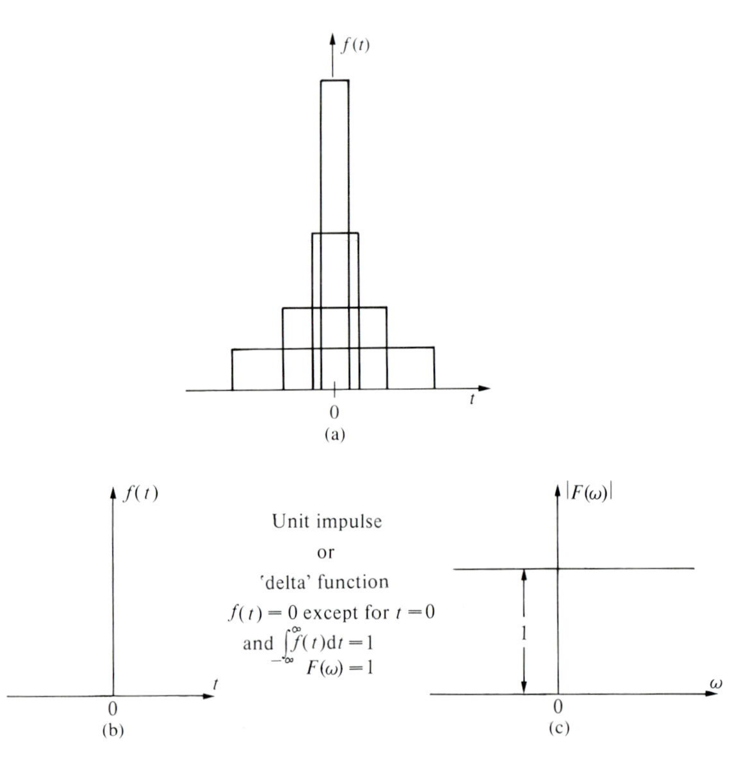

Fig. 6.9 The 'unit impulse' and its frequency spectrum. (a) Top hat functions of unit area but decreasing duration. (b) The delta function. (c) The Fourier transform of the unit impulse.

As will be seen in Section 7.4, the response of a system to an excitation in the form of a unit impulse (the so-called impulse response of the circuit) provides much information about the characteristics of the system and so is of great engineering interest.

Example 6.5 The Fourier transform of a cosine wave of finite duration, e.g. a radio-frequency pulse.

Referring to Fig. 6.10(a), let the angular frequency of the cosine wave be p, so that $f(t)$ is defined by:

$$\begin{aligned} f(t) &= \hat{V} \cos pt \qquad -\lambda/2 < t < \lambda/2 \\ &= 0 \qquad \text{elsewhere.} \end{aligned}$$

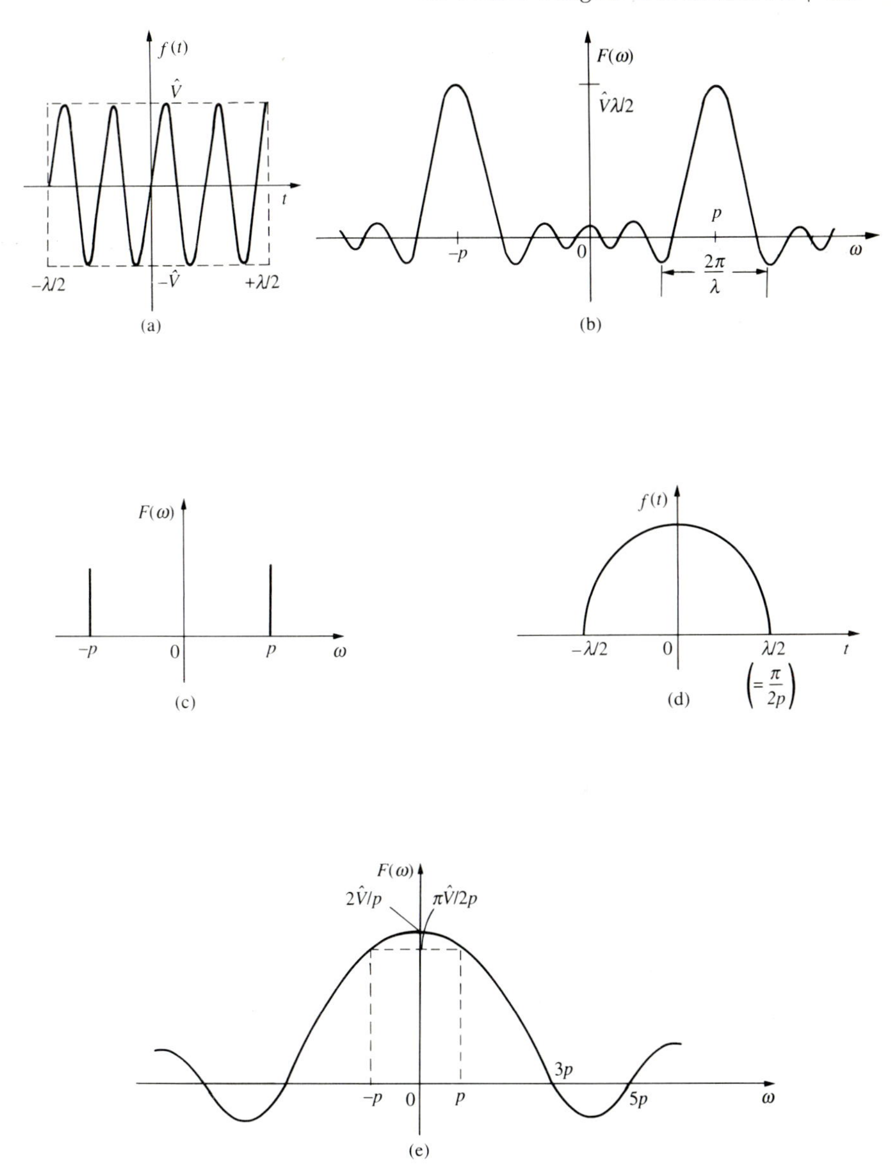

Fig. 6.10 (a) A cosine wave of duration λ. (b) The spectrum of the cosine wave of limited duration. (c) The spectrum of a cosine wave of angular frequency p and of unlimited duration, i.e. a mathematically pure cosine wave. (d) A cosine wave of half period duration. (e) The spectrum of the cosine wave of (d).

The Fourier transform $F(\omega)$ of $f(t)$ is given by

$$F(\omega) = \hat{V}\int_{-\infty}^{+\infty} f(t)\,\mathrm{e}^{-\mathrm{j}\omega t}\,\mathrm{d}t$$

$$= \hat{V}\int_{-\lambda/2}^{+\lambda/2} \cos pt\cdot\mathrm{e}^{-\mathrm{j}\omega t}\,\mathrm{d}t$$

$$= \frac{\hat{V}}{2}\int_{-\lambda/2}^{+\lambda/2} (\mathrm{e}^{\mathrm{j}pt} + \mathrm{e}^{-\mathrm{j}pt})\,\mathrm{e}^{-\mathrm{j}\omega t}\,\mathrm{d}t.$$

Thus

$$\frac{2F(\omega)}{\hat{V}} = \frac{1}{\mathrm{j}(p-\omega)}\int_{t=-\lambda/2}^{+\lambda/2} \mathrm{e}^{\mathrm{j}(p-\omega)t}\,\mathrm{d}[\mathrm{j}(p-\omega)t]$$

$$-\frac{1}{\mathrm{j}(p+\omega)}\int_{t=-\lambda/2}^{+\lambda/2} \mathrm{e}^{-\mathrm{j}(p+\omega)t}\,\mathrm{d}[-\mathrm{j}(p+\omega)]$$

$$= \frac{-\mathrm{j}}{(p-\omega)}\left[\mathrm{e}^{\mathrm{j}(p-\omega)t}\right]_{-\lambda/2}^{+\lambda/2} + \frac{1}{(p+\omega)}\left[\mathrm{e}^{-\mathrm{j}(p+\omega)t}\right]_{-\lambda/2}^{+\lambda/2},$$

which gives

$$F(\omega) = \frac{\hat{V}}{2}\left\{\frac{2}{(p-\omega)}\sin(p-\omega)\frac{\lambda}{2}\right\} + \frac{\hat{V}}{2}\left\{\frac{2}{(p+\omega)}\sin(p+\omega)\frac{\lambda}{2}\right\}$$

or

$$F(\omega) = \frac{\hat{V}\lambda}{2}\,\mathrm{sinc}\{(\omega-p)\lambda/2\} + \frac{\hat{V}\lambda}{2}\,\mathrm{sinc}\{(\omega+p)\lambda/2\}, \tag{6.26}$$

i.e. $F(\omega)$ consists of two sinc functions centred on $\omega = \pm p$ respectively (see Fig. 6.10).

The number of cycles of $\cos pt$ in the pulse is $\lambda/(2\pi/p) = p\lambda/2\pi$. If $p\lambda \gg 2\pi$, then the two sinc functions are well separated, as sketched in Fig. 6.10(b) and, in the limit of $p \to \infty$, the spectrum consists of two 'sharp' lines at $\omega = \pm p$, as was seen earlier (Fig. 6.6).

In a repetitively pulsed radar system, p could be 10 GHz (10^{10} Hz) and λ could be 0.1 μs. The number of cycles in a pulse would then be 1000 and the frequency spread about the carrier frequency (see Fig. 6.10(b)) would be $\pm(\pi/\lambda) = 10\pi$ MHz. This again demonstrates reciprocal spreading: $\Delta\omega\cdot\lambda \approx 2\pi$.

At the other extreme, if the pulse contains only one half cycle of cos Pt, then $p\lambda/2 = \pi/2$ and it follows from eqn (6.26) that

$$F(\omega) = \frac{-\cos\left(\dfrac{\pi\omega}{2p}\right)}{(\omega-p)} + \frac{\cos\left(\dfrac{\pi\omega}{2p}\right)}{(\omega+p)}$$

or

$$F(\omega) = \frac{-2p\hat{V}}{(\omega^2 - p^2)} \cos\left(\frac{\pi\omega}{2p}\right). \tag{6.27}$$

This expression for $F(\omega)$ is sketched in Fig. 6.10(e).

Notice that for $\omega = 0$, then, from eqn (6.27), $F(0) = 2\hat{V}/p$.

Also, $F(\omega) = 0$ for $\omega = (2n - 1)p$ (*except* $n = 1$, since then $\cos(\pi\omega/2p) = 0$ and both the numerator and the denominator of eqn (6.27) are zero).

For 'n' = 1 (i.e. $\omega = p$), then $\pi\omega/2p = \pi/2$ and both the numerator and the denominator of the right-hand side of eqn (6.27) are zero. In this case the limiting value of $F(\omega)$ as $\omega \to p$ is equal to the *quotient of the derivatives* of the numerator and the denominator, respectively, with respect to ω, evaluated at $\omega = p$ (L'Hospital's rule):

$$\lim_{\omega \to p} F(\omega) = (-2p\hat{V}) \frac{\left(-\dfrac{\pi}{2p}\right)\sin\left(\dfrac{\pi\omega}{2p}\right)}{2\omega}$$

or

$$F(p) = \pi\hat{V}/2p. \qquad \square$$

An important general point to note emerges from eqn (6.18), which can be rewritten in the form

$$a_n = \frac{2\hat{V}\lambda}{T} \operatorname{sinc}\left(\frac{n\pi\lambda}{T}\right).$$

This emphasizes that the 'envelope' function that describes the amplitudes of the Fourier components of any periodic function is the Fourier transform of the 'repeat element' of the periodic function; this is a useful point to remember.

Most transient signals in engineering situations cannot be represented by analytical mathematical functions, and resort must be made to signal processing procedures.

Any real signal of interest is 'band-limited' (to a frequency range 0–ω_b, say) either for intrinsic reasons or as a result of filtering. The signal is sampled at regular intervals $2\pi/\omega_s$ over a total time T, where ω_s is the sampling frequency; the 'record time', T, will be assumed to be longer than the effective duration of the transient signal (note that, for fundamental reasons, ω_s must be greater than $2\omega_b$; this is Nyquist's criterion). The number, N, of samples is equal to $\omega_s T/2\pi$. A detail discussion of the theory and practice of sampled data systems is outside the scope of this text, but it should be noted that the so-called discrete Fourier transform (DFT) of the signal can be computed from the sampled data. The considerable amount of computing required to process the data from the samples (1024 is a fairly typical number of samples) can be very significantly reduced by using the Fast Fourier Transform (FFT) algorithm; the reduction is in the ratio $N\log_2(N/N^2)$. In the frequency domain the range of the spectrum obtained is $0 \rightarrow \omega_s/2$ and the resolution is $\omega_s/2nN$ Hz.

6.7 The application of Laplace transforms

Any physical system and, in particular, any electrical system can be modelled at a fundamental level by a differential equation (or an integro-differential equation) that relates the output (or 'response') to the input (or 'excitation'. For example, CR and LR networks have been met already (see eqns (3.1) and (3.6)), and some of the general properties of oscillatory mechanical and electrical systems were outlined in Section 4.1.

The solution of the integro-differential equations of circuit analysis in the case of non-sinusoidal excitation functions, and transient excitations in particular, is greatly facilitated by the use of the Laplace transform, which, as will be seen very shortly, is generically related to the Fourier transform encountered in the previous section. The range of application of the Fourier transform is restricted to excitations $f(t)$ that satisfy the condition that $\int_{-\infty}^{\infty} |f(t)|^2\, dt$ is finite, which excludes many very important excitations (e.g. unit step, ramp; see Fig. 6.11), whereas the Laplace transform can be used in the contexts of most of the excitations that are met in engineering situations.

In considering an LCR network in Section 4.1, a second-order differential equation was obtained (eqn (4.3)) with the instantaneous value of the charge, q, on the capacitor as the dependent variable. To recapitulate, it was found that, for the stipulated boundary conditions, the solution of the differential equation was of the form

$$q(t) = \hat{q}\cdot e^{st},$$

where $s \equiv -\alpha \pm j\omega_r$, with $\alpha \equiv R/2L$ and $\omega_r^2 \equiv 1/(LC) - \alpha^2$. That is, q is in the form of decaying oscillations of angular frequency ω_r, the time

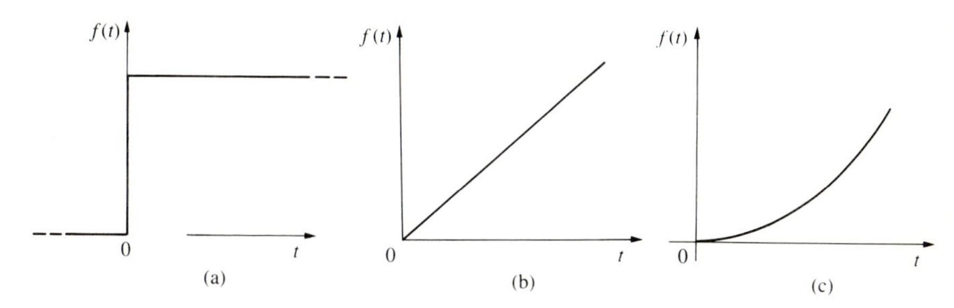

Fig. 6.11 Examples of functions that cannot be handled by the Fourier transform technique but for which Laplace transforms exist. (a) Unit step function: $u(t) = 0$, $t < 0_-$; $u(t) = 1$, $t \geqslant 0_+$. (b) Ramp function. (c) Increasing exponential function.

constant of the exponential decay of the amplitude of the oscillations being $1/\alpha$ seconds (see Fig. 4.1):

$$q(t) = \hat{q} \cdot \mathrm{e}^{-\alpha t} \cdot \mathrm{e}^{\pm \mathrm{j}\omega_r t}.$$

Notice that the original differential equation was solved by transforming it to an *algebraic* equation (the auxiliary equation) in which the variable was the **complex frequency**, s.

Generally speaking, algebraic equations are much easier to solve than differential equations.

If an excitation is in the form of a pure sinusoid (i.e. a 'steady state' sinusoid; $\alpha = 0$), then the response can be found by using steady-state a.c. circuit analysis as described in Chapter 3. In the case of a series LCR network, for example, if the required response is the current $i(t)$, then this can be found by solving the following integro-differential equation (obtained by applying KVL to the network in question):

$$R\, i(t) + L \frac{\mathrm{d}i(t)}{\mathrm{d}t} + \frac{1}{C} \int i(t)\, \mathrm{d}t = e(t),$$

where $e(t)$ is the excitation. Since it is assumed that $e(t)$ is in the steady-state sinusoidal form, $e(t) = \hat{E}\, \mathrm{e}^{\mathrm{j}\omega t}$, and that $i(t) = \hat{I}\, \mathrm{e}^{(\mathrm{j}\omega t + \phi)}$, then $\mathrm{d}i(t)/\mathrm{d}t = \mathrm{j}\omega\, i(t)$ and $\int i(t)\, \mathrm{d}t = i(t)/\mathrm{j}\omega$. So, under these circumstances, the integro-differential equation has been transformed into the *algebraic* equation

$$R\hat{I} + \mathrm{j}\omega L\hat{I} - \frac{\mathrm{j}\hat{I}}{\omega C} = \hat{E}\, \mathrm{e}^{-\mathrm{j}\phi},$$

which can easily be solved for $\hat{I}$.

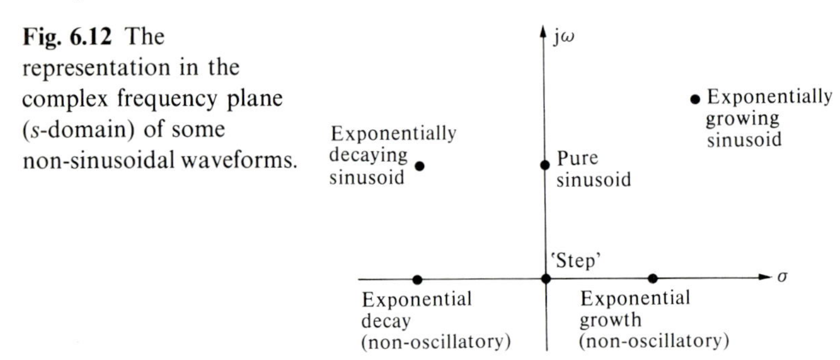

Fig. 6.12 The representation in the complex frequency plane (s-domain) of some non-sinusoidal waveforms.

The point of this discussion so far has been to introduce the concept of a complex frequency to represent a non-sinusoidal waveform (in the illustrative example, a decaying oscillation) and to emphasize the conveni-ence, in solving a differetial equation, of transforming it to an algebraic equation.

Since it is the *applications* of the Laplace transform to the solution of circuit problems that are of interest here, most of the properties of the transform will be stated without proof. A very wide range of problems can be solved by exploiting a small number of general properties of the Laplace transform in conjunction with the use of tables of transform pairs, supplemented if necessary by the routine application of the algebaic technique of partial fractions (see the Appendix).

Before giving the formal definition of the Laplace transform, it is useful to consider the representation in the 'complex frequency domain', or s-domain, of some commonly encountered transient functions (see Table 6.1 and Fig. 6.12).

Table 6.1. Representation in the complex frequency domain of some transient functions

Function	Complex frequency, s	
	real part, σ	imaginary part, ω
$e^{\sigma t}$ (increasing exponential)	σ	0
$e^{-\sigma t}$ (decreasing exponential)	$-\sigma$	0
step	0	0
$e^{j\omega t}$ (sinusoid)	0	ω
$e^{\sigma t}\,e^{j\omega t}$ (increasing amplitude of oscillation)	σ	ω
$e^{-\sigma t}\,e^{j\omega t}$ (decreasing amplitude of oscillation)	$-\sigma$	ω

The Laplace transform $F(s)$ of a function $f(t)$ is defined through

(6.28)
$$\mathscr{L}[f(t)] = F(s) \equiv \int_0^\infty f(t)\,\mathrm{e}^{-st}\,\mathrm{d}t.$$

Although the generic resemblance to the Fourier transform can be seen, there are two crucial differences. First, the range of integration is from $t = 0$ to $t = \infty$, which means that the excitation functions in question are treated as being 'switched on' at the instant $t = 0$. This allows functions that have a discontinuity at $t = 0$ to be treated, e.g. the 'step' function and the unit impulse function. Second, the integral contains the factor e^{-st} rather than the factor $\mathrm{e}^{-\mathrm{j}\omega t}$ of the Fourier integral.

Note that the usual convention is to denote functions in the time domain by lower-case letters ($f(t)$, $v(t)$, $i(t)$, ...) and their partners in the complex frequency domain (s-domain) by upper-case letters ($F(s)$, $V(s)$, $I(s)$, ...).

$F(s)$ and $f(t)$ are known as a Laplace transform pair and, as will be seen, there are very extensive libraries of transform pairs of practical interest.

In essence, the steps in the solution of a problem relating to the dynamic behaviour of a circuit or system will be:

(1) model the problem in the time domain (differential equation(s));
(2) transform the differential equation(s) into the s-domain;
(3) solve the (algebraic) s-domain equation(s) for the dependent variable(s) of interest;
(4) use an inverse Laplace transform to transform the s-domain solution to the time domain.

For completeness, the inverse Laplace transform is

(6.29)
$$f(t) = \mathscr{L}^{-1}[F(s)] = \frac{1}{2\pi \mathrm{j}} \oint_0^\infty \mathrm{F}(s)\,\mathrm{e}^{st}\,\mathrm{d}s.$$

The operation of taking the inverse Laplace transform of a function $F(s)$ involves an integration round a closed path in the s-plane; a proper consideration of this would necessitate a long diversion into the mathematical realm of functions of a complex variable. Fortunately, this is unnecessary here since, in a very wide range of practical situations, step 2 and step 4 can each be implemented by taking advantage of tabulated Laplace transform pairs, as will be seen later (see Table 6.3).

Example 6.6 Determine the Laplace transform of the unit step function $u(t)$.

The unit step function is defined by

$$u(t) = 1, \qquad t \geqslant 0$$
$$u(t) = 0, \qquad t < 0 \tag{6.30}$$

(Note that the unit step function performs the mathematical operation of 'switching on' a function $f(t)$, say, at $t = 0$. i.e. $f(t) \cdot u(t)$ is zero from $t = -\infty$ to $t = 0$ and thereafter assumes the values of $f(t)$.)

$$\mathscr{L}[u(t)] = \int_0^\infty u(t)\,\mathrm{e}^{-st}\,\mathrm{d}t \rightarrow \int_0^\infty 1 \cdot \mathrm{e}^{-st}\,\mathrm{d}t$$

$$= \left| \frac{\mathrm{e}^{-st}}{-s} \right|_0^\infty$$

Remembering that $s = \sigma + \mathrm{j}\omega$,

$$F(s) \equiv \mathscr{L}[u(t)] = \left| \frac{-\mathrm{e}^{-\sigma t}\,\mathrm{e}^{-\mathrm{j}\omega t}}{\sigma + \mathrm{j}\omega} \right|_0^\infty$$

Provided that $\sigma > 0$, then at the upper limit $F(s)$ is zero even though $\mathrm{e}^{\mathrm{j}\omega t}$ ($= \cos \omega t + \mathrm{j} \sin \omega t$, remember) is undefined. Hence

$$\mathscr{L}[u(t)] = \frac{1}{(\sigma + \mathrm{j}\omega)} \rightarrow \frac{1}{s}. \tag{6.31}$$

The derivations of the Laplace transforms of some of the commonly occurring functions are left as exercises.

Some important properties of Laplace transforms will now be stated without proof (the derivations are straightforward and can be treated as exercises). For the moment simply note the listed properties and refer back to the list as required; examples of their application will arise in the Examples.

The shifting theorem

It will be found that shifted (i.e. delayed) functions of time are of common

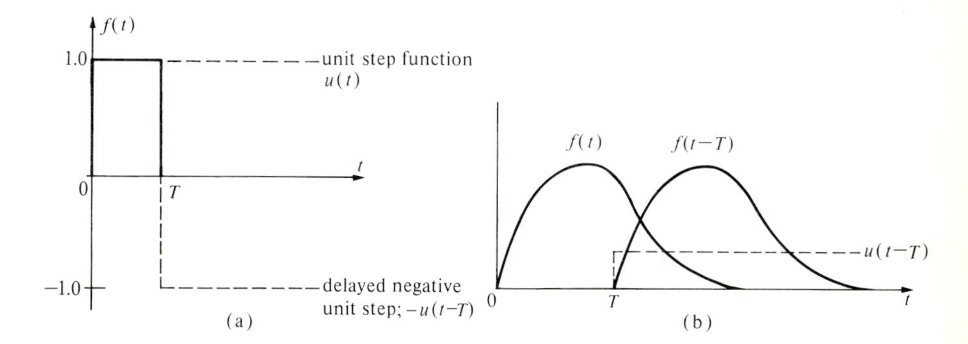

Fig. 6.13 (a) A square pulse generated by a unit step function $u(t)$ and a delayed negative unit step function $u(t-T)$. (b) A delayed version $f(t-T)$ of a function $f(T)$ switched on by a delayed unit step function $u(t-T)$.

occurrence in the analysis of the response of networks to transient excitations, e.g. due to propagation delays. Also, for instance, a rectangular pulse can be represented as the resultant of a step function and a delayed negative-going step function of the same amplitude (see Fig. 6.13).

If $F(s) = \mathscr{L}[f(t)]$, then the shifting theorem states:

$$\mathscr{L}[f(t-T)\cdot u(t-T)] = \mathrm{e}^{-sT}\cdot F(s). \tag{6.32}$$

(Note that the delayed step function, $u(t-T)$, has the effect of switching on $f(t)$ at the time $t = T$).

The Laplace transform of a derivative

$$\mathscr{L}[\mathrm{d}f(t)/\mathrm{d}t] = s\,F(s) - f(0), \tag{6.33}$$

where $f(0)$ is the value of $f(t)$ at $t = 0$.

$$\mathscr{L}[\mathrm{d}^2 f(t)/\mathrm{d}t^2] = s^2 F(s) - s\,f(0) - \left.\frac{\mathrm{d}f(t)}{\mathrm{d}t}\right|_{t=0}. \tag{6.34}$$

In general,

$$\mathscr{L}\left[\frac{\mathrm{d}^n f(t)}{\mathrm{d}t^n}\right] = s^n F(s) - \sum_{k=0}^{n-1} s^{n-1-k} f_0^k, \tag{6.35}$$

where

$$f_0^k \equiv \left.\frac{\mathrm{d}^k f(t)}{\mathrm{d}t^k}\right|_{t=0}.$$

The Laplace transform of an integral

$$\mathscr{L}\left[\int_0^t f(t)\,\mathrm{d}t\right] = \frac{F(s)}{s} + \frac{f^{-1}(0)}{s}, \tag{6.36}$$

where

$$f^{-1}(0) \equiv \int f(t)\,\mathrm{d}t\bigg|_{t=0}.$$

Frequency shifting

(6.37)
$$\mathscr{L}[\mathrm{e}^{-at} f(t)] = F(s + a).$$

Initial value theorem

(6.38)
$$\lim_{t \to 0} f(t) = \lim_{s \to \infty} s\, F(s).$$

Final value theorem

(6.39)
$$\lim_{t \to \infty} f(t) = \lim_{s \to 0} s\, F(s).$$

Notice that the Laplace transform is linear, i.e. $\mathscr{L}[a\, f(t)] = a\, F(s)$, but that $\mathscr{L}[f_1(t) \cdot f_2(t)] \neq F(s)_1 \cdot F(s)_2$.

The best way to illustrate the practical significance of these definitions, transformations, and theorems is to consider some examples in the context of electrical circuits.

Example 6.7 Consider the CR circuit shown in Fig. 6.14; the switch S is closed at time $t = 0$, so that the e.m.f. in the circuit can be represented by a step function of amplitude E_0. The problem is to determined the current $i(t)$.

In the time domain, KVL for the circuit gives:

(6.40) (step 1)
$$R\, i(t) + \frac{1}{C}\int i(t)\,\mathrm{d}t = E_0\, u(t),$$

and on transforming these functions to the s-domain,

(6.41) (step 2)
$$R\, I(s) + \frac{1}{C}\cdot\frac{I(s)}{s} + \frac{i^{-1}(0)}{Cs} = \frac{E_0}{s}.$$

Now $i^{-1}(0)$ is equal to $\int i(t)\,\mathrm{d}t$ evaluated at $t = 0$, which is the initial charge on the capacitor, say q_0. So on rearranging eqn (6.41),

(6.42) (step 3)
$$I(s) = \frac{\left(E_0 - \dfrac{q_0}{C}\right)}{R\left(s + \dfrac{1}{CR}\right)}.$$

Fig. 6.14

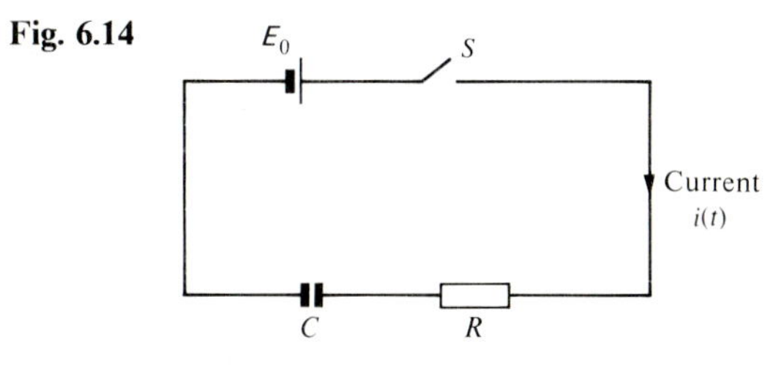

In order to obtain an expression for $i(t)$, the inverse transform of $I(s)$ is required:

(step 4) $$\mathscr{L}^{-1}\left[\left(s+\frac{1}{CR}\right)^{-1}\right]$$

is of the form $\mathscr{L}^{-1}[(s+a)^{-1}]$ and is listed in Table 6.3. Thus, on transforming eqn (6.42) back to the time domain,

$$i(t) = \frac{\left(E_0 - \dfrac{q_0}{C}\right) e^{-t/CR}}{R}. \tag{6.43}$$

This was a relatively simple problem. Even so, the solution using the Laplace transform technique is arguably simpler than solving eqn (6.40) by the method used in Section 3.1. □

In many problems it will be possible with experience, for the reader to visualize the s-domain algebraic equation without the necessity of first writing down the time domain equation. The voltage–current relationships for circuit elements are summarized in Table 6.2.

Table 6.2. Voltage–current relationships for circuit elements

Element	Time domain	Steady-state a.c.	s-domain
Resistor R	$e(t) = R\,i(t)$	$E = RI$	$E(s) = R\,I(s)$; $Z(s) = R$
Capacitor C	$e(t) = \dfrac{1}{C}\displaystyle\int i(t)\,\mathrm{d}t$	$E = \dfrac{I}{\mathrm{j}\omega C}$	$E(s) = \dfrac{I(s)}{sC} + \dfrac{i^{-1}(0)}{sC}$
			$Z(s) = \dfrac{1}{sC} + \dfrac{i^{-1}(0)}{sC}$
Inductor L	$e(t) = L\dfrac{\mathrm{d}i(t)}{\mathrm{d}t}$	$E = \mathrm{j}\omega L$	$E(s) = sLI(s) - sL\,i(0)$
			$Z(s) = sL - sL\,i(0)$

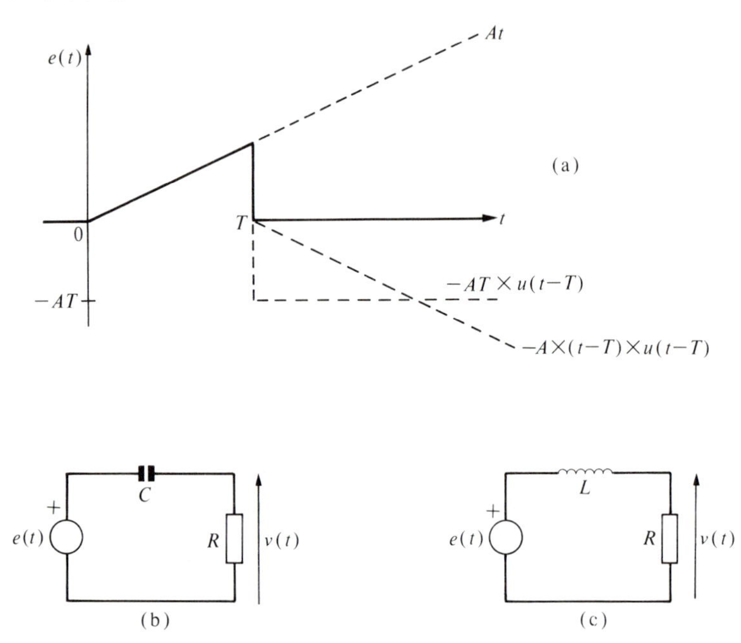

Fig. 6.15 (a) The decomposition of a ramp function excitation for the purpose of determining the response of a network. (b) and (c): See text.

Example 6.8 A rather more complicated problem is that of obtaining an expression for the voltage $v(t)$ across the resistor R in the circuit of Fig. 6.15 when the excitation $e(t)$ is a voltage in the form of a ramp function of duration T.

Assuming that the capacitor C is uncharged at $t = 0$, then, by reference to Table 6.2, step 2 can be proceeded to immediately:

$$V(s) = E(s)\frac{R}{\left(R + \dfrac{1}{sC}\right)}.$$

It is convenient to convert the denominator into a standard form that can be found in tables such as Table 6.3, i.e.

$$V(s) = \frac{s\,E(s)}{\left(s + \dfrac{1}{CR}\right)}. \tag{6.44}$$

Table 6.3 does not contain a 'terminated' ramp function, although it does contain the *mathematical* ramp function, $f(t) = At$, which continues to infinity, of course. However, it can be seen from Fig. 6.15 that the terminated ramp function can be analysed into the algebraic sum of a ramp, a delayed negative-going step, and a delayed negative ramp. Hence by using Table 6.3, and exploiting the shifting theorem, it follows that

$$E(s) = \underset{+\text{ve ramp}}{\frac{A}{s^2}} - \underset{-\text{ve step}}{\frac{AT\,\mathrm{e}^{-sT}}{s}} - \underset{-\text{ve ramp}}{\frac{A\,\mathrm{e}^{-sT}}{s^2}} \tag{6.45}$$

On substituting for $E(s)$ in eqn (6.44),

$$V(s) = A\left\{\frac{1}{s\left(s+\dfrac{1}{CR}\right)} - \frac{T\,\mathrm{e}^{-sT}}{\left(s+\dfrac{1}{CR}\right)} - \frac{\mathrm{e}^{-sT}}{s\left(s+\dfrac{1}{CR}\right)}\right\}.$$

Using Table 6.3,

$$\mathscr{L}^{-1}\left[\left(s+\frac{1}{CR}\right)^{-1}\right] = \mathrm{e}^{-t/CR}$$

and, using the shifting theorem,

$$\mathscr{L}^{-1}\left[\left(s+\frac{1}{CR}\right)^{-1}\mathrm{e}^{-sT}\right] = \mathrm{e}^{-(t-T)/CR}\cdot u(t-T).$$

Also, from Table 6.3,

$$\mathscr{L}^{-1}[\{(s+a)(s+b)\}^{-1}] = (\mathrm{e}^{-at} - \mathrm{e}^{-bt})/(b-a)$$

and so

$$\mathscr{L}^{-1}\left[\left\{s\left(s+\frac{1}{CR}\right)\right\}^{-1}\right] = CR\{1-\mathrm{e}^{-t/CR}\}$$

and

$$\mathscr{L}^{-1}\left[\frac{\mathrm{e}^{-sT}}{s\left(s+\dfrac{1}{CR}\right)}\right] = CR\{1-\mathrm{e}^{-(t-T)/CR}\}\cdot u(t-T).$$

So, finally,

$$v(t) = ACR\left\{1 - \mathrm{e}^{-t/CR} - u(t-T) - \frac{T}{CR}\mathrm{e}^{-(t-T)/CR}\cdot u(t-T) + \mathrm{e}^{-(t-T)/CR}\cdot u(t-T)\right\}. \tag{6.46}$$

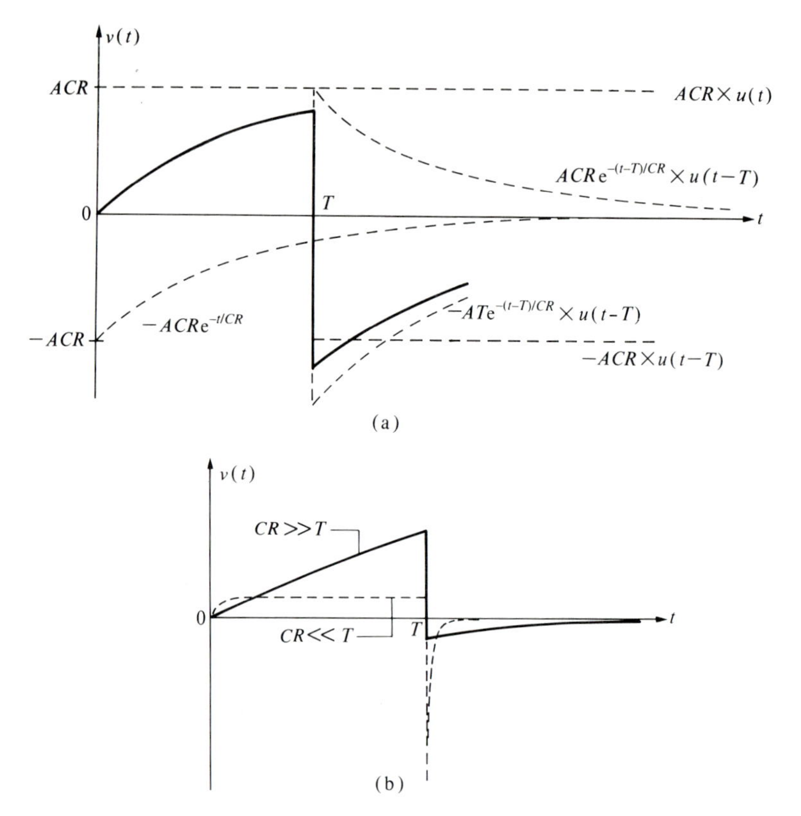

Fig. 6.16 The response $v(t)$ of the network of Fig. 6.15(b) (a 'high pass' CR network) to a ramp function excitation. (a) The general response (eqn (6.46)). (b) The response is reasonably faithful to the excitation if $CR \gg T$, but a good approximation to the time derivative, $(CR)\,\mathrm{d}e(t)/\mathrm{d}t$, is obtained if $CR \ll T$.

This expression for $v(t)$ is drawn in Fig. 6.16(a). □

It is left as an exercise to show that if the ramp voltage of Fig. 6.15(a) is applied to an LR network the expression for the voltage across the inductor L has the same mathematical form as the expression of eqn (6.46) apart from the replacement of the product CR by the quotient L/R.

In the situation depicted in Fig. 6.15(c) a ramp voltage is applied to an LR network. The method for determining $v(t)$ involves a simple example of the use of the **partial fractions** technique to resolve an s-domain function into a *sum* of functions that are tabulated. The s-domain voltage $V(s)$

corresponding to $v(t)$ is given by

$$V(s) = \frac{E(s)R}{(R + sL)}$$

$$= \frac{R}{L\left(s + \frac{R}{L}\right)} E(s).$$

Using eqn (6.45) for $E(s)$ it is found that

$$V(s) = \frac{AR}{L}\left\{\frac{1}{s^2\left(s + \frac{R}{L}\right)} - \frac{T\mathrm{e}^{-sT}}{s\left(s + \frac{R}{L}\right)} - \frac{\mathrm{e}^{-sT}}{s^2\left(s + \frac{R}{L}\right)}\right\}. \tag{6.47}$$

Again $v(t)$ is found by obtaining the inverse Laplace transforms of each of the terms in eqn (6.47).

Although, as it happens, functions in the forms of all of the types present in eqn (6.47) are to be found in Table 6.3, it is nevertheless useful to indicate how the method of partial fractions (see the Appendix) can be used to obtain 'simpler' terms. For example, the term $\{s^2(s + R/L)\}^{-1}$ can be expanded as shown below:

$$\frac{1}{s^2\left(s + \frac{R}{L}\right)} = \frac{F}{\left(s + \frac{R}{L}\right)} + \frac{Gs + H}{s^2}$$

$$= \frac{s^2(F + G) + s(GR/L + H) + HR/L}{s^2\left(s + \frac{R}{L}\right)}.$$

Equating the coefficients of the terms in s^2, s^1, s^0 in the numerators of the two sides of this equation, it follows that

$$s^2\text{:} \quad 0 = F + G$$

$$s^1\text{:} \quad 0 = GR/L + H$$

$$s^0\text{:} \quad 1 = HR/L.$$

On solving these simultaneous equations,

$$F = L^2/R^2; \qquad G = -L^2/R^2; \qquad H = L/R$$

and

$$\frac{1}{s^2\left(s + \frac{R}{L}\right)} = \frac{L^2/R^2}{\left(s + \frac{R}{L}\right)} - \frac{L^2}{R^2 s} + \frac{L}{Rs^2}.$$

Table 6.3. Laplace transform pairs of some functions and operations

	$f(t)$ (switched on at $t = 0$, i.e. $u(t)$ is implicit)	$F(s)$
Functions:		
Step	$A \cdot u(t)$	A/s
Unit impulse	$\delta(t)$	1
Ramp	$A \cdot t$	A/s^2
	t^n	$n!/s^{n+1}$
	e^{-at}	$1/(s+a)$
	$t^n e^{-at}$	$\dfrac{n!}{(s+a)^{n+1}}$
	$\dfrac{e^{-at} - e^{-bt}}{(b-a)}$	$\dfrac{1}{(s+a)(s+b)}$
	$e^{-at}(1 - at)$	$\dfrac{s}{(s+a)^2}$
	$\dfrac{1}{ab}\left(1 + \dfrac{b e^{-at} - a e^{-bt}}{(a-b)}\right)$	$\dfrac{1}{s(s+a)(s+b)}$
	$\dfrac{1}{a^2}(e^{-at} + at - a e^{-at})$	$\dfrac{1}{s^2(s+a)}$
	$\dfrac{1 - e^{-at} - at e^{-at}}{a^2}$	$\dfrac{1}{s(s+a)^2}$
	$\sin(\omega t + \phi)$	$\dfrac{s \sin\phi + \omega\cos\phi}{s^2 + \omega^2}$
	$\cos(\omega t + \phi)$	$\dfrac{s\cos\phi - \omega\sin\phi}{s^2 + \omega^2}$
	$\dfrac{(a-b) e^{-at} - (a-c) e^{-ct}}{(c-b)}$	$\dfrac{(s+a)}{(s+b)(s+c)}$
	$e^{-at}\sin\omega t$	$\dfrac{\omega}{(s+a)^2 + \omega^2}$
	$e^{-at}\cos\omega t$	$\dfrac{(s+a)}{(s+a)^2 + \omega^2}$

continued . . .

Table 6.3. *Continued.*

	$f(t)$	$F(s)$
	$\sinh \omega t$	$\dfrac{\omega}{(s^2 - \omega^2)}$
	$\cosh \omega t$	$\dfrac{s}{(s^2 - \omega^2)}$
Periodic waveform of period T;		
	$f(t) = f(t \pm nT)$	$\dfrac{1}{(1 - \mathrm{e}^{-Ts})} \displaystyle\int_0^T f(t)\,\mathrm{e}^{-st}\,\mathrm{d}t$
Operations:		
	$f'(t)$	$s\,F(s) - f(0)$
	$\dfrac{\mathrm{d}^n f(t)}{\mathrm{d}t^n}$	$s^n F(s) - \displaystyle\sum_{k=0}^{n-1} s^{n-1} \cdot f_0^k,$
		where $f_0^k \equiv \left.\dfrac{\mathrm{d}^k f(t)}{\mathrm{d}t^k}\right\rvert_{t=0+}$
	$\displaystyle\int_0^t f(t)\,\mathrm{d}t$	$\dfrac{F(s)}{s} + \dfrac{1}{s}\left[\displaystyle\int f(t)\,\mathrm{d}t\right]_{t=0}$
	$f(t - T)$	$\mathrm{e}^{-Ts}\,F(s)$
	$-t\,f(t)$	$\dfrac{\mathrm{d}F(s)}{\mathrm{d}s}$
	$\dfrac{1}{t}\,f(t)$	$\displaystyle\int_s^\infty F(s)\,\mathrm{d}s$
	$f(at)$	$\dfrac{1}{a}\,F(s/a)$
	$\mathrm{e}^{at}\,f(t)$	$F(s - a)$
Initial value theorem:		
	$\lim\limits_{t \to 0} f(t)$	$\lim\limits_{s \to \infty} s\,F(s)$
Final value theorem:		
	$\lim\limits_{t \to \infty} f(t)$	$\lim\limits_{s \to 0} s\,F(s)$

Although this simple exercise was not strictly necessary in the problem just considered, the partial-fractions technique should always be borne in mind as a possible means of simplifying the problem of obtaining the inverse Laplace transform of a complicated s-domain function.

The techniques of mesh and nodal analysis and the concepts of Thévenin and Norton equivalent circuits can be carried over from time-domain circuit analysis.

6.8 Some practical aspects of pulse propagation

The practical problems associated with the generation, transmission, and shaping of pulses constitute in themselves a very large and important sector of the subject of electronic engineering, for instance in digital systems, data communications, telecommunications, and radar. Hence it will be possible only to touch on a few of the most important aspects here. The bandwidth problem associated with the transmission of a train of pulses was discussed in Section 6.4. Here the distortion of transient and pulsed signals that occurs when they are transmitted through CR networks will now be discussed.

It is obvious from eqn (6.46) and Fig. 6.16(a) that the response of a CR network to a ramp excitation exhibits considerable distortion. Let us consider two extreme situations in the case of the network of Fig. 6.15(b).

(i) $CR \gg T$. For $t \leqslant T$, eqn (6.46) becomes

$$v(t) = ACR(1 - e^{-t/CR}) \tag{6.48}$$

and, using the series expansion for an exponential function,

$$v(t) \approx ACR\left[1 - \left(1 - \frac{t}{CR} + \frac{t^2}{2C^2R^2} - \cdots\right)\right]$$

and since if $T \ll CR$, then, in this region, $t \ll CR$ also,

$$v(t) \approx At. \tag{6.49}$$

For $t \geqslant T$, eqn (6.46) becomes

$$v(t) = ACR\,e^{-(t-T)/CR}\left[1 - \frac{T}{CR} - e^{-t/CR}\right] \tag{6.50}$$

or, again using the series expansion for an exponential function,

$$v(t) \approx -\frac{AT^2}{2CR}\,e^{-(t-T)/CR}. \tag{6.51}$$

The response described by eqns (6.49) and (6.51) is sketched in Fig. 6.16(b).

(ii) $CR \ll T$. For $t \gg CR$ but with $t \leqslant T$, eqn (6.46) becomes

$$v(t) \approx ACR \tag{6.52}$$

and for $t \geqslant T$, eqn (6.46) becomes

$$v(t) \approx -AT\,\mathrm{e}^{-(t-T)/CR}. \tag{6.53}$$

The response described by eqns (6.52) and (6.53) is also sketched in Fig. 6.16(b).

It will have been noticed, no doubt, from the foregoing analysis and/or from Fig. 6.16(b) that in case (i) the form of $v(t)$ is a reasonable facsimile of the ramp voltage $e(t)$, bearing in mind that $AT \gg AT^2/2CR$. In case (ii), $v(t)$ has a form that approximates closely to a scaled version of the time derivative of the ramp voltage. Of course it is impossible to attain the infinite value of the *mathematical* derivative at time $t = T$ in a real physical network, but there is a negative-going spike of comparatively large amplitude. The differentiating property that this 'high pass'* CR network possesses under the condition $CR \ll T$ applies to any form of excitation, of course.

Although eqn (6.46) is the exact solution to the specified problem (and this underlines the usefulness of Laplace transforms in this context), it must not be overlooked that inevitably the specification is itself only an approximation to the real situation. On the one hand the representation of a network is only approximate; e.g. 'stray' and/or 'parasitic' resistance, inductance and capacitance may not be recognized or may not be known accurately. Also the mathematical representation of an excitation is in practice an approximation, since a real excitation does not have discontinuities in either its value or the derivatives of its value, e.g. right-angled 'corners', instantaneous changes in slope, etc.

Example 6.8 As another example of the determination of the response of a network to a transient excitation, consider a ramp voltage applied to a low-pass RC network as shown in Fig. 6.17(a). In the s-domain the response $v(t)$ is given by

$$V(s) = \frac{E(s)}{sC\left(R + \dfrac{1}{sC}\right)} \quad \text{or} \quad V(s) = \frac{E(s)}{CR\left(s + \dfrac{1}{CR}\right)}.$$

* A CR network in the configuration shown in Fig. 6.15(b) is an example of a 'high pass' network, since if $e(t)$ is sinusoidal, then $v(t)/e(t)$ increases from zero towards a value of unity as the frequency of $e(t)$ is increased from zero to infinity. Conversely, for a 'low pass' network (e.g. interchange the positions of C and R in the network of Fig. 6.15(b)) the response decreases from its value at d.c. towards zero as the frequency is increased from zero towards infinity..

Using eqn (6.45) for $E(s)$,

$$V(s) = \frac{A}{CRs^2\left(s + \dfrac{1}{CR}\right)} - \frac{AT\,\mathrm{e}^{-sT}}{CRs\left(s + \dfrac{1}{CR}\right)} - \frac{A\,\mathrm{e}^{-sT}}{CRs^2\left(1 + \dfrac{1}{CR}\right)},$$

and using eqn (6.47) with CR in place of L/R,

$$V(s) = ACR\cdot\left[\frac{1}{\left(s + \dfrac{1}{CR}\right)} - \frac{1}{s} + \frac{1}{CRs^2} - \frac{T\,\mathrm{e}^{-sT}}{(CR)^2 s\left(s + \dfrac{1}{CR}\right)} - \frac{\mathrm{e}^{-sT}}{\left(s + \dfrac{1}{CR}\right)} + \frac{\mathrm{e}^{-sT}}{s} - \frac{\mathrm{e}^{-sT}}{CRs^2}\right].$$

The inverse Laplace transform of the terms of this s-domain equation can be obtained from Table 6.3, yielding

$$v(t) = ACR\cdot\left[\mathrm{e}^{-t/CR} - u(t) + \frac{t}{CR} - \frac{T}{CR}\{1 - \mathrm{e}^{-(t-T)/CR}\}\cdot u(t-T) - \mathrm{e}^{-(t-T)/CR}\cdot u(t-T) + u(t-T) - \frac{(t-T)}{CR}\cdot u(t-T)\right]. \tag{6.54}$$

This response is sketched in Fig. 6.17(b).

As with the high-pass CR network discussed previously, it is useful to consider two extreme situations.

(i) $CR \gg T$. For $t \leqslant T$, eqn (6.54) becomes

$$v(t) = ACR\cdot\left(\mathrm{e}^{-t/CR} - 1 + \frac{1}{CR}\right), \tag{6.55}$$

and exploiting the series expansion of $\mathrm{e}^{-t/CR}$,

$$v(t) \approx \frac{1}{CR}\cdot\frac{At^2}{2}. \tag{6.56}$$

For $t \geqslant T$, eqn (6.54) becomes

$$v(t) = ACR\cdot\mathrm{e}^{-(t-T)/CR}\left(\frac{T}{CR} - 1 + \mathrm{e}^{-T/CR}\right) \tag{6.57}$$

and since $T \ll CR$,

$$v(t) \approx \frac{AT^2}{2CR}\,\mathrm{e}^{-(t-T)/CR}. \tag{6.58}$$

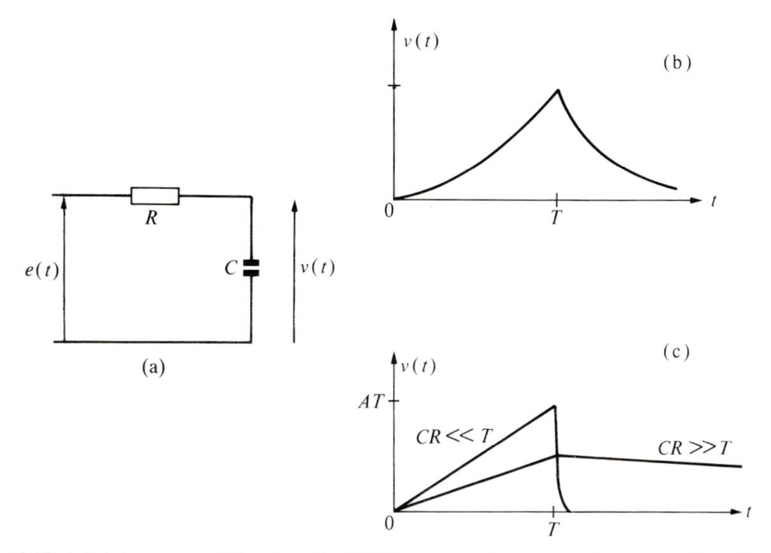

Fig. 6.17 (a) A low-pass RC network. (b) The general response to a ramp function excitation. (c) The response is reasonably faithful to the excitation if $CR \ll T$, but a good approximation to $(CR)^{-1} \int e(t)\,dt$ is obtained if $CR \gg T$.

The form of the response described by eqns (6.55) and (6.57) is sketched in Fig. 6.17(c).

(ii) $CR \ll T$. In the region $t \leqslant T$, but with $t \gg CR$, eqn (6.54) becomes

$$v(t) \approx At. \tag{6.59}$$

For $t \geqslant T$ we use eqn (6.57) again, but since $T \gg CR$,

$$v(t) \approx AT \cdot e^{-(t-T)/CR}. \tag{6.60}$$

The form of this response is also sketched in Fig. 6.17(c). It will be noticed that in these circumstances a reasonable facsimile of the ramp excitation $e(t)$ is obtained if $CR \ll T$ and that if on the other hand $CR \gg T$, then a good approximation to a scaled version of $\int e(t)\,d(t)$ is obtained. Again, this integrating property of a low-pass RC network applies to any form of excitation.

For an excitation in the form of an ideal rectangular pulse of amplitude A and duration λ,

$$e(t) = A \cdot u(t) - A \cdot u(t - \lambda),$$

i.e. the sum of a positive step and a delayed negative step and

$$E(s) = \frac{A}{s} - \frac{A\,e^{-s\lambda}}{s}. \tag{6.61}$$

If this excitation is applied to a high-pass CR network, then using

$$V(s) = \frac{E(s)R}{\left(R + \frac{1}{sC}\right)},$$

the response is

$$V(s) = \frac{A}{\left(s + \frac{1}{CR}\right)} - \frac{A\,\mathrm{e}^{-s\lambda}}{\left(s + \frac{1}{CR}\right)}$$

or

$$v(t) = A\,\mathrm{e}^{-t/CR} - A\,\mathrm{e}^{-(t-\lambda)/CR} \cdot u(t-\lambda). \tag{6.62}$$

In the case of a train of rectangular pulses, if there is time for the transients to decay to negligible proportions between successive pulses, then the analysis for a single pulse can be used. The waveform of eqn (6.62) is shown in Fig. 6.18, together with the differentiated and the integrated forms. The high-pass CR network is met very often as an interstage coupling network in electronic systems, and this means, in the case of transient and pulsed signals, that some distortion is introduced as indicated in Fig. 6.18(a) and (b).

A typical interstage coupling situation is shown in Fig. 6.19(a), where C_2 represents a shunt capacitance arising at the input to the 'next' stage of the system. It is left as an exercise to show that $v(t)$ has the same general form as eqn (6.62) apart from a multiplying factor $C_1/(C_1 + C_2)$ and with $(C_1 + C_2)R_2$ replacing CR.

It is interesting to analyse the network shown in Fig. 6.19(b); it can easily be shown that

$$\frac{V(s)}{E(s)} = \frac{R_2}{(R_1 + R_2)} \cdot \frac{(1 + sC_1R_1)}{\left[1 + s(C_1 + C_2)\frac{R_1R_2}{(R_1 + R_2)}\right]}. \tag{6.63}$$

Hence, if

$$sC_1R_1 = s(C_1 + C_2)\frac{R_1R_2}{(R_1 + R_2)},$$

then $V(s)/E(s)$ is independent of s, and $v(t)$ will be an undistorted version of $e(t)$. This equation reduces to $C_1R_1 = C_2R_2$; a network satisfying this condition is usually called a **compensated attenuator**. Networks of this kind are often used in probes for oscilloscopes, in which case R_2 and C_2 would be the input resistance and capacitance, respectively, of the oscilloscope itself. Typical values for the circuit elements in the case of a general-purpose laboratory CRO would be $R_2 = 1\ \mathrm{M\Omega}$, $C_2 \approx 30\ \mathrm{pF}$,

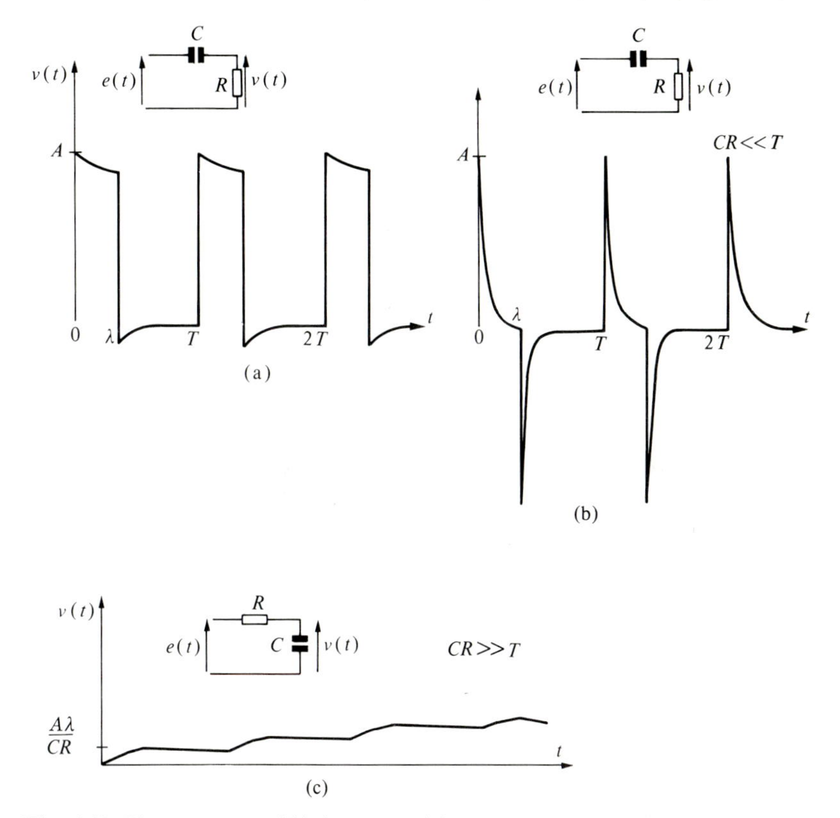

Fig. 6.18 The response of high-pass and low-pass *CR* networks to an excitation in the form of a train of repetitive rectangular pulses of amplitude A, duration λ, and period T. (a) The general case for a high-pass network. (b) An extreme case (differentiation) for a high-pass network. (c) Another extreme case (integration) for a low-pass network.

$R_1 = 9\text{M}\Omega$. C_1 would be an adjustable capacitor integral with the probe unit itself. For these component values the signal amplitude would be attenuated by a factor of ten (a '$10\times$' probe), but an ancilliary advantage is that the input impedance of the probe is ten times that at the terminals of the CRO, i.e. 10 MΩ (shunted by a capacitance of value slightly less than C_1).

So far, the characteristics of the source of pulses have not been considered. The output impedance of a source inevitably contains a reactive component that reflects the fact that it is impossible in practice to generate ideal mathematically rectangular pulses; to put it another way, real pulses cannot be obtained that have mathematical discontinuities in $e(t)$ and its derivatives. For reasonably 'clean' rectangular pulses, the rise

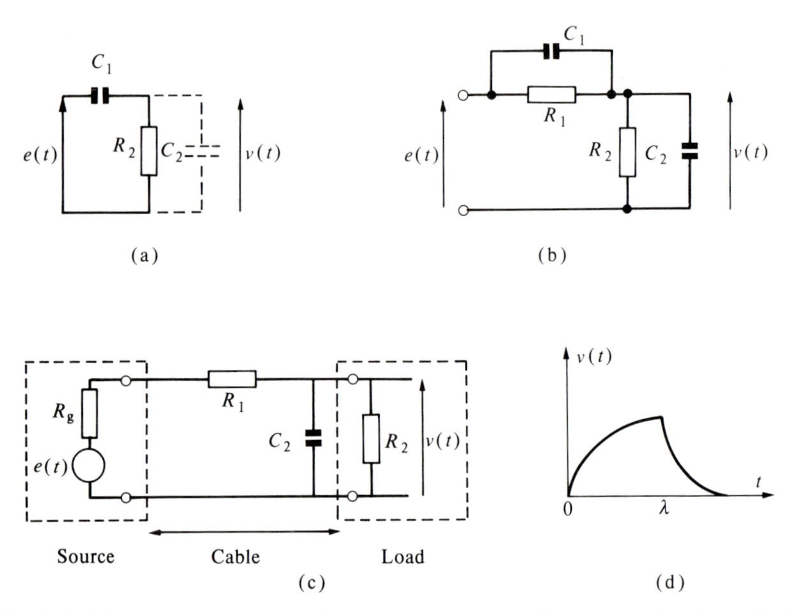

Fig. 6.19 (a) A representation of an inter-stage coupling network where C_2 is the shunt input capacitance of the succeeding stage. (b) This network becomes a so-called compensated attenuator if $C_1R_1 = C_2R_2$. (c) The equivalent circuit for a voltage source and a load connected by a cable having an equivalent lumped series resistance R_1; C_2 includes the shunt capacitance of the load. (d) See the text.

and fall times are usually taken to be the time for the signal to rise, or fall, as the case may be, between the limits 10 and 90 per cent of the amplitude of the pulse. Pulse generators are commonly available that give 'rectangular' pulses having durations as short as 15 ns with rise times of 5 ns at a pulse repetition rate as high as 50 MHz.

A situation that may often by encountered in practice is that in which the source of pulses is connected to an amplifier via a coaxial cable. If the cable is short enough so that the propagation time along the cable is much less than the rise time of the pulses, then the cable can be represented by an equivalent lumped series resistance and lumped shunt capacitance.* The circuit diagram for such a situation is shown in Fig. 6.19(c), where C_2 includes the shunt capacitance at the input to the amplifier (commonly that of a CRO) as well as the equivalent lumped capacitance of the cable ($\sim$50 to 100 pF m^{-1}) and R_1 is the lumped series resistance of the cable. For simplicity, the impedance of the source is shown as being purely

* If the propagation time is of the order of, or greater than, the rise time, then the cable must be treated as a 'distributed element. The propagation time for pulses on typical cable is about 4 ns per metre.

resistive. It can be shown easily in this case that

$$\frac{V(s)}{E(s)} = -1\bigg/\left[C_2(R_g + R_1)\left\{s + \frac{(R_2 + R_g + R_1)}{C_2R_2(R_g + R_1)}\right\}\right],$$

and if $E(s)$ corresponds to a rectangular pulse, as specified by eqn (6.61), then

$$v(t) = \frac{AR_2}{(R_2 + R_g + R_1)}\{[1 - \exp(-t(R_2 + R_g + R_1)/C_2R_2(R_g + R_1))] - [1 - \exp(-(t - \lambda)(R_2 + R_g + R_1)/C_2R_2(R_g + R_1))]\},$$

which is sketched in Fig. 6.19(d). So it can be seen that the effect of the shunt capacitance is to distort the input waveform with a tendency towards integration. If the value of $(R_g + R_1)$ can be reduced, then

$$e(t) \approx A\{[1 - \exp(-t/C_2(R_g + R_1))] - [1 - \exp(-(t - \lambda)/C_2(R_g + R_1))]\}\cdot u(t - \lambda),$$

and the exponential decays are much sharper, which gives a closer approximation to a rectangular pulse. This can be achieved in practice by connecting a buffer amplifier of low output impedance (e.g. an 'emitter follower' or a 'source follower') directly between the source and the cable.

For longer lengths of cable and/or shorter pulse rise times, the cables have to be considered as (distributed) transmission lines (see Chapter 8).

In practical situations there are always present stray capacitances and inductances that cause distortions of pulses; indeed only brief attention has been paid in this chapter to networks containing inductance. The general effect of inductance, in combination with capacitance and resistance, is to cause 'ringing'. However, *LCR* networks were treated in Chapter 4, albeit not by Laplace transform methods, so they will not be discussed further here.

Problems

6.1 (a) Derive the Fourier series in eqn (6.4) for the square wave shown in Fig. 6.2(d). (b) Derive the Fourier series in eqn (6.3) for the sawtooth wave shown in Fig. 6.2(g).

6.2 Show that if the wave $v(t) = \hat{V}\cos\omega t$ is full-wave rectified as depicted in Fig. 6.20, then the Fourier series $v(t)$ for this waveform is given by

$$v(t) = a_0 + \sum_{n=1}^{\infty} a_n \cos 2n\omega t$$

where

$$a_0 = \frac{2\hat{V}}{\pi}$$

$$a_n = -\frac{4\hat{V}}{\pi(4n^2 - 1)} \quad (n \text{ even})$$

$$a_n = \frac{4\hat{V}}{\pi(4n^2 - 1)} \quad (n \text{ odd}).$$

Fig. 6.20

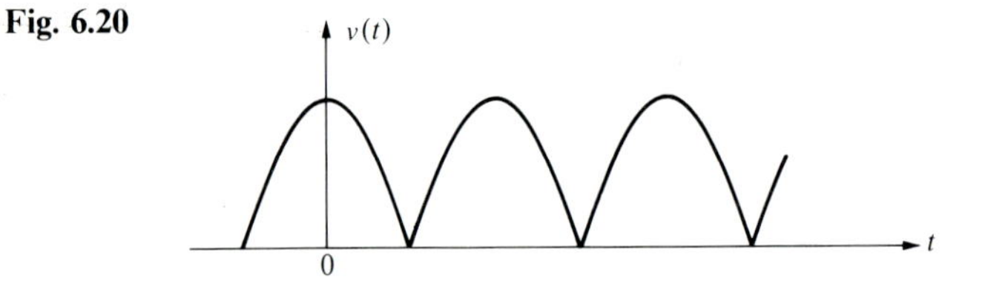

6.3 Show that the Fourier series that represent the square wave of Fig. 6.2(e) and the half-wave-rectified waveform of Fig. 6.2(f) are, respectively,

$$v(t) = \frac{\hat{V}}{2} + \frac{2\hat{V}}{\pi}\left\{\sin \omega t + \frac{\sin 3\omega t}{3} + \frac{\sin 5\omega t}{5} + \frac{\sin 7\omega t}{7} + \cdots\right\}$$

$$v(t) = \frac{\hat{V}}{\pi} + \frac{\hat{V}}{2}\cos \omega t + \frac{2\hat{V}}{\pi}\left\{\frac{\cos 2\omega t}{2} - \frac{\cos 4\omega t}{15} + \frac{\cos 6\omega t}{35} - \cdots\right\}.$$

6.4 What are the r.m.s. values of the non-sinusoidal periodic waveforms shown in Fig. 6.2(b), (c), and (g)?

6.5 Show that the Fourier series for the train of triangular pulses shown in Fig. 6.21 is given by

$$v(t) = \hat{V}\left[\frac{\omega\tau}{2\pi} + \frac{4}{\pi\omega\tau}\left\{\sum_{n=1}^{\infty} \frac{\left(1 - \cos \frac{n\omega\tau}{2}\right)}{n^2}\right\}\right],$$

where $\omega \equiv 2\pi/T$.

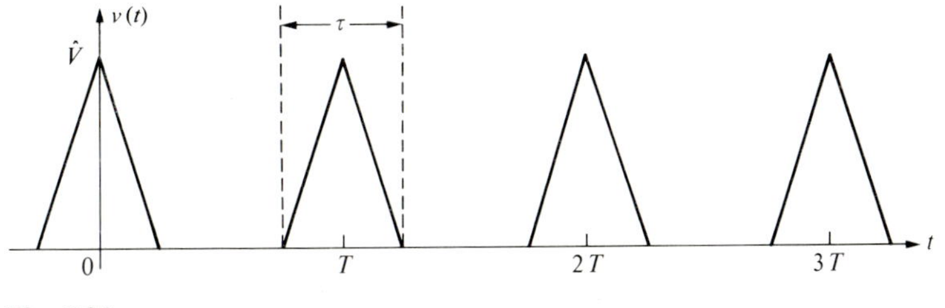

Fig. 6.21

6.6 For a train of rectangular pulses as shown in Fig. 6.7(a) obtain the expressions for the Fourier coefficients given in eqns (6.18).

6.7 Derive an expression for the time domain voltage across the inductance in the circuit shown in Fig. 6.22 if the excitation is a ramp voltage $e(t) = A \cdot t$.

Fig. 6.22

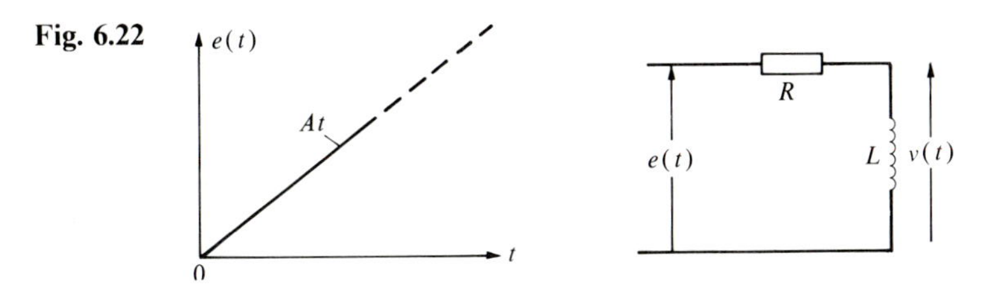

6.8 For the network and voltage excitation $e(t)$ shown in Fig. 6.23 show that the voltage $V(s)$ is the s-domain is given by

$$V(s) = \frac{A}{CR}\frac{1}{s\left(s+\frac{1}{CR}\right)} - \frac{A}{CR}\frac{e^{-Ts}}{s\left(s+\frac{1}{CR}\right)}.$$

Fig. 6.23

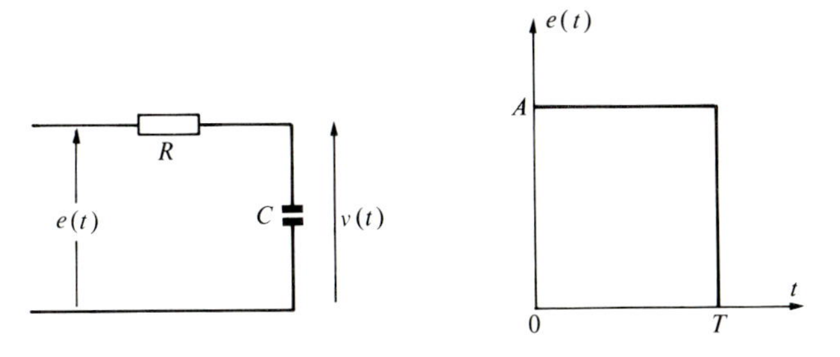

(You may assume that the capacitor is uncharged at time $t = 0$.)

Hence show that the time domain voltage $v(t)$ is given by

$$v(t) = A\{1 - u(t - T) - e^{-t/CR} + e^{-(t-T)/CR} \cdot u(t - T)\}.$$

Sketch the general form of $v(t)$ and also the form of $v(t)$ for $CR \gg T$.

6.9 A non-periodic signal voltage $e(t)$ of the form shown in Fig. 6.24 is applied to the high-pass CR network. Show that

$$v(t) = ACR\{1 - e^{-t/CR}\} - 2ACR\{1 - e^{-(t-T/2)/CR}\} \cdot u(t - T/2) + ACR\{1 - e^{-(t-T)/CR}\} \cdot u(t - T).$$

Fig. 6.24

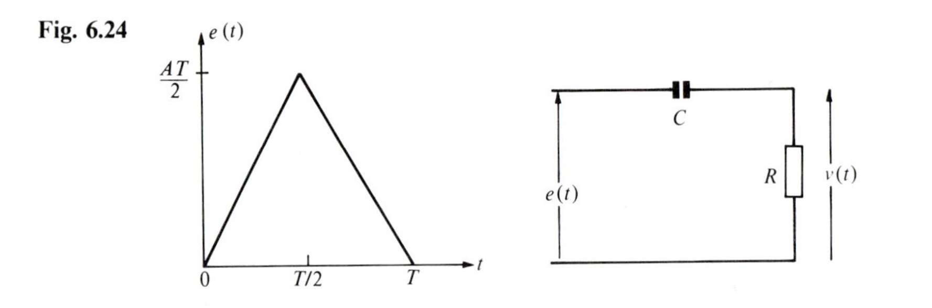

6.10 A voltage pulse $e(t)$ of the form shown in Fig. 6.25 is applied to the low-pass CR network. Show that for the pulse

$$E(s) = \frac{A}{\tau}\left\{\frac{1}{s^2} - \frac{e^{-s\tau}}{s^2} - \frac{e^{-s(\tau+T)}}{s^2} + \frac{e^{-s(2\tau+T)}}{s^2}\right\}$$

and hence show that

$$\begin{aligned} v(t) = \frac{ACR}{\tau}\Bigg\{&\left[-1 + \frac{t}{CR} + e^{-t/CR}\right] \\ &-\left[-1 + \frac{(t-T)}{CR} + e^{-(t-\tau)/CR}\right] \cdot u(t-\tau) \\ &-\left[-1 + \frac{(t-(\tau+T))}{CR} + e^{-(t-(\tau+T))/CR}\right] \cdot u(t-(\tau+T)) \\ &+\left[-1 + \frac{(t-(2\tau+T))}{CR} + e^{-(t-(2\tau+T))/CR}\right] \cdot u(t-(2\tau+T))\Bigg\}. \end{aligned}$$

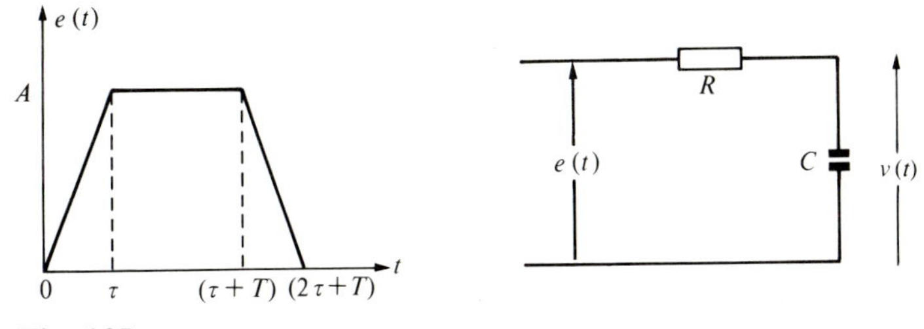

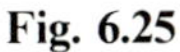
Fig. 6.25

6.11 Show that the response $v(t)$ of the high-pass CR network to a voltage excitation $e(t)$ of the form shown in Fig. 6.26 is given by

$$v(t) = A\cdot\left\{\frac{CR}{T}(1 - \mathrm{e}^{-t/CR}) - 2\mathrm{e}^{-(t-T)/CR}\cdot u(t-T) - \frac{CR}{T}(1 - \mathrm{e}^{-(t-2T)/CR}\cdot u(t-2T)\right\}.$$

Fig. 6.26

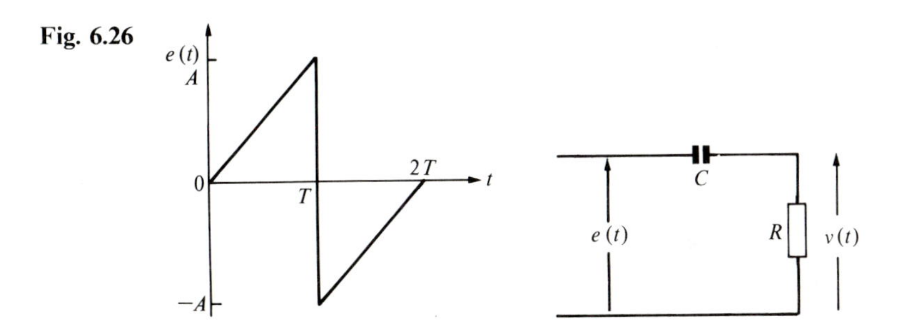

6.12 For the transient voltage excitation $e(t)$ shown in Fig. 6.27 show that the s-domain voltage $V(s)$ is given by

$$V(s) = \frac{A}{Ts^2}\{1 - 3\mathrm{e}^{-Ts} + 3\mathrm{e}^{-2Ts} - \mathrm{e}^{-3Ts}\} - \frac{A}{s}\{\mathrm{e}^{-Ts} - \mathrm{e}^{-2Ts}\}.$$

Fig. 6.27

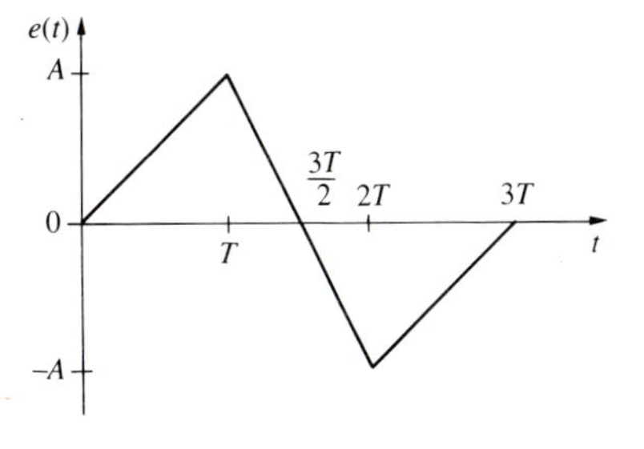

Two-port networks

7

7.1 Introduction

From an operational point of view the interest in a network or, more generally, a system is focused on the **response** at the pair of output terminals in relation to a specified **excitation** applied to the input terminals; detailed knowledge of the circuit elements, and their interconnections, inside the 'box' is not required. Such a representation of a network or system is often called a 'black box'. The pairs of input terminals and output terminals are usually referred to as the input port and the output port, respectively, and the network is a **two-port network**. Commonly encountered examples are coupling and matching networks, filters, attenuators, and phase-shifters, as well as amplifiers in general.

The functional relation between the response and the excitation is specified by the **system function**, H. The situation can be summarized as follows:

$$\text{Response} = \text{System function} \times \text{Excitation}$$
$$\mathscr{R} = \mathscr{H} \times \mathscr{E}$$

Analysis

To determine the response given the excitation and the network.

Synthesis

To design the network given the excitation and the required response.

In this chapter the following assumptions will be made about the networks and components.

(a) *Linear*. If the excitation is multiplied by a constant factor, then the response is multiplied by the same factor.

(b) *Lumped*. A component is assumed to be of small enough physical size so that an applied alternating electric field can be assumed, to an acceptable degree of accuracy, to have the same value at all points along the length of the component at any particular instant of time. This means that a component can be represented by a combination of a pure resistor and/or a pure capacitor and/or a pure inductor (if the assumption does

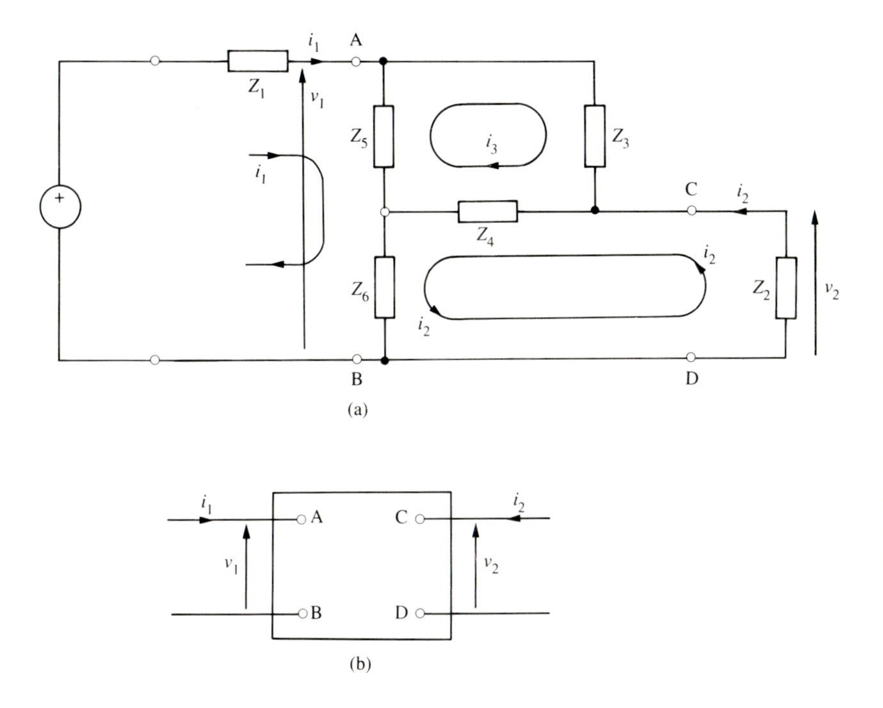

Fig. 7.1 (a) A general bridge network. (b) A representation of a two-port network.

not hold to a sufficient degree of accuracy, then a component must be treated as a 'distributed' circuit element; see Chapter 8). Furthermore, if the physical dimensions of a component or of a circuit as a whole are comparable with, or greater than, the wavelength of electromagnetic waves of the frequency of the electric field, then there will be significant radiation of power. This means, in turn, that energy is not conserved within the circuit and Kirchhoff's laws cannot be applied.

(c) *Passive*. The circuits considered will not contain energy sources.

As an example of a two-port network consider the network of Fig. 7.1(a), which is similar to that of Fig. 2.6 but with general impedances $Z_1 \cdots Z_6$ as the passive circuit elements instead of pure resistors, and with the source of e.m.f. E_2 removed; e_1 is now a source of a sinusoidal e.m.f. In this illustration one of the input terminals and one of the output terminals are connected together inside the 'black box' and form a 'common' connection between the input port and the output port; this occurs quite often in practice and such a network is often referred to as a **three-terminal network**. As far as the relation between the source e_1 and the load Z_2 is concerned, Z_3, Z_4, Z_5, and Z_6 are located within a two-port black box, as in Fig. 7.1(b), and it will now be shown how the properties of the two-port network can be represented by a set of three parameters.

The set of mesh equations that are equivalent to eqns (2.7) (but check the changes in sign) are

$$\left.\begin{aligned} i_1(Z_5 + Z_6) + (i_2 Z_6 - i_3 Z_5) &= e_1 - i_1 Z_1 = v_1 \\ i_1 Z_6 + i_2(Z_4 + Z_6) + i_3 Z_4 &= -i_2 Z_2 = v_2 \\ -i_1 Z_5 + i_2 Z_4 + i_3(Z_3 + Z_4 + Z_5) &= 0 \end{aligned}\right\} \tag{7.1}$$

By using the procedure outlined in Section 2.4 (see eqns (2.10)) it follows that

$$i_1 = v_1 C_{11}/D + v_2 C_{21}/D, \qquad i_2 = v_1 C_{12}/D + v_2 C_{22}/D, \tag{7.2}$$

where $C_{12} = C_{21}$ in the case of a passive network such as this.

Notice that the convention of using lower-case symbols to represent time-dependent currents and voltages has been retained.

The determinant D and its cofactors can be calculated, since the structure of the network is known, but more general importance lies in the fact that eqns (7.2) may be written in the form:

$$i_1 = y_{11} v_1 + y_{12} v_2, \qquad i_2 = y_{21} v_1 + y_{22} v_2, \tag{7.3}$$

where $y_{11} \equiv C_{11}/D$, etc. The quantities y_{11}, y_{12}, y_{21}, y_{22} are known as **y-parameters**, or **admittance parameters**, since they have dimensions of $(\text{ohm})^{-1}$; $y_{21} = y_{12}$, of course. The sign conventions for the currents and voltages at the input and output ports of a two-port network are shown in Fig. 7.1(b).

It follows from eqn (7.3) that v_1, v_2 can be made the dependent variables; i.e.

$$v_1 = z_{11} i_1 + z_{12} i_2, \qquad v_2 = z_{21} i_1 + z_{22} i_2, \tag{7.4}$$

where z_{11}, z_{12}, z_{21}, z_{22} are known as **z-parameters** or **impedance parameters**. The derivation of the explicit relationships between the y- and

z-parameters is left as a problem; for instance

$$z_{12} = -y_{12}/(y_{11}y_{22} - y_{12}y_{21}).$$

Finally, v_1, i_1, v_2, i_2 can be related by

$$v_1 = h_{11}i_1 + h_{12}v_2, \qquad i_2 = h_{21}i_1 + h_{22}v_2, \tag{7.5}$$

where h_{11}, h_{12}, h_{21}, h_{22} are known as ***h*-parameters** or **hybrid parameters**, since, while h_{12}, h_{21} are dimensionless, h_{11} has the dimensions of impedance and h_{22} has the dimensions of admittance.

It is very important to note that, although a specific network has been used as an illustration, eqns (7.3), (7.4) and (7.5) can be used to describe any linear, passive,* two-port network.

Note also that at most four, and in the case of passive networks three, parameters are required to specify one pair of currents or voltages (e.g. at the output port) if the other pair of currents or voltages is known.

The y-, z-, and h-parameters of a network are sometimes known collectively as **matrix parameters**.

It is very important to note that the matrix parameters of a network can be determined from *measurements* made at the ports of the network, without the necessity to analyse the networks contained within the black box. From eqns (7.3) it can be seen that the y-parameters can be determined by short-circuiting the input terminals and then the output terminals ($v_1 = 0$ and $v_2 = 0$, respectively) and measuring i_1, v_2, i_2 and then i_1, v_1, i_2 respectively:

$$v_1 = 0: \qquad y_{12} = i_1/v_2 \qquad y_{22} = i_2/v_2 \tag{7.6}$$

$$v_2 = 0: \qquad y_{11} = i_1/v_1 \qquad y_{21} = i_2/v_1. \tag{7.7}$$

Hence the y-parameters are sometimes known as 'short-circuit admittance parameters'.

The z-parameters can be determined by open-circuiting the input terminals and the output terminals successively and measuring the

* Actually the equations can be used to describe linear, *active*, two-port networks also (e.g. bipolar transistors), but in this case $h_{21} \neq h_{12}$.

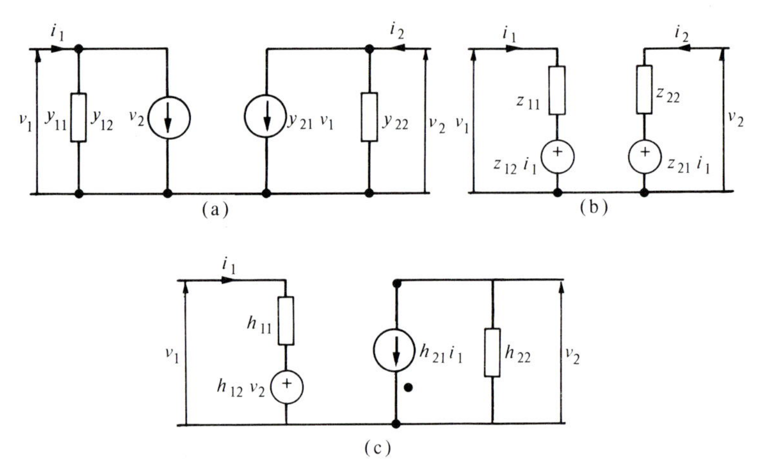

Fig. 7.2 Equivalent circuits for a two-port network in terms of (a) its *y*-parameters, (b) its *z*-parameters, and (c) its *h*-parameters.

appropriate voltages and currents:

$$i_1 = 0: \qquad z_{12} = v_1/i_2 \qquad z_{22} = v_2/i_2 \tag{7.8}$$

$$i_2 = 0: \qquad z_{11} = v_1/i_1 \qquad z_{21} = v_2/i_1. \tag{7.9}$$

Hence the *z*-parameters are sometimes known as 'open-circuit impedance parameters'.

Similarly the *h*-parameters can be determined by successively open-circuiting the input terminals and then short-circuiting the output terminals:

$$i_1 = 0: \qquad h_{12} = v_1/v_2 \qquad h_{22} = i_2/v_2 \tag{7.10}$$

$$v_2 = 0: \qquad h_{11} = v_1/i_1 \qquad h_{21} = i_2/i_1. \tag{7.11}$$

A cautionary note: in the foregoing, 'short circuit' does not necessarily imply a d.c. short circuit, since a large-enough capacitor, with a reactance much less than the magnitude of the other impedances of interest, may constitute a good enough approximation to an a.c. short circuit. In any case, caution should be exercised when short-circuiting a pair of terminals, in case damage is caused to the circuit components.

The equivalent circuits for a two-port network in terms of *y*-, *z*-, and *h*-parameters, as defined by eqns (7.3), (7.4) and (7.5), are shown in Figs. 7.2(a), (b), and (c) respectively.

The possibility of measuring matrix parameters is especially useful in the case of active networks, since quite often the equivalent circuit model for an active device contains elements whose values are not accurately

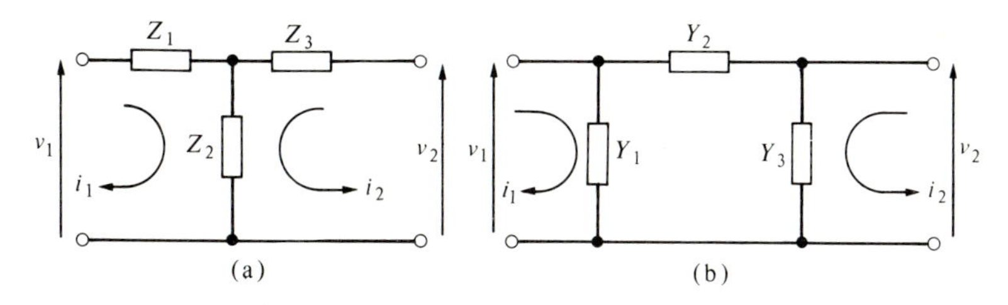

Fig. 7.3 (a) T-form and (b) Π-form of equivalent network for a passive two-port network.

known. Further, for circuits operating at high frequency (>10 MHz, say) the effects of 'stray' reactances, and of reactances inherent in the packaging of devices, are difficult to calculate. Sophisticated network analysers are available commercially and are invaluable tools for the designer.

The h-parameters, as defined above, are widely used to describe the circuit properties of bipolar transistors at frequencies up to about 30 kHz. In the case of such transistors, h_{11} ($\equiv h_i$) is the input resistance, h_{12} ($\equiv h_r$) is the reverse voltage transfer ratio, h_{21} ($\equiv h_f$) is the forward current gain, and h_{22} ($\equiv h_o$) is the output admittance.

For frequencies up to about 100 MHz, y-parameters have more utility than h-parameters, because the characteristics of circuits are increasingly influenced by shunt capacitances as the frequency increases. At even higher frequencies the so-called 's-parameters' are used to characterize passive and active circuit elements of all kinds.

7.2 T- and Π-networks

Since there are only three independent elements in the set of z-, or y-, or h-parameters in the case of a passive network, the actual network can be represented by an equivalent network containing only three circuit elements. Further, the equivalent network can be expressed in the form of a T-network or a Π-network as shown in Fig. 7.3; practical considerations will determine which form is the most convenient to use in particular contexts.

It is useful to determine the relations between the circuit elements in the T- and Π-forms of an equivalent circuit. The mesh equations for the network of Fig. 7.3 are

$$\begin{aligned} i_1(Z_1 + Z_2) + i_2 Z_2 &= v_1 \\ i_1 Z_2 + i_2(Z_3 + Z_2) &= v_2. \end{aligned} \tag{7.12}$$

So

$$z_{11} = Z_1 + Z_2 \qquad z_{12} = Z_2$$
$$z_{21} = Z_2 \qquad z_{22} = Z_3 + Z_2. \tag{7.13}$$

Following the procedures for solving simultaneous equations given in Section 2.4 (eqn (2.10)),

$$i_1 = \frac{v_1 C_{11}}{D_z} + \frac{v_2 C_{21}}{D_z}$$
$$i_2 = \frac{v_1 C_{12}}{D_z} + \frac{v_2 C_{22}}{D_z}, \tag{7.14}$$

where

$$C_{11} = (-1)^2(Z_3 + Z_2),\ C_{12} = C_{21} = (-1)^3 Z_2,\ C_{22} = (-1)^4(Z_1 + Z_2),$$

and

$$D_z = (Z_1 + Z_2)(Z_3 + Z_2) - Z_2^2, \qquad \text{or} \qquad D_z = Z_1 Z_2 + Z_1 Z_3 + Z_2 Z_3.$$

Now the nodal equations for the Π-network of Fig. 7.3(b) are

$$v_1(Y_1 + Y_2) - v_2 Y_2 = i_1$$
$$-v_1 Y_2 + v_2(Y_2 + Y_3) = i_2. \tag{7.15}$$

So

$$y_{11} = Y_1 + Y_2 \qquad y_{12} = y_{21} = -Y_2 \qquad y_{22} = Y_2 + Y_3. \tag{7.16}$$

From eqn (7.15) it follows that

$$v_1 = \frac{i_1 C_{11}}{D_y} + \frac{i_2 C_{21}}{D_y}$$
$$v_2 = \frac{i_1 C_{12}}{D_y} + \frac{i_2 C_{22}}{D_y}, \tag{7.17}$$

where

$$C_{11} = (-1)^2(Y_2 + Y_3), \qquad C_{12} = C_{21} = (-1)^3(-Y_2),$$
$$C_{22} = (-1)^4(Y_1 + Y_2),$$

and

$$D_y = (Y_1 + Y_2)(Y_2 + Y_3) - Y_2^2 \qquad \text{or} \qquad D_y = Y_1 Y_2 + Y_1 Y_3 + Y_2 Y_3.$$

On comparing eqns (7.12) and (7.17) it can be seen that the elements

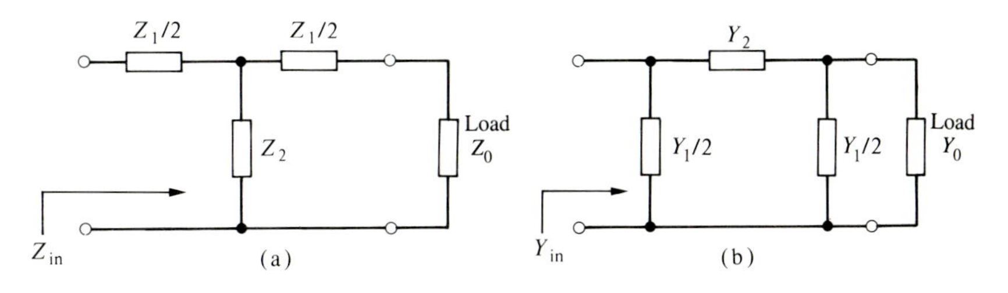

Fig. 7.4 (a), (b) T- and Π-forms, respectively, of a symmetrical two-port network.

of the T- and Π-forms of the equivalent network are related as follows:

$$Z_1 + Z_2 = (Y_2 + Y_3)/D_y \qquad Z_2 = Y_2/D_y \qquad Z_3 + Z_2 = (Y_1 + Y_2)/D_y,$$

whence

$$Z_1 = Y_3/D_y \qquad Z_3 = Y_1/D_y \qquad Z_2 = Y_2/D_y. \tag{7.18}$$

Similarly, the admittances Y_1, Y_2, Y_3 can be expressed in terms of the impedances Z_1, Z_2, Z_3 and D_z; you should be able to show that

$$Y_1 = Z_3/D_z \qquad Y_3 = Z_1/D_z \qquad Y_2 = Z_2/D_z. \tag{7.19}$$

A two-port network is said to be symmetrical if open- or short-circuit measurements, as the case may be, yield identical values for the z-, y-, h-parameters irrespective of which pair of terminals is designated as the input port; i.e. the network 'looks' identical when looked into from either port. In the case of an equivalent T-network, or an actual T-network, this implies that $Z_1 = Z_3$ or that $Y_1 = Y_3$. It is common practice to designate each of the series elements in a symmetrical T-network by $Z_1/2$ and the shunt elements in a Π-network by $Y_1/2$, as shown in Fig. 7.4.

In high-frequency systems (frequency >100 MHz, say) it is often necessary for a two-port network, for example an attenuator (see Example 7.4), to be matched to a source connected to its input port and to a load connected to its output port, where the source and load have the same impedance. If the source and the load each have impedance Z_0, or admittance Y_0, then, in Fig. 7.4, the requirement is that $Z_{in} = Z_0$ or $Y_{in} = Y_0$, as the case may be.

For the T-network of Fig. 7.4(a) it can be seen, by inspection, that

$$Z_{in} = \frac{Z_1}{2} + \frac{Z_2(Z_1/2 + Z_0)}{(Z_1/2 + Z_2 + Z_0)}.$$

To meet the requirement $Z_{in} = Z_0$, it follows from this equation that

$$Z_0 \equiv Z_{0T} = \sqrt{(Z_1 Z_2 + Z_1^2)}. \tag{7.20}$$

Z_{0T} is called the **characteristic impedance** of the T-network.

The open-circuit input impedance (Z_{oc}) is the input impedance with the output port open-circuited ($Z_L = \infty$) and, by inspection, is given by

$$Z_{oc} = Z_1/2 + Z_2. \tag{7.21}$$

The short-circuit input impedance ($Z_L = 0$) is given by

$$Z_{sc} = Z_1/2 + \frac{\frac{1}{2}Z_1 Z_2}{\{\frac{1}{2}Z_1 + Z_2\}}$$

or

$$Z_{sc} = \frac{Z_1 Z_2 + \frac{1}{4}Z_1^2}{\{Z_2 + \frac{1}{2}Z_1\}}. \tag{7.22}$$

The practical utility of these expressions for Z_{oc} and Z_{sc} is that (as you can easily show for yourself)

$$Z_{0T} = \sqrt{(Z_{oc} Z_{sc})}. \tag{7.23}$$

It follows from this relationship that the characteristic impedance (or the characteristic admittance; see immediately below) of a symmetrical two-port network can be obtained from measurements made at the input port with the output port open-circuited and short-circuited in turn.

It is left as a problem to show tht for the Π equivalent circuit the characteristic admittance $Y_{o\Pi}$ is given by the dual expression

$$Y_{0\Pi} = \sqrt{(Y_1 Y_2 + \frac{1}{4}Y_1^2)} \tag{7.24}$$

and that

$$Y_{0\Pi} = \sqrt{(Y_{oc} Y_{sc})}. \tag{7.25}$$

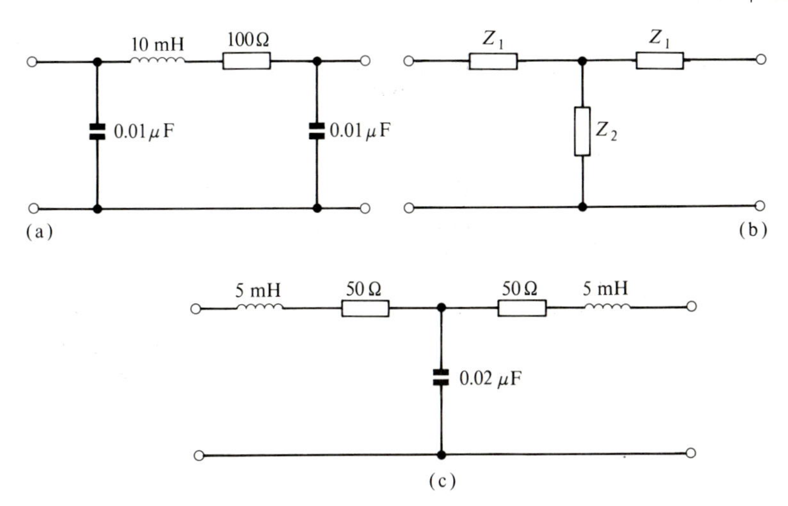

Fig. 7.5 (a) A symmetrical Π-network. (b) A general symmetrical T-network. (c) The symmetrical T-network equivalent to the network of (a).

Example 7.1 As a simple example of the above type of analysis, consider the problem of finding the T-network that is equivalent to the Π-network shown in Fig. 7.5(a).

For simplicity, assume that the operating frequency is $10^4/2\pi$ Hz, so that $\omega = 10^4$ rad s^{-1}.

$$Y_1 = Y_3 = \mathrm{j}10^4 10^{-8} = \mathrm{j}10^{-4}\ \mathrm{S}$$

$$Y_2 = (100 + \mathrm{j}10^4 10^{-2})^{-1} = \frac{(1-\mathrm{j})}{2\cdot 10^2}\ \mathrm{S}$$

$$D_y = 2Y_1Y_2 + Y_1^2 \approx 10^{-6}(1+\mathrm{j})\ \mathrm{S}^2.$$

So, using eqns (7.18),

$$Z_1 = \frac{\mathrm{j}10^{-4}}{10^{-6}(1+\mathrm{j})} \qquad Z_2 = \frac{(1-\mathrm{j})}{2\cdot 10^2\cdot 10^{-6}(1+\mathrm{j})}$$

or

$$Z_1 = 50(1+\mathrm{j})\ \Omega \qquad Z_2 = -\mathrm{j}5\cdot 10^3\ \Omega.$$

Remembering that the operating frequency is $10^4/2\pi$ Hz, the component values shown in Fig. 7.5(c) are obtained. □

7.3 The interconnection of two-port networks

7.3.1 Matrix parameters

Many electrical and electronic systems are composed of interconnected two-port networks, and it is important to consider the various ways in which the interconnections can be made. The five modes of interconnection are shown in Fig. 7.6, but before they are discussed further, it is necessary to introduce another set of matrix parameters in addition to the *z*-, *y*-, and *h*-parameters already met. **Transmission parameters** (*a*-parameters) express the relation between the input variables v_1, i_1 and the output variables v_2, i_2, and are defined (see Fig. 7.7) as follows:

$$\begin{aligned} v_1 &= a_{11}v_2 + a_{12}i_2 \\ i_1 &= a_{21}v_2 + a_{22}i_2 \end{aligned} \tag{7.26}$$

Note carefully that the sense of i_2 is *opposite* to that which was adopted in the context of *z*-, *y*-, and *h*-parameters; the reason for adopting this convention will be discussed below.

The definitive equations (eqns (7.3), (7.4), (7.5), and (7.26)) for the *y*-, *z*-, *h*-, and *a*-parameters can be written in the form:

$$\begin{bmatrix} i_1 \\ i_2 \end{bmatrix} = \begin{bmatrix} y_{11} & y_{12} \\ y_{21} & y_{22} \end{bmatrix} \begin{bmatrix} v_1 \\ v_2 \end{bmatrix} \quad \begin{bmatrix} v_1 \\ v_2 \end{bmatrix} = \begin{bmatrix} z_{11} & z_{12} \\ z_{21} & z_{22} \end{bmatrix} \begin{bmatrix} i_1 \\ i_2 \end{bmatrix}$$

$$\text{`}[y]\text{'} \qquad\qquad\qquad \text{`}[z]\text{'}$$

$$\begin{bmatrix} v_1 \\ i_2 \end{bmatrix} = \begin{bmatrix} h_{11} & h_{12} \\ h_{21} & h_{22} \end{bmatrix} \begin{bmatrix} i_1 \\ v_2 \end{bmatrix} \quad \begin{bmatrix} v_1 \\ i_1 \end{bmatrix} = \begin{bmatrix} a_{11} & a_{12} \\ a_{21} & a_{22} \end{bmatrix} \begin{bmatrix} v_2 \\ i_2 \end{bmatrix} \tag{7.27}$$

$$\text{`}[h]\text{'} \qquad\qquad\qquad \text{`}[a]\text{'}$$

Although matrix algebra will not be introduced in this text, except at a very elementary level, it is worth making some comments on the last one, say, of the above equations. A matrix equation in this form can be expressed in words as: column vector $\begin{bmatrix} v_1 \\ i_1 \end{bmatrix}$ equals column vector $\begin{bmatrix} v_2 \\ i_2 \end{bmatrix}$ multiplied (or operated on) by the square (2 × 2) matrix $\begin{bmatrix} a_{11} & a_{12} \\ a_{21} & a_{22} \end{bmatrix}$.

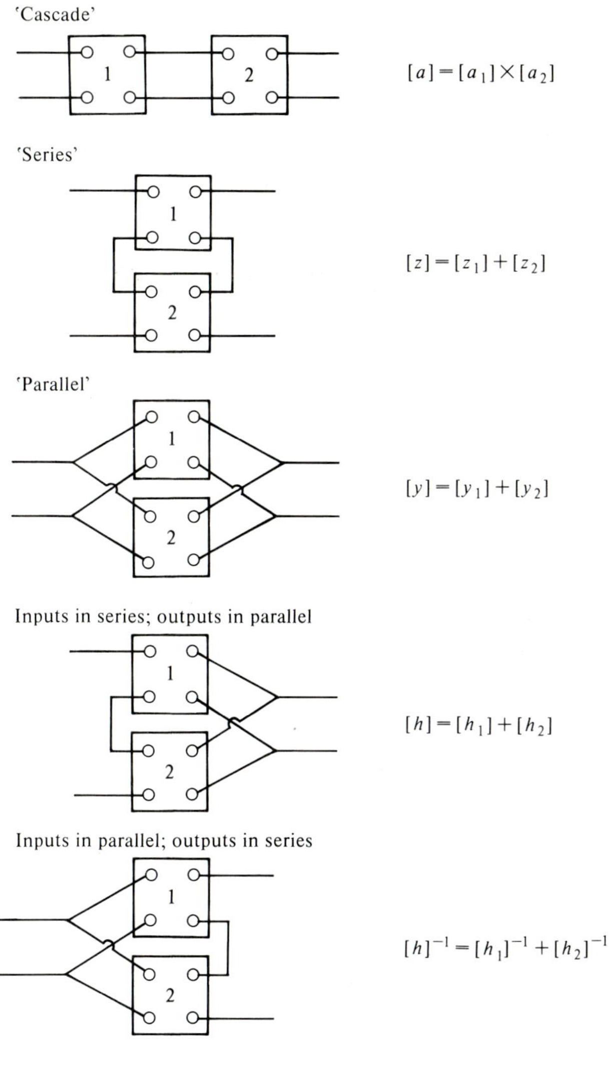

Fig. 7.6 Interconnection of two-port networks.

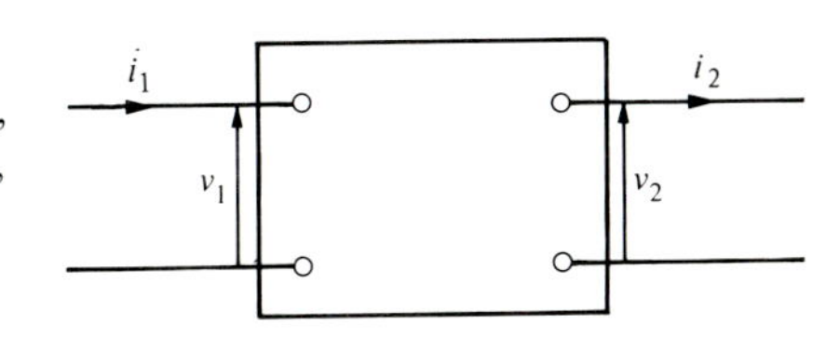

Fig. 7.7 The definitions, in relation to the transmission parameters, of the input variables v_1, i_1 and the output variables v_2, i_2 of a two-port network.

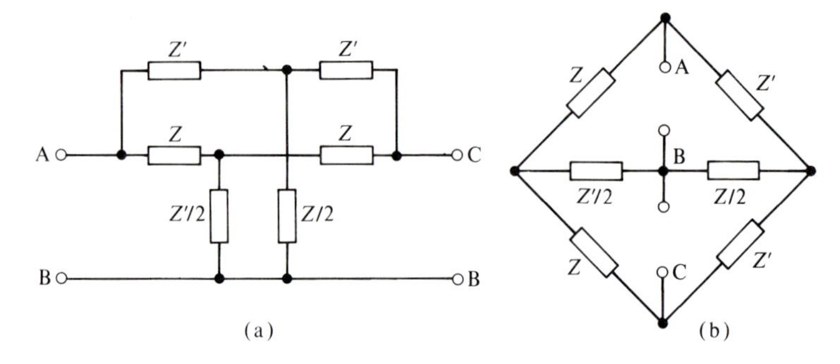

Fig. 7.8 Two ways of drawing the circuit diagram for a twin-T network.

A comparison of this matrix equation and eqns (7.26) (or a look ahead to eqn (7.35)) reveals the rule to be followed in multiplying together a 2×2 matrix and a column vector having two rows (a 2×1 matrix).

The utility of the matrix representation is that when networks are interconnected (as shown in Fig. 7.6), the matrices combine in a mathematically simple fashion to give a resultant matrix for the combined network.

(7.28)

- (a) Series connection $\quad [z] = [z_1] + [z_2]$
- (b) Parallel connection $\quad [y] = [y_1] + [y_2]$
- (c) Inputs in series; outputs in parallel $\quad [h] = [h_1] + [h_2]$
- (d) Inputs in parallel; outputs in series $\quad [h]^{-1} = [h_1]^{-1} + [h_2]^{-1}$
- (e) Cascade connection $\quad [a] = [a_1] \times [a_2]$

The cascade interconnection of two-port networks will be of most interest here. However, one example of the parallel connection of two networks that is of practical interest is the 'twin-T' network (see Fig. 7.8), which consists of two symmetrical T-networks connected in parallel.

Example 7.2 Obtain the y-matrix for the twin-T network of Fig. 7.8 and hence obtain an expression for the rejection frequency (or 'notch' frequency) of the network.

Using the results (eqns (7.16) for the y-parameters of a T-network together with eqns (7.19) and the relevant equation from eqns (7.28),

$$[y] = \begin{bmatrix} \dfrac{\frac{1}{2}Z' + Z}{D_1} & \dfrac{-\frac{1}{2}Z'}{D_1} \\ \dfrac{-\frac{1}{2}Z'}{D_1} & \dfrac{Z + \frac{1}{2}Z'}{D_1} \end{bmatrix} + \begin{bmatrix} \dfrac{\frac{1}{2}Z + Z'}{D_2} & \dfrac{-\frac{1}{2}Z}{D_2} \\ \dfrac{-\frac{1}{2}Z}{D_2} & \dfrac{Z' + \frac{1}{2}Z}{D_2} \end{bmatrix},$$

where $D_1 \equiv Z(Z + Z')$ and $D_2 \equiv Z'(Z + Z')$.

In the addition of two matrices (which must be of the same dimensions), each matrix element of the resultant matrix is simply the sum of the corresponding elements in the two component matrices. Hence

$$[y] = \{ZZ'(Z + Z')\}^{-1} \begin{bmatrix} (\frac{1}{2}Z'^2 + 2ZZ' + \frac{1}{2}Z^2) & -\frac{1}{2}(Z'^2 + Z^2) \\ -\frac{1}{2}(Z'^2 + Z^2) & (\frac{1}{2}Z'^2 + 2ZZ' + \frac{1}{2}Z^2) \end{bmatrix}. \tag{7.29}$$

The twin-T network is said to be 'balanced' (see Fig. 7.8(b)) when the current flowing through a load connected to the output terminals BC is zero. Since under these circumstances v_2 is also zero, it follows from eqn (7.3) that the value of y_{21} must be zero. For this to be so it can be seen from eqn (7.29) that $Z'^2 = -Z^2$.

It is usual to construct a twin-T network from resistors and capacitors, since it is easy to make such components variable; for $Z = R$, and $Z' = -j/\omega C$, then $Z'^2 = -Z^2$ if $\omega RC = 1$. In this situation the network acts as a rejection filter (a 'band-stop' or 'notch' filter) at the angular frequency $\omega = (CR)^{-1}$. A useful practical feature of a twin-T network is that there is one terminal that is common to both the input and output ports, so that the source and the load can have a common ground. In contrast, in a conventional bridge network the output port does not share a common terminal with the input port. □

It must not be assumed automatically that two two-port networks retain their individual two-port characteristics after being interconnected. There exist general rules, or validity tests, for interconnected two-port networks, but it would not be very profitable to discuss them in detail here, for the following reasons:

(a) Many of the two-port networks of interest have a common connection between an input and an output terminal (so-called 'three-terminal' networks), in which case the interconnections shown in Fig. 7.9 satisfy the validity tests; the twin-T network falls into this category.

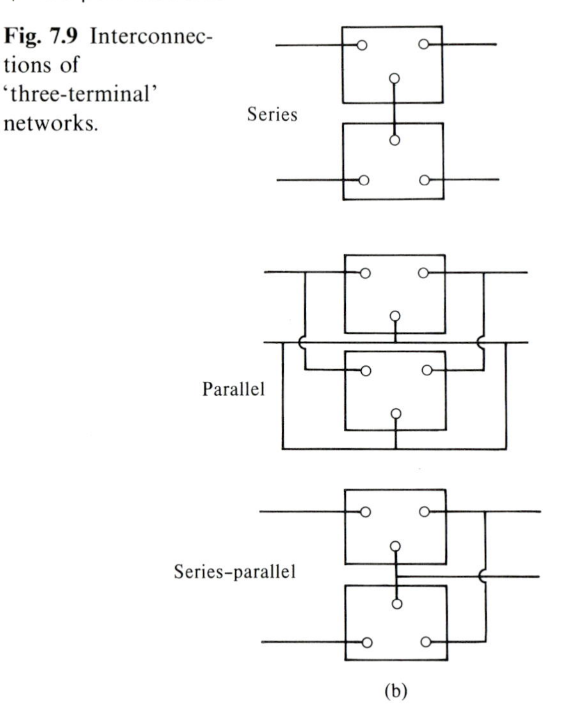

Fig. 7.9 Interconnections of 'three-terminal' networks.

(b) In the case of cascade, or tandem, connection of two-port networks, on which most interest is focused in this text, the above-mentioned problem does not arise.

The relationships between the elements of the z-, y-, h-, and a-matrices of a two-port network can be derived easily, if somewhat tediously. For example, consider the equations that define the z- and y-parameters (eqns (7.3) and (7.4)). From eqns (7.3),

$$v_1 = \frac{i_2}{y_{21}} - \frac{y_{22}v_2}{y_{21}} \qquad \text{and} \qquad v_2 = \frac{i_1 - y_{11}v_1}{y_{12}}.$$

Hence,

$$v_1\left\{1 - \frac{y_{22}y_{11}}{y_{21}y_{12}}\right\} = \frac{i_2}{y_{21}} - \frac{y_{22}i_1}{y_{21}y_{12}}$$

or

$$v_1\frac{\{y_{12}y_{21} - y_{22}y_{11}\}}{y_{21}y_{12}} = \frac{-y_{22}i_1}{y_{21}y_{12}} + \frac{i_2}{y_{21}}.$$

Thus

$$v_1 = \frac{y_{22}i_1}{D_y} - \frac{y_{12}i_2}{D_y}, \tag{7.30}$$

where $D_y \equiv y_{11}y_{22} - y_{12}y_{21}$.

On comparing eqn (7.30) with eqns (7.4) it can be seen that

$$z_{11} = \frac{y_{22}}{D_y} \qquad z_{12} = \frac{-y_{12}}{D_y}.$$

By using similar arguments, the matrix parameter conversion table (Table 7.1) can be constructed.

Table 7.1. Matrix parameter conversion table.

	z		y		h		a	
z	z_{11}	z_{12}	$\frac{y_{22}}{D_y}$	$\frac{-y_{12}}{D_y}$	$\frac{D_h}{h_{22}}$	$\frac{h_{12}}{h_{22}}$	$\frac{a_{11}}{a_{21}}$	$\frac{D_a}{a_{21}}$
	z_{21}	z_{22}	$\frac{-y_{21}}{D_y}$	$\frac{y_{11}}{D_y}$	$\frac{-h_{21}}{h_{22}}$	$\frac{1}{h_{22}}$	$\frac{1}{a_{21}}$	$\frac{a_{22}}{a_{21}}$
y	$\frac{z_{22}}{D_z}$	$\frac{-z_{12}}{D_z}$	y_{11}	y_{12}	$\frac{1}{h_{11}}$	$\frac{-h_{12}}{h_{11}}$	$\frac{a_{22}}{a_{12}}$	$\frac{-D_a}{a_{12}}$
	$\frac{-z_{21}}{D_z}$	$\frac{z_{11}}{D_z}$	y_{21}	y_{22}	$\frac{h_{21}}{h_{11}}$	$\frac{D_h}{h_{11}}$	$\frac{-1}{a_{12}}$	$\frac{a_{11}}{a_{12}}$
h	$\frac{D_z}{z_{22}}$	$\frac{z_{12}}{z_{22}}$	$\frac{1}{y_{11}}$	$\frac{-y_{12}}{y_{11}}$	h_{11}	h_{12}	$\frac{a_{12}}{a_{22}}$	$\frac{D_a}{a_{22}}$
	$\frac{-z_{21}}{z_{22}}$	$\frac{1}{z_{22}}$	$\frac{y_{21}}{y_{11}}$	$\frac{D_y}{y_{11}}$	h_{21}	h_{22}	$\frac{-1}{a_{22}}$	$\frac{a_{21}}{a_{22}}$
a	$\frac{z_{11}}{z_{21}}$	$\frac{D_z}{z_{21}}$	$\frac{-y_{22}}{y_{21}}$	$\frac{-1}{y_{21}}$	$\frac{-D_h}{h_{21}}$	$\frac{-h_{11}}{h_{21}}$	a_{11}	a_{12}
	$\frac{1}{z_{21}}$	$\frac{z_{22}}{z_{21}}$	$\frac{-D_y}{y_{21}}$	$\frac{-y_{11}}{y_{21}}$	$\frac{-h_{22}}{h_{21}}$	$\frac{-1}{h_{21}}$	a_{21}	a_{22}

NB $D_z = z_{11}z_{22} - z_{12}z_{21}$, etc.

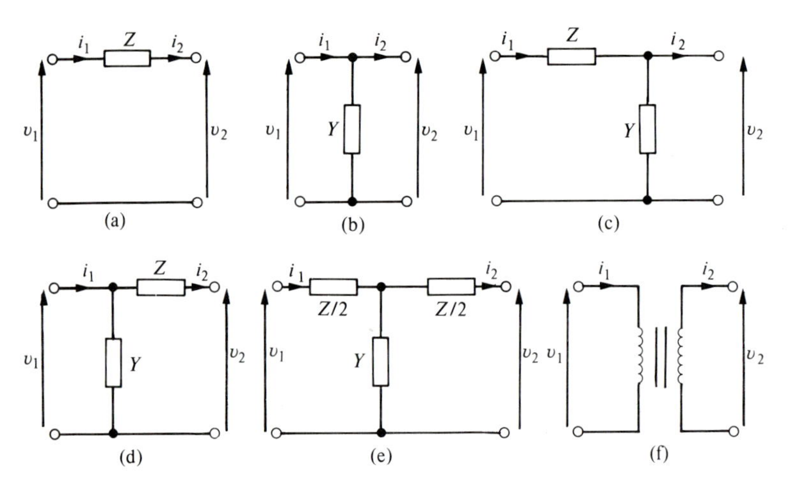

Fig. 7.10 Some two-port networks for which the transmission parameters are derived in the text.

The application of transmission parameters in the case of cascade-connected two-port networks can be illustrated usefully by some simple examples (see Fig. 7.10).

Example 7.3 Obtain the transmission matrices for the networks shown in Figs. 7.10(a), (b), and (c).

(a) For the network of Fig. 7.10(a),

$$\begin{aligned} v_1 &= v_2 + i_2 Z \\ i_1 &= i_2 \end{aligned} \qquad [a] = \begin{bmatrix} 1 & Z \\ 0 & 1 \end{bmatrix}. \tag{7.31}$$

(b) For the network of Fig. 7.10(b),

$$v_1 = v_2$$

$$\frac{(i_1 - i_2)}{Y} = v_2$$

or

$$\begin{aligned} v_1 &= v_2 \\ i_1 &= v_2 Y + i_2 \end{aligned} \qquad [a] = \begin{bmatrix} 1 & 0 \\ Y & 1 \end{bmatrix}. \tag{7.32}$$

(c) For the network of Fig. 7.10(c), which is the resultant of the cascade connection of the previous two networks,

$$v_1 = i_1 Z + \frac{(i_1 - i_2)}{Y}$$

$$v_2 = \frac{(i_1 - i_2)}{Y} \quad \text{or} \quad i_1 = v_2 Y + i_2.$$

So

$$\begin{aligned} v_1 &= v_2(1 + YZ) + i_2 Z \\ i_1 &= v_2 Y + i_2 \end{aligned} \qquad [a] = \begin{bmatrix} (1 + YZ) & Z \\ Y & 1 \end{bmatrix}. \tag{7.33}$$

It should be noticed that the matrix (eqn (7.33)) is equal to the product of the two previous matrices (eqns (7.31) and (7.32)) for the constituent networks, i.e.

$$\begin{bmatrix} 1 & Z \\ 0 & 1 \end{bmatrix} \times \begin{bmatrix} 1 & 0 \\ Y & 1 \end{bmatrix} = \begin{bmatrix} (1 + YZ) & Z \\ Y & 1 \end{bmatrix}, \tag{7.34}$$

where use has been made of the matrix multiplication rule

$$\begin{bmatrix} a_{11} & a_{12} \\ a_{21} & a_{22} \end{bmatrix} \times \begin{bmatrix} b_{11} & b_{12} \\ b_{21} & b_{22} \end{bmatrix} = \begin{bmatrix} (a_{11}b_{11} + a_{12}b_{21}) & (a_{11}b_{12} + a_{12}b_{22}) \\ (a_{21}b_{11} + a_{22}b_{21}) & (a_{21}b_{12} + a_{22}b_{22}) \end{bmatrix}. \tag{7.35}$$

□

With regard to the convention that has been adopted for the sense of 'i_2', it can easily be checked, using the above type of procedure, that if i_2 is defined in the opposite sense, then the matrices (7.31), (7.32) and (7.33) become

$$\begin{bmatrix} 1 & -Z \\ 0 & -1 \end{bmatrix}, \begin{bmatrix} 1 & 0 \\ Y & -1 \end{bmatrix}, \quad \text{and} \quad \begin{bmatrix} (1 + YZ) & -Z \\ Y & -1 \end{bmatrix} \text{respectively.}$$

Now the product of the first two of these matrices is

$$\begin{bmatrix} 1 & -Z \\ 0 & -1 \end{bmatrix} \times \begin{bmatrix} 1 & 0 \\ Y & -1 \end{bmatrix} = \begin{bmatrix} (1 + YZ) & Z \\ -Y & 1 \end{bmatrix} \neq \begin{bmatrix} (1 + YZ) & -Z \\ Y & -1 \end{bmatrix}.$$

However, if the matrix

$$\begin{bmatrix} 1 & 0 \\ 0 & -1 \end{bmatrix}$$

is interposed between the two matrices on the left-hand side of the above equation, then you will be able to confirm that the correct resultant matrix is obtained.

It is a matter of taste as to which convention is adopted with regard to the sense of 'i_2', but in this text the convention specified in Fig. 7.6 will continue to be used.

The transmission matrices for the networks of Fig. 7.10(d) and (e) can be derived by the methods used in Example 7.3 and are:

$$\begin{bmatrix} 1 & Z \\ Y & (1 + YZ) \end{bmatrix} \tag{7.36}$$

and

$$\begin{bmatrix} (1 + \frac{1}{2}YZ) & (Z + \frac{1}{4}YZ^2) \\ Y & (1 + \frac{1}{2}YZ) \end{bmatrix}. \tag{7.37}$$

Radiofrequency systems are generally characterized by a particular level of impedance (e.g. 75 Ω, 50 Ω), and so it is necessary that subsystems, such as attenuators and phase shifters, should be matched to the relevant impedance level. The use of transmission parameters in this context is illustrated in the next Example.

Example 7.4 Design an attenuator to work between a source of output impedance R_0 and a resistive load R_0 and which is to provide a *voltage* attenuation ratio of $1{:}n$.

Consider the symmetrical T-network of Fig. 7.11, where the problem is to achieve $v_2/v_1 = 1/n$ whilst simultaneously arranging that the impedance presented to the source is still equal to R_0.

If the circuit is redrawn as in Fig. 7.11(b), where $G_0 \equiv 1/R_0$, then it follows, by using eqns (7.32) and (7.37), tht the two-port network with terminals AA′, BB′ has a transmission matrix $[a]$ given by

$$[a] = \begin{bmatrix} (1 + \frac{1}{2}RG) & (R + \frac{1}{4}GR^2) \\ G & (1 + \frac{1}{2}GR) \end{bmatrix} \times \begin{bmatrix} 1 & 0 \\ G_0 & 0 \end{bmatrix}$$

Hence

$$[a] = \begin{bmatrix} (1 + \frac{1}{2}RG) + G_0R + \frac{1}{4}G_0GR^2) & 0 \\ (G + G_0 + \frac{1}{2}G_0GR) & 0 \end{bmatrix}. \tag{7.38}$$

Now, if eqn (7.26) is considered in the light of the problem as specified by the network in Fig. 7.11(b), then it can be seen that 'i_2' = 0, since the voltage v_2 across the load resistance is imagined to be measured by an ideal voltmeter of infinite internal resistance. Hence $v_2/v_1 = 1/a_{11}$, and from eqn (7.38) this means that for $v_2/v_1 = 1/n$,

$$1 + \tfrac{1}{2}RG + G_0R + \tfrac{1}{4}G_0GR^2 = n.$$

It is common practice to write such equations in terms of 'normalized'

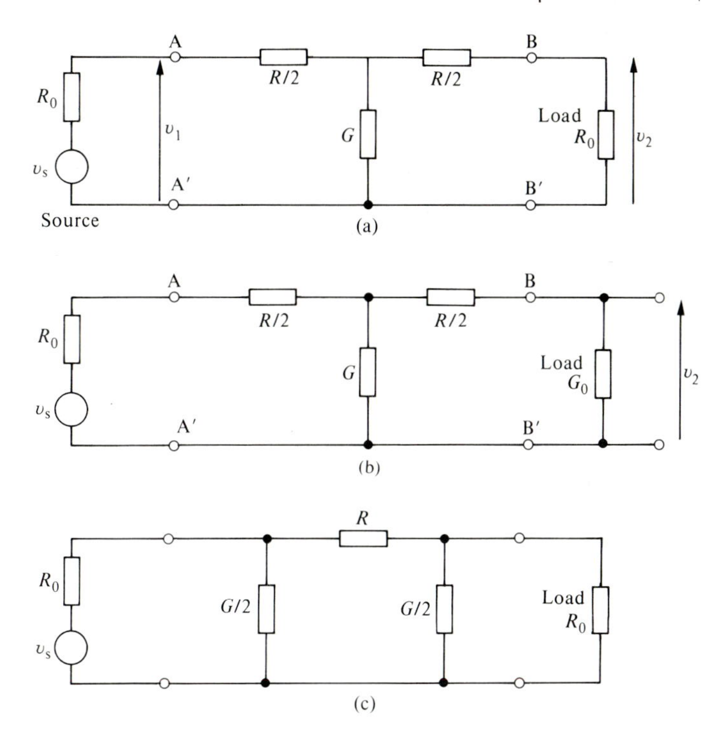

Fig. 7.11 A matched attenuator network.

resistances (or impedances or admittances) and, since $R_0 G_0 = 1$, it follows that

$$\frac{G}{G_0}\cdot\left(\frac{R}{R_0}\right)^2 + 2\frac{G}{G_0}\cdot\frac{R}{R_0} + 4\frac{R}{R_0} + 4(1-n) = 0. \tag{7.39}$$

Now the problem is to find a relation between R and G such that the characteristic impedance of the symmetrical T-network is equal to the source and load resistance, namely R_0. Using the defining equation for the characteristic impedance (eqn (7.20)) in the form

$$R_0^2 = \frac{R}{G} + \tfrac{1}{4}R^2 \qquad \text{then } \frac{G}{G_0}\left(\frac{R}{R_0}\right)^2 + 4\frac{R}{R_0} - 4\frac{G}{G_0} = 0. \tag{7.40}$$

On substituting for (G/G_0) in eqn (7.39) from eqn (7.40), it follows that

$$\left(\frac{R}{R_0}\right)^2(1+n) + 4\frac{R}{R_0} + 4(1-n) = 0.$$

The solution of this quadratic equation for (R/R_0) yields

$$\frac{R}{R_0} = \frac{2(n-1)}{(n+1)} \quad \text{and thence} \quad \frac{G}{G_0} = \frac{(n^2-1)}{2n}. \tag{7.41}$$

For example, for $n = 2$ (6 dB attenuation), then $R/R_0 = \frac{2}{3}$ and $G/G_0 = \frac{3}{4}$. If the attenuator is to be used in a 75 Ω antenna downlead, then the required value of R is 50 Ω, and to realize the required value of G, use a 100 Ω resistor. □

An alternative way of realizing a matched symmetrical network to obtain the required attenuation factor $1/n$ is with a symmetrical Π-network as shown in Fig. 7.11(c). It is left as a problem to show, by following an analysis analogous to that above, that the required relation between G, R, and R_0 are

$$\frac{G}{G_0} = \frac{2(n-1)}{(n+1)} \quad \text{and} \quad \frac{R}{R_0} = \frac{(n^2-1)}{2n}, \tag{7.42}$$

as would be expected from the principle of duality.

If identical T- or Π-networks are cascaded, then the combination is also matched to the characteristic impedance (and could be reduced to a single T- or Π-network). For the case of m cascaded, purely resistive networks of the type just considered above, the overall attenuation of the voltage, or the current, would be n^m. A transmission line, such as a coaxial cable, can be represented as a 'ladder network' of cascaded T- or Π-networks (see Section 8.3).

In the decibel notation (see Section 4.2.1) the power gain of a two-port network is given by (see Fig. 7.12):

$$\begin{aligned} \text{power gain} &= 10 \log\{(v_2^2/R_{out})/(v_1^2/R_{in})\} \\ &= 20 \log(v_2/v_1) + 10 \log(R_{in}/R_{out}) \text{ dB.} \end{aligned} \tag{7.43}$$

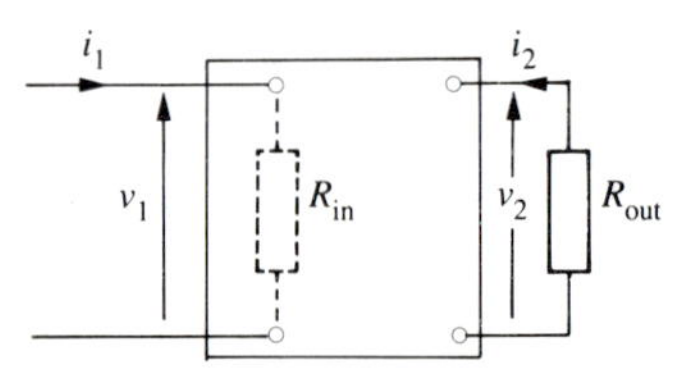

Fig. 7.12 A two-port network, showing the resistances for defining the power gain.

The **insertion loss** of a two-port network is the reciprocal of the power gain:

$$\text{insertion loss} = 10\log\left\{\frac{\text{input power}}{\text{output power}}\right\}\text{dB}$$

It is convenient in many practical situations to express a voltage, current or power gain in relation to some reference value of gain; then the term $10\log(R_{in}/R_{out})$ disappears. For example, in a situation where the gain of a two-port network is frequency-dependent,

$$\frac{\text{power gain at frequency } f}{\text{power gain at frequency } f_0} = \left[\frac{v_1^2 R_{out}}{R_{in} v_2^2}\right]_f \Bigg/ \left[\frac{v_1^2 R_{out}}{R_{in} v_2^2}\right]_{f_0}$$

$$= (v_1^2/v_2^2)_f/(v_1^2/v_2^2)_{f_0}.$$

If the power gain at frequency f is greater than that at f_0, then it would be said that the power (or voltage or current) gain at frequency f is N dB 'up' on that at f_0, where

$$N = 20\log\{(v_1/v_2)_f/(v_1/v_2)_{f_0}\} = 20\log\{(i_2/i_1)_f/(i_2/i_1)_{f_0}\}. \tag{7.44}$$

In practice, matrix parameters for a particular two-port circuit element may be given (e.g. the y-parameters in the case of a transistor for use at radio frequencies), or may be measured as indicated in Section 7.1, or may be calculated if the form of the network inside the 'black box' is known. In any case, if the network is to be connected in cascade as part of a multistage system it is necessary to be able to characterize the loaded network (see Fig. 7.13) by its input impedance, or admittance, and the equivalent Thévenin source or Norton source at the output port.

Example 7.5 Obtain expressions for the current gain i_2/i_s, the input admittance Y_{in} and the Norton equivalent source at the output port for the network of Fig. 7.13(b).

The equations defining the y-parameters are

$$i_1 = y_{11}v_1 + y_{12}v_2 \quad \text{and} \quad i_2 = y_{21}v_1 + y_{22}v_2.$$

In addition,

$$v_2 = -i_2/Y_L \quad \text{and} \quad i_1 = i_s - v_1 Y_s.$$

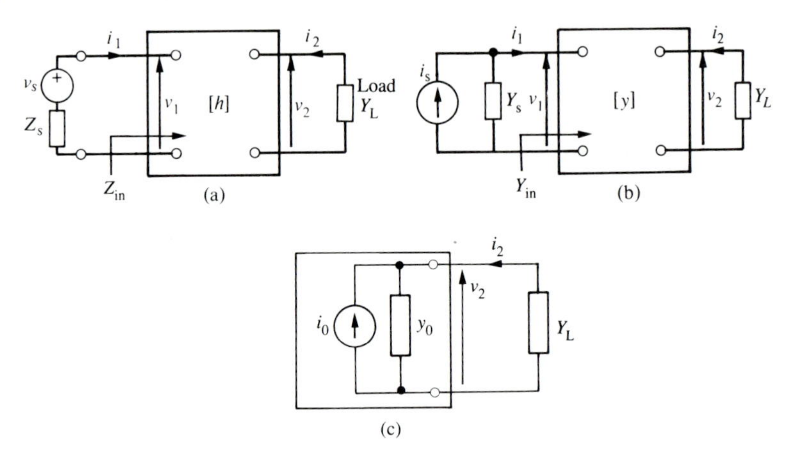

Fig. 7.13 A loaded two-port network: (a) voltage amplification; (b) current amplification; (c) Norton equivalent source at the output port.

Straightforward, but rather tedious, algebraic manipulation of these simultaneous equations yields the current gain:

$$\frac{i_2}{i_s} = \frac{-y_{22}/Y_s}{\dfrac{y_{12}y_{22}}{Y_L Y_s} - \left(1 + \dfrac{y_{22}}{Y_L}\right)\left(1 + \dfrac{y_{11}}{Y_s}\right)}. \tag{7.45}$$

Further algebra yields Y_{in} ($\equiv i_1/v_1$):

$$Y_{in} = y_{11} - \frac{y_{12}y_{21}}{Y_L\left(1 + \dfrac{y_{22}}{Y_L}\right)}. \tag{7.46}$$

Finally,

$$v_2 = \frac{-y_{21}i_s}{[y_{22}(y_{11} + Y_s) - y_{12}y_{21}]} + \frac{i_2(y_{11} + Y_s)}{[y_{22}(y_{11} + Y_s) - y_{12}y_{21}]}. \tag{7.47}$$

By comparing eqn (7.47) directly with eqn (2.1) or (2.2), it follows that the Norton equivalent source at the output port of the network is a current source i_0 of value

$$i_0 = \frac{-y_{21}i_s}{(y_{11} + Y_s)} \tag{7.48}$$

(the minus sign means that in Fig. 7.13(c) the current source is indicated in the 'wrong' sense), together with a shunt admittance

$$y_0 = \frac{y_{22}(y_{11} + Y_s) - y_{12}y_{21}}{(y_{11} + Y_s)}. \tag{7.49}$$

These equations may seem rather complicated, but they do illustrate the feedthrough interaction between the source and load admittances, and they are required in detailed analyses such as stability analyses of active circuits. However, in many situations, simplifying approximations can be made. For example, if $y_{11} \gg Y_s$ and $Y_L \gg y_{22}$ (i.e. good 'current coupling'; see Section 3.8.2), then

$$\frac{i_2}{i_s} \approx \frac{-y_{22}}{\dfrac{y_{12}y_{22}}{Y_L} - y_{11}}$$

and the short-circuit current gain ($Y_L = \infty$) is y_{22}/y_{21}.

The input admittance becomes

$$Y_{in} \approx y_{11} - \frac{y_{12}y_{21}}{Y_L},$$

and under short-circuited output conditions, $Y_{in} = y_{11}$.

For the Norton equivalent source the approximate relationships under these good current coupling conditions are:

$$i_0 \approx \frac{-y_{21}i_s}{y_{11}} \quad \text{and} \quad y_0 \approx \frac{y_{11}y_{22} - y_{12}y_{21}}{y_{11}}. \qquad \square$$

It is left as an exercise (Problem 7.12) to derive the analogous expressions in the case of the h-parameters for a two-port network.

7.3.2 Impedance matching

In telecommunication systems, the power levels at the receiving end of the system are often very low ($\sim 10^{-10}$ W, or even less, in free-space links), so it is imperative that there should be good power matching between the elements of the system, e.g. in the chain antenna → pre-amplifier → downlead → amplifier.

Two-element '⅂' or 'Γ' networks (see Fig. 7.14; usually referred to as 'L' networks) are one solution to the impedance matching problem, although with one particular disadvantage, as will be seen. Nevertheless, because of their simplicity they are quite widely employed.

Consider first the circuit shown in Fig. 7.14(a), where it is required to provide conjugate matching (see Section 3.5) between the source and the

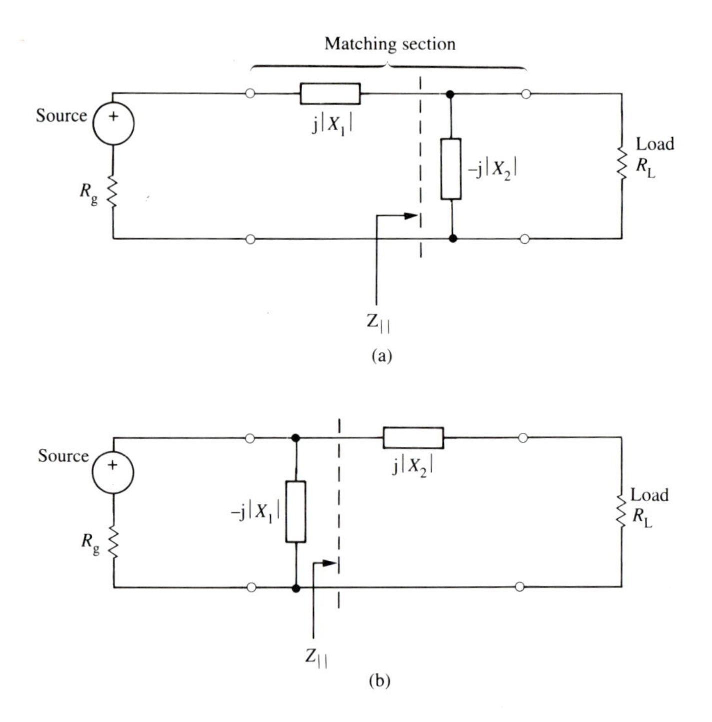

Fig. 7.14 (a) '⅂' and (b) '⌈' matching networks.

load. Since power losses are to be kept to a minimum, reactive elements are used in the matching section.

It turns out to be a matter of practical convenience as to whether the series element is made an inductance (L) and the shunt element a capacitance (C) or vice versa; in the present case the shunt element will be made a capacitance, i.e. a negative reactance.

For the network consisting of R_L in parallel with the shunt element of reactance $-j|X_2|$,

$$Z_{\|} = \frac{R_L(-j|X_2|)}{(R_L - j|X_2|)}$$

or

$$Z_{\|} = \frac{R_L|X_2|^2 - jR_L^2|X_2|}{(R_L^2 + |X_2|^2)}.$$

For conjugate matching it is required that

$$R_g = \frac{R_L|X_2|^2}{(R_L^2 + |X_2|^2)} \tag{7.50}$$

and

$$|X_1| = \frac{R_L^2|X_2|}{(R_L^2 + |X_2|^2)}. \tag{7.51}$$

From eqn (7.50),

$$\frac{1}{R_g} = \frac{R_L}{|X_2|^2} + \frac{1}{R_L}. \tag{7.52}$$

Two quality factors Q_g, Q_L are defined through

$$Q_g \equiv \frac{|X_1|}{R_g} \quad \left(\text{i.e. } Q_g = \frac{\omega_0 L}{R_g}\right) \quad \text{and} \quad Q_L \equiv \frac{R_L}{|X_2|} \quad (Q_L = \omega_0 R_L C),$$

where $\omega_0^2 \equiv (LC)^{-1}$. On substituting in eqn (7.52),

$$Q_L^2 = \frac{R_L}{R_g} - 1. \tag{7.53}$$

Notice that for eqn (7.53) to hold true it is necessary that $R_L > R_g$.
From eqn (7.51) it follows that

$$\frac{R_g}{|X_1|} = \frac{R_g}{|X_2|} + \frac{R_g|X_L|}{R_L^2}$$
$$= \frac{R_g R_L^2 + R_g|X_2|^2}{R_L^2|X_2|}.$$

Substituting for $R_g R_L^2$ from eqn (7.50) yields

$$\frac{R_g}{|X_1|} = \frac{|X_2|}{R_L}, \qquad \text{i.e. } Q_L = Q_g \equiv Q, \text{ say.}$$

So, if $R_L > R_g$, conjugate matching is realized at a frequency ω_0 (i.e. 'narrow band' matching) if

$$L = \frac{QR_g}{\omega_0} \quad \text{and} \quad C = \frac{Q}{\omega_0 R_L}.$$

(7.54) *Design equations*

$$\text{where } Q = \sqrt{\left(\frac{R_L}{R_g} - 1\right)}.$$

It is left as a problem to show that the design equations for the '⌈' network of Fig. 7.14(b) are

$$C = \frac{Q}{\omega_0 R_g} \qquad L = \frac{QR_L}{\omega_0}$$

Design equations

$$\text{where } Q = \sqrt{\left(\frac{R_g}{R_L} - 1\right)} \quad (\text{NB } R_g > R_L).$$

An advantage of choosing a capacitance as the shunt element is that stray shunt capacitance can be subsumed into the value required for C. On the other hand, if a d.c. path to ground is required for biasing purposes, then an inductor must be chosen as the shunt element.

A drawback to simple matching of elements in this form is that the operating Q-value of the network is likely to be low, since it is determined by the ratio R_L/R_g. Even for R_L/R_g as large as 10, $Q = 3$ only, and for $R_L/R_g = 2$, $Q = 1$. Such low values for Q mean that the matching network has poor selectivity.

It would be useful if the Q-value of a matching network could be assigned independently of the values of the souce and load impedances; this can be accomplished by using three-element networks (see Fig. 7.15(a)).

The Π-network can be resolved into two 'L'-sections and a virtual resistance (see Fig. 7.15(b)). 'Looking' from BB′ towards R_L, then if the matching criteria established above for an L-section are applied, the input impedance is purely resistive (R', say). If the Q-value of the network to the right of CC′ is denoted by Q_2, then, from eqn (7.53),

$$Q_2 = \sqrt{\left(\frac{R_L}{R'} - 1\right)}.$$

Similarly, if the Q-value of the L-section that matches R' to the source is denoted by Q_1, then

$$Q_1 = \sqrt{\left(\frac{R_g}{R'} - 1\right)}.$$

Note that R' must be smaller than both R_L and R_g.

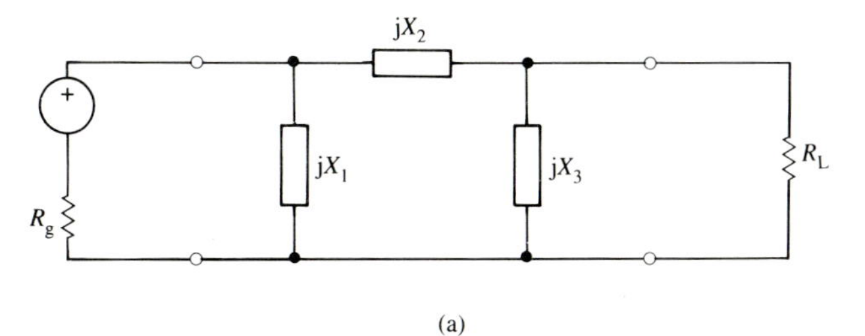

(a)

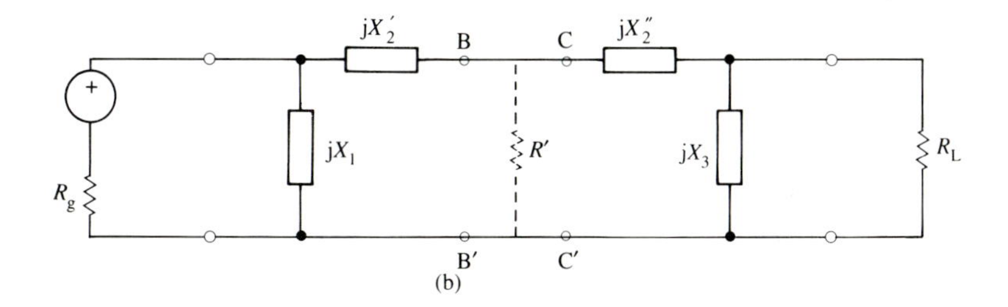

(b)

Fig. 7.15 (a) A three-element matching network (Π-form). (b) The Π-network resolved into two L-sections and a virtual resistance R'.

The effective operational Q-value for the matching circuit is determined by the section having the highest Q-value (i.e. by whichever is the greater of Q_1, Q_2), and can be assigned by a suitable choice of the value of R'. In this circumstance,

$$|X_1| = \frac{R_g}{Q_1}, \qquad |X'_2| = Q_1 R', \qquad |X''_2| = Q_2 R', \qquad |X_3| = \frac{R_L}{Q_2}.$$

If it is required that R' be greater than R_g, R_L, then a T-network must be used (see Fig. 7.16(a)); the analysis proceeds analogously to that for the Π-network, and it is left as a problem to show that

$$|X_1| = Q_1 R_g, \qquad |X'_2| = \frac{R'}{Q_1}, \qquad |X''_2| = \frac{R'}{Q_2}, \qquad |X_3| = Q_2 R_L, \tag{7.55}$$

where

$$Q_1 = \sqrt{\left(\frac{R'}{R_g} - 1\right)}, \qquad Q_2 = \sqrt{\left(\frac{R'}{R_L} - 1\right)}.$$

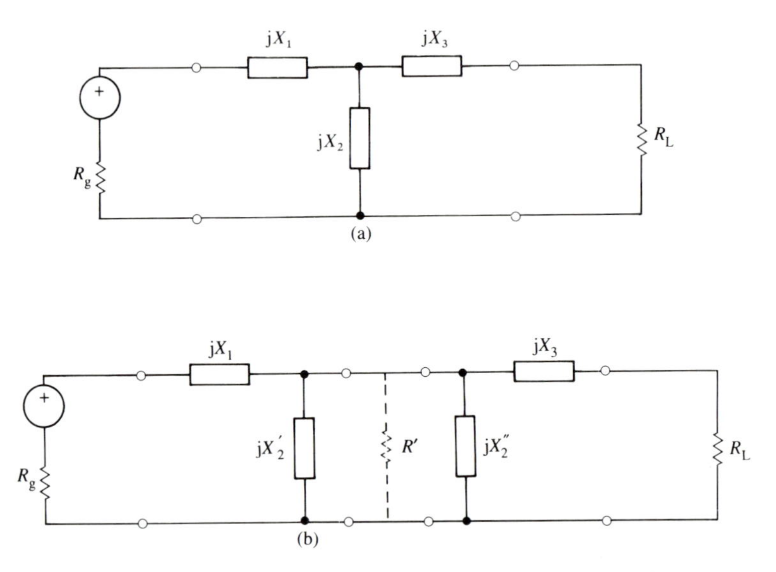

Fig. 7.16 (a) A three-element matching network (T-form). (b) The T-network resolved into two ⅂ sections and a virtual resistance R'.

Example 7.6 A signal souce at a frequency of 500 MHz and of output resistance 50 Ω is to be matched to a load of resistance 100 Ω. Determine the elements of a T-network to realise the matching condition if the operating Q-factor is to have a value of 15 (with shunt capacitors, series inductors).

Use design eqns (7.55). Since $R_L > R_g$, $Q_1 > Q_2$, and so $Q_1 = 15$. Hence $R' = 11.3$ kΩ and therefore $Q_2 = 10.58$. So

$$X_1: \quad L = 1.20\ \mu\text{H} \qquad X_2'': \quad C = 1.49\ \text{pF}$$

$$X_2': \quad C = 2.11\ \text{pf} \qquad X_3: \quad L = 1.68\ \mu\text{H}.$$

□

In practice the source and load impedances will be complex (e.g. transistor output and input impedances), but the analyses are essentially the same as above. The design equations for the various combinations of series and shunt reactive elements can easily be entered as listings to a computer program with R_g, R_L as variables and the operating Q-value as a parameter.

7.4 System functions

One of the very important features of the theory of electrical networks is that it is possible, generally speaking, to *synthesize* a network that will have a desired response to a specified excitation. Network synthesis, particuarly with regard to the design of filters (although all networks may be thought of as filters in a general sense), is an exceedingly important aspect of electronics in general and in telecommunications and radar in particular. Many computer-aided design packages are now available to assist the designer, but it is important that the underlying principles are appreciated. The principal aim of this section is to describe some of the basic features of system functions and the responses of networks. The general theory is rather formal, but it is first necessary to say something very briefly about the properties of the mathematical functions that represent the excitations and responses of networks before turning to illustrative practical examples.

7.4.1 Poles, zeros, and the time-domain response

Since the excitations and responses of interest are generally not simply sinusoidal, they are expressed as functions of the complex frequency s that was introduced in Section 6.7; hence system functions are expressed as functions of s also. If an excitation is written as an s-domain function $\mathscr{E}(s)$ and the response of a system as $\mathscr{R}(s)$, then the system function $\mathscr{H}(s)$ is defined through

$$\mathscr{R}(s) = \mathscr{H}(s) \times \mathscr{E}(s) \tag{7.56}$$

Now, although the term has not been used explicitly, many system functions have been encountered earlier in the text. For example, consider the networks shown in Fig. 7.17. For the network of Fig. 7.17(a) it follows, using the notation displayed in Table 6.2, that

$$V(s) = I(s)\frac{R/sC}{\left(R + \dfrac{1}{sC}\right)},$$

or, the system function $\mathscr{H}(s)$ is given by

$$\mathscr{H}(s) \equiv \frac{V(s)}{I(s)} = \frac{1}{C\left(s + \dfrac{1}{CR}\right)}. \tag{7.57a}$$

It is left as an exercise to show that for the networks (b), (c) and (d) of

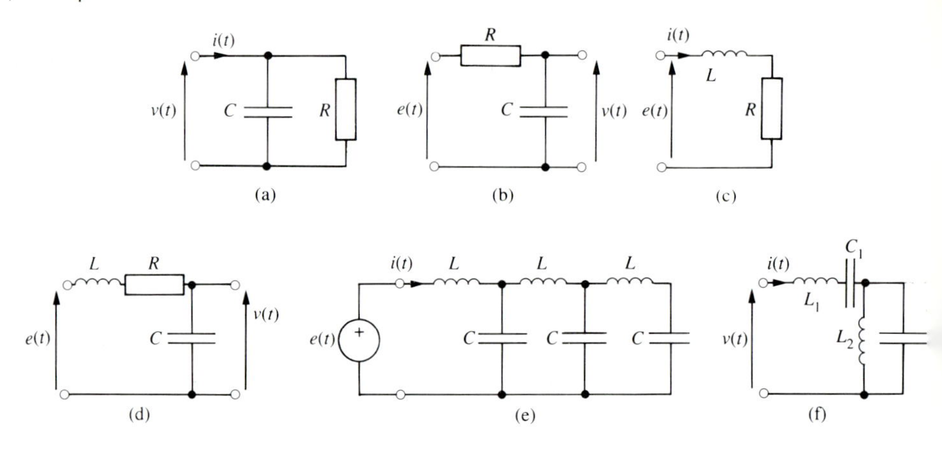

Fig. 7.17 Some simple networks, showing the excitations and responses defined in eqns (7.57).

Fig. 7.17, the system functions are respectively

(7.57b)
$$\mathscr{H}(s) \equiv \frac{V(s)}{E(s)} = \frac{1}{CR\left(s + \dfrac{1}{CR}\right)}$$

(7.57c)
$$\mathscr{H}(s) \equiv \frac{I(s)}{E(s)} = \frac{1}{L\left(s + \dfrac{R}{L}\right)}$$

(7.57d)
$$\mathscr{H}(s) \equiv \frac{V(s)}{E(s)} = \frac{1}{LC\left(s^2 + \dfrac{sR}{L} + \dfrac{1}{LC}\right)}.$$

For the one-port network of Fig. 7.17(e) it can be shown, rather tediously, that

(7.57e)
$$\frac{E(s)}{I(s)} = -\frac{s^6(LC)^3 - 4s^4(LC)^2 + 2s^2LC + 2}{sC\{s^4(LC)^2 - 4s^2LC + 3\}},$$

and, for the network of Fig. 7.17(f),

(7.57f)
$$\mathscr{H}(s) \equiv \frac{V(s)}{I(s)} = \frac{[L_1C_2s^4 + (C_2/C_1 + L_1/L_2 + 1)s^2 + 1/L_2C_1]}{(C_2s^3 + s/L_2)}.$$

The reader will have noticed that system functions can be defined in a number of different ways, depending on whether the specified responses and excitations are voltages or currents.

If the excitation is a current source and the response is a voltage, with both the excitation and the response measured between the same pair of terminals, then the system function is a 'driving point' impedance (e.g. Fig. 7.17(a)). If the current and voltage are interchanged, then the system function is a 'driving point' admittance (e.g. Fig. 7.17(c)). Since it turns out that the mathematical descriptions of dual networks are identical, with voltages and currents interchanged, it is common practice to use the embracing term driving-point 'immittance'.

A system function in which the excitation and the response are measured at different ports is called a 'transmittance'; it could be a transfer admittance (e.g. Fig. 7.17(e)) or a transfer impedance (i.e. output voltage/input current). Additionally, a system function could be a 'voltage ratio transfer function' or a 'current ratio transfer function'.

Now the reason for giving the rather cumbersome expressions of eqns (7.57e) and (7.57f) was to illustrate the point, which can be proved quite generally, that the system functions for physically realizable networks are rational functions that can be written as the ratio of two polynomials in s. So $\mathscr{H}(s)$ can be written generally as

$$\mathscr{H}(s) = \frac{a_p s^p + a_{p-1} s^{p-1} + \cdots + a_1 s + a_0}{b_q s^q + b_{q-1} s^{q-1} + \cdots + b_1 s + b_0}$$

or

$$\mathscr{H}(s) = \frac{N_H(s)}{b_q s^q + b_{q-1} s^{q-1} + \cdots + b_1 s + b_0},$$

where $N_H(s)$ denotes the numerator polynomial. Now such polynomials can be factorized; in particular, the polynomial in the denominator of the last equation can be written as the product of its factors:

$$\mathscr{H}(s) = \frac{N_H(s)}{(s + s_{H_0})(s + s_{H_1})(s + s_{H_2})\cdots(s + s_{H_q})}. \tag{7.58}$$

The quantities $-s_{H_0}, -s_{H_1}, \ldots$ are called the **poles** of $\mathscr{H}(s)$, since $\mathscr{H}(s) \to \infty$ for $s \to -s_{H_0}, -s_{H_1}, \ldots$

The values of s that make the numerator equal to zero are called the **zeros** of $\mathscr{H}(s)$.

A quotient such as the right-hand side of eqn (7.58) can be expanded by the method of partial fractions (see the Appendix), so that $\mathscr{H}(s)$ can be expressed as a *sum* of terms:

$$\mathscr{H}(s) = \frac{K_{H_0}}{(s + s_{H_0})} + \frac{K_{H_1}}{(s + s_{H_1})} + \cdots,$$

where K_{H_0}, $K_{H_1}, \ldots$ are constants. Further, the Laplace transformations of excitation functions are also ratios of polynomials in s (see Table 6.3 or more extensive tables), and these can also be written in terms of their roots:

$$\mathscr{E}(s) = \frac{N_E(s)}{(s + s_{E_1})(s + s_{E_1})(s + s_{E_3})\cdots}.$$

If $\mathscr{E}(s)$ is expanded into a sum of terms by the method of partial fractions, then

$$\begin{aligned}\mathscr{R}(s) = \mathscr{H}(s) \times \mathscr{E}(s) = {} & \underbrace{\frac{K_{H_0}}{(s + s_{H_0})} + \frac{K_{H_1}}{(s+s_{H_1})}}_{\text{(natural response)}} \\ & + \cdots + \underbrace{\frac{K_{E_1}}{(s + s_{E_1})} + \frac{K_{E_2}}{(s + s_{E_2})}}_{\text{(forced response)}} + \cdots \end{aligned} \tag{7.59}$$

The sum of the inverse Laplace transformations of each of the terms on the right-hand side of eqn (7.59) yields the time-domain response $r(t)$ of the network or system, i.e. the response contains a term for each pole of the system function and for each pole of the excitation. As the two examples given below will indicate, the terms arising from the system function represent the 'natural' response of the system and the terms from the excitation represent the 'forced' response. The free oscillations of a damped system, as discussed in Section 4.1, are an example of the natural response of a system (a second-order system in that instance).

Although the foregoing general discussion has been kept very brief, it has been rather formal, and it is more interesting if the general points are illustrated by examining some relatively simple practical situations.

Consider the circuit shown in Fig. 7.18(a), where the network is excited by a step-function current source of amplitude A, i.e. $i(t) = A \cdot u(t)$ and $I(s) = A/s$. The indicated response (the voltage $v(t)$) is, in the s-domain, given by $V(s) = I(s)Z(s)$, where

$$Z(s) = \frac{R\left(R + \dfrac{1}{sC}\right)}{\left(2R + \dfrac{1}{sC}\right)}.$$

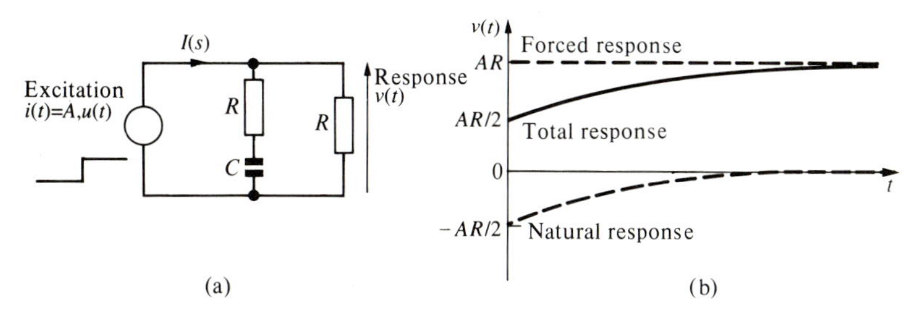

Fig. 7.18 The response (b) to the step-function excitation of the network shown in (a).

Thus

$$V(s) = \frac{A}{s} \cdot \frac{R\left(s + \dfrac{1}{CR}\right)}{2\left(s + \dfrac{1}{2CR}\right)}; \tag{7.60}$$

i.e. the response is given by the product of the excitation (A/s) and the system function (in this case a transfer impedance). Notice that the system function has poles at $s = -(1/2CR)$ and $s = 0$, and a zero at $s = -(1/CR)$.

The partial-fractions technique (see the Appendix) is now used to convert the right-hand side of eqn (7.60) into a sum of functions that are given in tables of Laplace transforms:

$$\frac{s + \dfrac{1}{CR}}{s\left(s + \dfrac{1}{2CR}\right)} = \frac{B}{s} + \frac{D}{\left(s + \dfrac{1}{2CR}\right)}.$$

On bringing the right-hand side to the common denominator, then

$$s + \frac{1}{CR} = Bs + \frac{B}{2CR} + Ds.$$

Equating terms in s^1: $1 = B + D$, hence $D = 1 - B$.

Equating terms in s^0: $\dfrac{1}{CR} = \dfrac{B}{2CR}$, hence $B = 2$ and so $D = -1$.

So

$$V(s) = \frac{AR}{2}\left\{\frac{2}{s} - \frac{1}{\left(s + \dfrac{1}{2CR}\right)}\right\}.$$

On transforming back to the time-domain (use Table 6.3 to obtain the inverse Laplace transformations):

$$v(t) = AR\,u(t) - \frac{AR}{2}\,\mathrm{e}^{-t/2CR} \tag{7.61}$$

step (forced response) — damped exponential (natural response)

This analysis, although elementary, is a reminder (refer to Section 6.7) of the general procedure to be followed in all problems where Laplace transforms are used to transform functions to the s-domain as an aid to eventually obtaining the time-domain response of a system.

Eqn (7.61) demonstrates the general feature that the time-domain response is the sum of a term arising from the excitation (the forced response) and a term arising from the system function, which is the natural response of the system alone.

As another example worth analysing in detail, because incidentally of the information it reveals about resonant systems in general, consider the parallel LCR circuit illustrated in Fig. 7.19. What is the response $v(t)$ if the excitation is a square-wave signal? In practice the signal source will usually be a voltage generator of relatively low output impedance (e.g. 60 Ω), which must be decoupled from the LCR network by a large-value resistor R_g ($\gg Q^2R\ \Omega$) in order not to damp out the resonance (see Section 4.3). Further, the square waveform can be decomposed into a positive-going step followed by a delayed negative-going step of equal and opposite amplitude. So, in determining the response of the network, only a step-function excitation need be considered.

In the circuit of Fig. 7.19 it is assumed that the input impedance (Z_{in}) of the measurement circuit (buffer amplifier/CRO) is so much greater than that of the LCR network at resonance (Q^2R) that it can be discounted as a significant source of damping of the resonant network; this is easily accomplished if an FET amplifier is used for which $Z_{in} \approx 10^8\ \Omega$. If the above approximation cannot be satisfied, then the input impedance of the measurement circuit is effectively in parallel with R_g and can be allowed for.

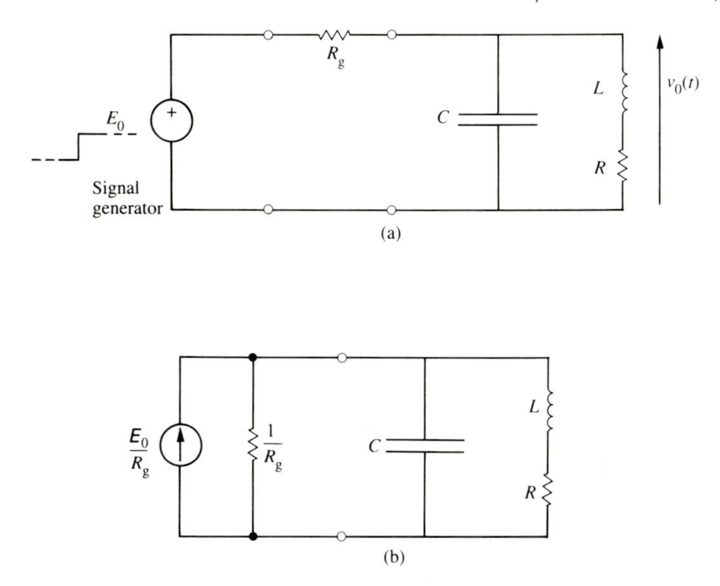

Fig. 7.19 An *LCR* network shock-excited (a) by a voltage source and (b) by the equivalent current source.

Referring to Fig. 7.19,

$$\frac{V_0(s)}{E_0(s)} = \frac{\dfrac{(R + sL)/sC}{\left(R + sL + \dfrac{1}{sC}\right)}}{\left\{R_g + \dfrac{(R + sL)/sC}{\left(R + sL + \dfrac{1}{sC}\right)}\right\}}. \tag{7.62}$$

Assuming, as outlined above, that only a step-function excitation need be considered, then, putting $E_0(s) = A/s$, say, it follows that

$$V(s) = \frac{A}{CR_g} \cdot \frac{(s + R/L)}{s\left\{s^2 + s\left(\dfrac{R}{L} + \dfrac{1}{CR_g}\right) + \dfrac{(R + R_g)}{LCR_g}\right\}}. \tag{7.63}$$

If the quality factor (see Sections 4.1, 4.2) of the unloaded *LCR* network (i.e. for an *isolated LCR* network; $R_g \to \infty$) is denoted by Q_u, where $Q_u = \omega_0 L/R$ and ω_0 is the (natural) resonance frequency, then eqn (7.63) becomes

$$V(s) = \frac{A}{CR_g} \cdot \frac{(s + \omega_0/Q_u)}{s\left\{s^2 + s\omega_0\left(\frac{1}{Q_u} + \frac{1}{\omega_0 CR_g}\right) + \omega_0^2\left(1 + \frac{R}{R_g}\right)\right\}}. \tag{7.64}$$

Equation (7.64) is in the form

$$V(s) = \frac{A}{CR_g} \cdot \frac{(s + \omega_0/Q_u)}{s\left(s^2 + \frac{s\omega_0}{Q_L} + \omega_r^2\right)}, \tag{7.65}$$

where

$$\omega_r^2 \equiv \omega_0^2(1 + R/R_g) \tag{7.66}$$

and

$$\frac{1}{Q_L} \equiv \frac{1}{Q_u} + \frac{1}{\omega_0 CR_g}. \tag{7.67}$$

The frequency ω_r is the resonance frequency of the loaded network (i.e. loaded by R_g), whereas ω_0 is the natural frequency of the unloaded network.

The factor Q_L is the **loaded Q-factor** and it is common practice to write

$$\frac{1}{Q_L} = \frac{1}{Q_u}(1 + \beta), \tag{7.68}$$

where

$$\beta \equiv \frac{Q_u}{\omega_0 CR_g} = \frac{Q_u}{R_g}\sqrt{\left(\frac{L}{C}\right)} \tag{7.69}$$

is called the **coupling factor** (coupling to the 'outside world' via R_g).

The denominator of eqn (7.65) can be factorized:

$$V(s) = \frac{A}{CR_g} \frac{(s + \omega_0/Q_u)}{s(s + \alpha_1)(s + \alpha_2)}, \tag{7.70}$$

where

$$\alpha_1 + \alpha_2 = \frac{\omega_0}{Q_L} \quad \text{and} \quad \alpha_1\alpha_2 = \omega_r^2. \tag{7.71}$$

Note that the poles of the system function (in this case the voltage transfer function $V(s)/(A/s)$) are $s = -\alpha_1$, $s = -\alpha_2$.

From eqns (7.71),

$$\begin{matrix}\alpha_1\\\alpha_2\end{matrix} = \frac{\omega_0}{2Q_L} \pm \frac{\omega_0}{2Q_L}\sqrt{\left(1 - \frac{4Q_L^2\omega_r^2}{\omega_0^2}\right)}. \tag{7.72}$$

Even for Q-values as low as 5 (and Q will be much larger than 5 in most practical cases), $4Q_L^2\omega_r^2/\omega_0^2 \gg 1$, since $\omega_r \approx \omega_0$.

So for the majority of practical situations (with an error of 0.5 per cent at most),

$$\begin{matrix}\alpha_1\\\alpha_2\end{matrix} = \frac{\omega_0}{2Q_L} \pm j\omega_r. \tag{7.73}$$

(NB in this situation the poles are unequal in value and, in fact, form a *complex conjugate pair.*) So eqn (7.70) can be written

$$V(s)\cdot\frac{CR_g}{A} = \frac{(s + \omega_0/Q_u)}{s(s + \alpha_1)(s + \alpha_2)},$$

and by using the partial-fractions technique,

$$\frac{(s + \omega_0/Q_u)}{s(s + \alpha_1)(s + \alpha_2)} = \frac{B}{s} + \frac{D}{(s + \alpha_1)} + \frac{F}{(s + \alpha_2)}. \tag{7.74}$$

This equation gives

$$B = \frac{\omega_0}{\alpha_1\alpha_2 Q_u}, \qquad D = \frac{(1 - B\alpha_2)}{(\alpha_2 - \alpha_1)}, \qquad F = -\frac{(1 - B\alpha_1)}{(\alpha_2 - \alpha_1)}.$$

After substituting for D and F in eqn (7.74) and performing the inverse Laplace transform (tabulated), it follows that

$$v(t)\frac{CR_g}{A} = B\cdot u(t) + \frac{(1 - B\alpha_2)}{(\alpha_2 - \alpha_1)}\cdot e^{-\alpha_1 t} - \frac{(1 - B\alpha_1)}{(\alpha_2 - \alpha_1)}\cdot e^{-\alpha_2 t}. \tag{7.75}$$

On substituting for B, α_1, α_2, it follows that

$$v(t) = \frac{A}{CR_g\omega_r}\cdot\left\{\frac{\omega_0}{Q_u\omega_r}u(t) + \left(1 - \frac{Q_L}{Q_u\left(1 + \frac{4Q_L^2\omega_r^2}{\omega_0^2}\right)}\right)e^{-\omega_0 t/2Q_L}\cdot\sin\omega_r t\right.$$
$$\left. + \frac{2\omega_r Q_L}{\omega_0 Q_u\left(1 + \frac{4Q_L^2\omega_r^2}{\omega_0^2}\right)}e^{-\omega_0 t/2Q_L}\cdot\cos\omega_r t\right\}. \tag{7.76}$$

The first term in this equation constitutes the forced response (a step function) and the second and third terms constitute the natural response.

As was noted earlier, $(4Q_L^2\omega_r^2/\omega_0^2) \gg 1$, usually, and so eqn (7.76) becomes approximately

$$v(t) = \frac{A}{CR_g\omega_r} \cdot \left\{ \frac{\omega_0}{Q_u\omega_r} u(t) + \left(1 - \frac{\omega_0^2}{4Q_uQ_L\omega_r^2}\right) e^{-\omega_0 t/2Q_L} \cdot \sin \omega_r t + \frac{\omega_0}{2Q_uQ_L\omega_r} e^{-\omega_0 t/2Q_L} \cdot \cos \omega_r t \right\}. \tag{7.77}$$

Again, since $(Q_uQ_L)^{-1} \approx 0.04$, or very much less, in most practical situations eqn (7.77) can be approximated by

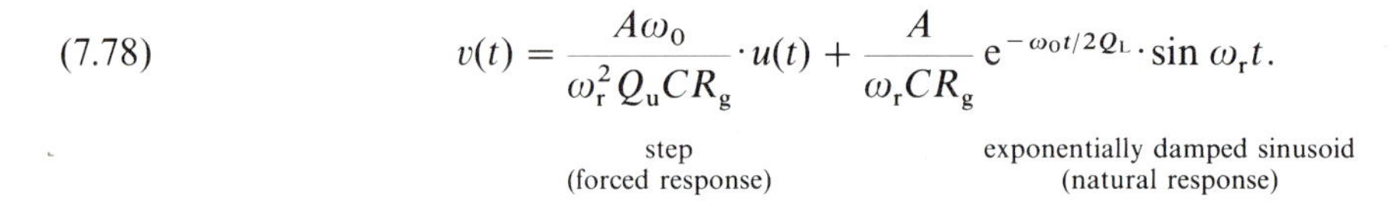

$$v(t) = \underbrace{\frac{A\omega_0}{\omega_r^2 Q_u CR_g} \cdot u(t)}_{\substack{\text{step} \\ \text{(forced response)}}} + \underbrace{\frac{A}{\omega_r CR_g} e^{-\omega_0 t/2Q_L} \cdot \sin \omega_r t}_{\substack{\text{exponentially damped sinusoid} \\ \text{(natural response)}}}. \tag{7.78}$$

The situation in which the poles of the system function are complex (see eqn (7.73)) and which led to this oscillatory solution is called the **underdamped case**.

To recapitulate (see Section 4.1), three situations are distinguished:

(i) Overdamped: In eqn (7.72), $4Q_L^2\omega_r^2/\omega_0^2 < 1$, so that α_1, α_2 are real. This means that the terms $e^{-\alpha_1 t}$, $e^{-\alpha_2 t}$ in eqn (7.75) are non-oscillatory.

(ii) Critically damped: $4Q_L^2\omega_r^2/\omega_0^2 = 1$ (i.e. $Q_L \approx \frac{1}{2}$), which means that α_1, α_2 are real and equal to each other, so the response is again non-oscillatory.

(iii) Underdamped: $4Q_L^2\omega_r^2/\omega_0^2 > 1$, so that α_1, α_2 are complex and the terms $e^{-\alpha_1 t}$, $e^{-\alpha_2 t}$ are oscillatory. This is the case treated in detail above.

It is left as an exercise to obtain the time-domain responses in cases (i) and (ii).

The critically damped condition gives the fastest approach to equilibrium and is often exploited in mechanical and electromechanical systems such as servo systems and moving-coil meters. In such systems the Q-value is rather low (< 1) and it is common practice to use the **damping factor** ζ $(\equiv 1/2Q)$.

This detailed analysis of the response of an *LCR* network has revealed the great power of the technique of exploiting Laplace transforms; with experience, the steps of the process can become purely routine. The precise

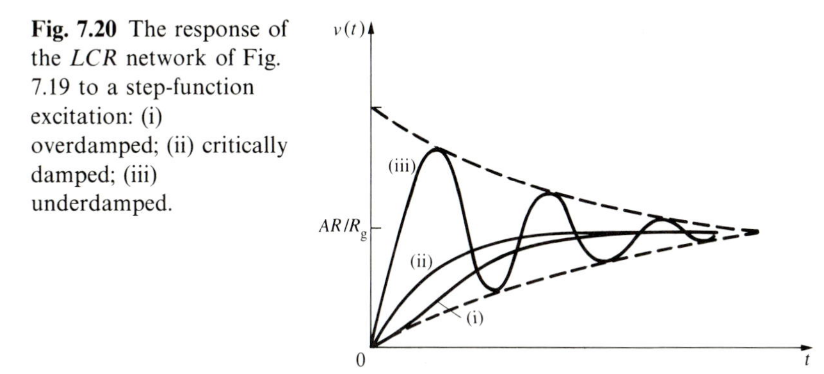

Fig. 7.20 The response of the *LCR* network of Fig. 7.19 to a step-function excitation: (i) overdamped; (ii) critically damped; (iii) underdamped.

response of a system from $t = 0$ to infinity can be obtained, if necessary, although in particular cases simplifying approximations can make for significant reductions in the length of the analysis; for instance, exponential terms can be replaced by their series expansions and only the significant terms retained.

Of course the accuracy of a calculated response depends on the accuracy of the model used to represent the system in question and on the accuracy of the mathematical representation of the excitation. For example, no circuit model can completely account for all the 'stray' elements (e.g. stray and parasitic capacitances), and also it is not possible, in practice, to obtain mathematically 'clean' steps, square waves, ramps, etc. as excitation functions.

Some of the quantities that are used to characterize the transient response of first- and second-order systems are shown in Fig. 7.21. In the case of the first-order response it is useful to note that:

Time to reach 95% $= 3\tau_c$ closely.
Time to reach 99% $= 5\tau_c$ closely.
Time to reach 99.9% $= 7\tau_c$ closely.

For a second-order system,

$$\omega_r \tau_p = \frac{\pi}{\sqrt{(1 - 1/4Q^2)}}.$$

In the case of an amplifier of bandwith B, it is important and useful to note that the product $(B \cdot \tau_r) \approx 1$ (see 'reciprocal spreading', Section 6.6). A commonly encountered practical situation where this constraint is significant is in the observation of short rise-time signals using the CRO.

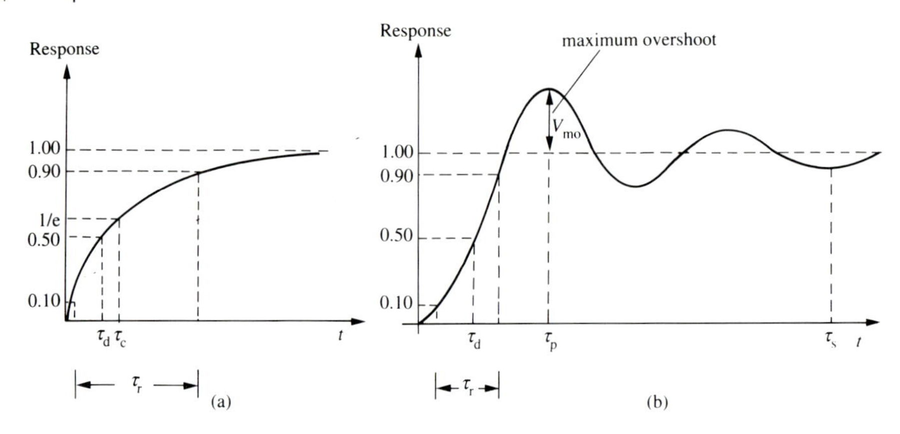

Fig. 7.21 The transient response of (a) a first-order system such as a *CR* network, and (b) a second-order system such as an *LCR* network. Symbols: τ_r, 'rise time'; τ_c, 'time constant'; τ_d, 'delay time', τ_s, 'settling time'.

In order to display faithfully the shape of a pulse, the rise time of the CRO amplifier must be much less than the rise time of the pulse. Many standard laboratory oscilloscopes have a bandwidth of 20 MHz, which limits the observable rise times to ~ 10 ns, roughly, whereas a 100 MHz bandwidth CRO can cope with rise times of the order of a nonosecond.

In a wide range of practical situations the transient natural response term in eqn (7.78) decays so quickly that it is not noticed; e.g. for f_0 $(=\omega_0/2\pi) = 10$ kHz and $Q = 10$, then $Rt/2L$ $(=\omega_0 t/2Q) = 1$ for $t \approx 0.3$ ms. For $f_0 = 1$ MHz and $Q = 100$, then $Rt/2L = 1$ for $t \approx 30$ μs. However, in low-Q systems, particularly in electromechanical situations, the transient term may be very noticeable.

The properties of a system function are represented in a compact way by its pole–zero diagram in the s-domain. As a matter of fact, a system function for a realizable network is specified completely by its poles and zeros (apart from a scale factor). For example, the poles and zeros of the functions of eqns (7.60) and (7.62) are displayed in Fig. 7.22(a) and (b) respectively. Two very important points are demonstrated in these two diagrams. First, complex poles always occur in pairs, one pole being the complex conjugate of the other. Second, for passive networks, the poles lie in the left-hand half-plane; this reflects the fact that, in terms like e^{st}, the negative real part of s means that the response is decaying in magnitude with time. A positive real part for S indicates a response that is growing in magnitude, which is physically impossible in a passive network. For active systems the existence of poles in the right-hand half-plane is an indication of instability. Notice also that the further a pole is away from the 'imaginary' axis, the more rapidly damped is the

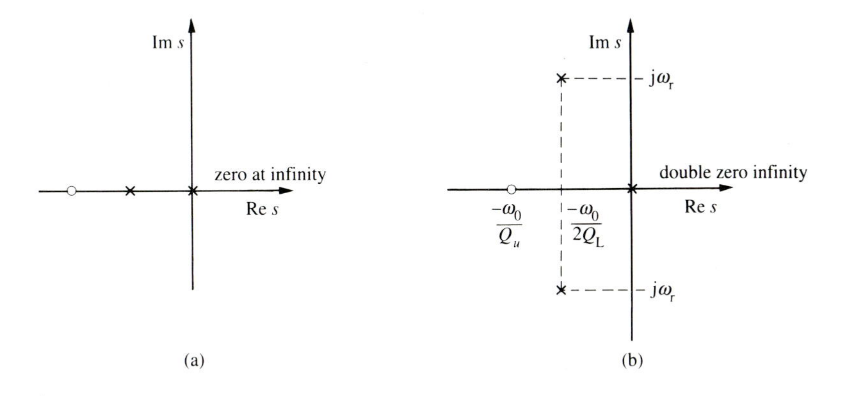

Fig. 7.22 The poles (×) and zeros (○) of (a) the function $V(s)$ of eqn (7.60) and (b) the function $V(s)$ of eqn (7.70).

component of the natural response attributed to that pole. So the behaviour of a multi-pole system is often dominated by the single pole that is closest to the 'imaginary' axis, i.e. the **dominant pole**.

The relation between a system function and the response of the system to an excitation in the form of a unit impulse $\delta(t)$ is of considerable practical interest (see Section 6.6 for the definition of the unit impulse function). Since $\mathscr{R}(s) = \mathscr{E}(s)\mathscr{H}(s)$, and $\mathscr{E}(s) = \mathscr{L}[\delta(t)] = 1$ in this case, it follows that the time-domain **impulse response** $r(t)$ is equal to $\mathscr{L}^{-1}[\mathscr{H}(s)]$. Hence in practice the system function for a real system can be determined from the measurement of its impulse response. Of course a true mathematical impulse cannot be generated but a 'sharp enough' pulse will suffice in practice.

7.4.2 Impedance and frequency normalization

A particular form of network (e.g. a simple CR coupling network) may be utilized in a variety of systems operating in different regions of the spectrum of 'engineering' frequencies; for example, instrumentation systems operating at a fraction of a hertz at one extreme, and telecommunication systems operating at tens of gigahertz at the other extreme. Obviously the values of capacitance and inductance used in the network will depend on the range of operating frequencies of the system in question. Also, the levels of impedance that are required in practice vary over many orders of magnitude. Hence, for the purpose of tabulating the properties of commonly used networks through their system functions, compactness is achieved by 'normalizing' the impedance level to 1 Ω and the (angular) frequency to 1 radian per second.

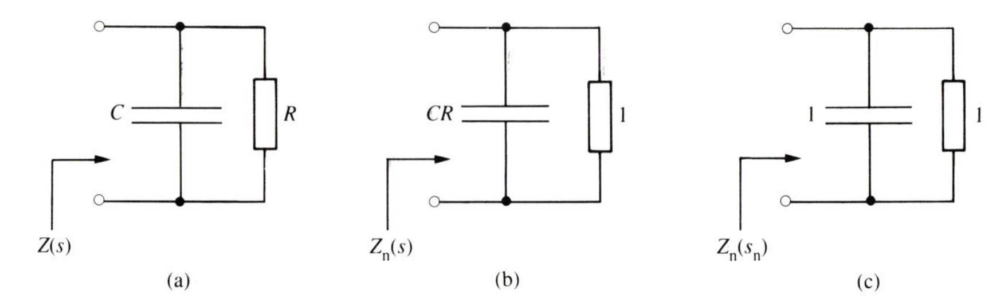

Fig. 7.23 An illustration of the normalization of impedance and frequency for a network.

A simple *CR* network can be used to illustrate the procedures involved in normalizing a network; see Fig. 7.23. For this network, the driving point impedance is given by

$$Z(s) = \frac{R \cdot \dfrac{1}{sC}}{R + \dfrac{1}{sC}} \qquad \text{or} \qquad Z(s) = \frac{R}{sCR + 1}.$$

Notice that this is an example of a 'first order' system function, since the complex frequency *s* occurs only to the first power. In a 'second order' system function, *s* appears to the second power, e.g. in resonant *LCR* networks (see Section 7.4.1).

Dividing through by *R*, the value of the resistor in the network is reduced (normalized) to 1 Ω (see Fig. 7.23(b) and notice that the value of the capacitor has been increased by the factor *R* in order to reduce its reactance by the same factor *R*):

$$Z_n(s) = \frac{1}{sCR + 1}.$$

Frequency is normalized with reference to a 'critical' frequency: the '*corner*' frequency $(CR)^{-1}$ in the case of a first-order network such as this (see Bode diagrams in the following section) or the resonance (or 'mid-band' frequency) in the case of second-order system functions.

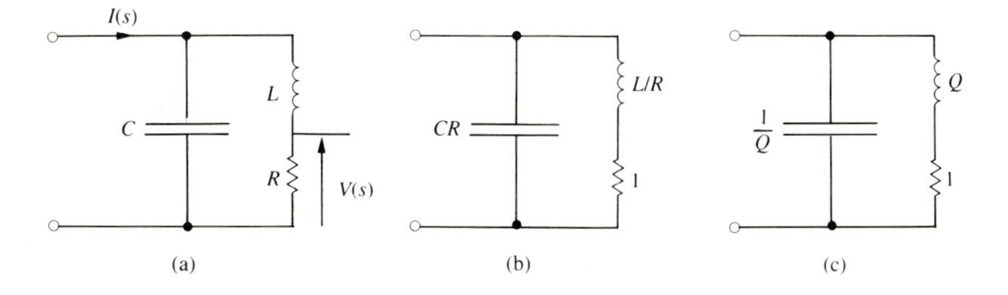

Fig. 7.24

So in this case the normalized frequency s_n is defined through $s_n \equiv s/(1/CR)$ or $s_n = sCR$:

$$Z_n(s_n) = \frac{1}{s_n + 1},$$

and the normalized network takes the form shown in Fig. 7.23(c). This expression is in the general form:

$$Z(s) = \frac{1}{(s + 1)}. \tag{7.79}$$

If a network with the properties of the normalized network of Fig. 7.23(c) should happen to be required in a particular practical situation, then rules for 'denormalization' have to be followed in order to transform the normalized network to the 'real' world. These rules will be given shortly, but first consider the parallel LCR (resonant) network shown in Fig. 7.24, discussed in the previous section.

The system function of interest is the 'transimpedance' $V(s)/I(s) \equiv Z(s)$. Now

$$V(s) = R \cdot I(s) \cdot \frac{\dfrac{1}{(R + sL)}}{\dfrac{1}{(R + sL)} + sC},$$

and so

$$Z(s) = \frac{\omega_0^2 R}{s^2 + \dfrac{sR}{L} + \omega_0^2},$$

where the resonance frequency is given by $\omega_0^2 \equiv 1/LC$, and $Q \equiv \omega_0 L/R = 1/\omega_0 CR$.

Normalization to an impedance level of 1 Ω is again accomplished by dividing the expression for $Z(s)$ by R, yielding

$$Z_n(s) = \frac{1}{\left(\dfrac{s}{\omega_0}\right)^2 + \left(\dfrac{s}{\omega_0}\right)Q + 1} \quad \text{(see Fig. 7.24(b))}.$$

If the normalized frequency is again defined in relation to a critical frequency (in this case the resonance frequency, or 'mid-band' frequency, ω_0) by $s_n \equiv s/\omega_0$, then

$$Z_n(s_n) = \frac{1}{\left(s_n^2 + \dfrac{s_n}{Q} + 1\right)} \quad \text{(see Fig. 7.24(c))},$$

which is of the general form

$$Z(s) = \frac{1}{\left(s^2 + \dfrac{s}{Q} + 1\right)}. \tag{7.80}$$

The expression on the right-hand side of eqn (7.80) is often known as the '*universal resonance curve*'.

Example 7.6 Normalize the driving-point impedance function for the network shown in Fig. 7.25.

$$Z(s) = R + \frac{R \cdot \dfrac{1}{sC}}{\left(R + \dfrac{1}{sC}\right)} = \frac{R\left(R + \dfrac{2}{sC}\right)}{\left(R + \dfrac{1}{sC}\right)}.$$

Normalizing with respect to resistance (i.e. dividing through by R) gives

$$Z_n(s) = \frac{R + \dfrac{2}{sC}}{R + \dfrac{1}{sC}} \quad \text{(see Fig. 7.25 (b))}$$

Defining the normalized frequency s_n through $s_n \equiv sCR$ leads to

$$Z_n(s_n) = \frac{s_n + 2}{s_n + 1}$$

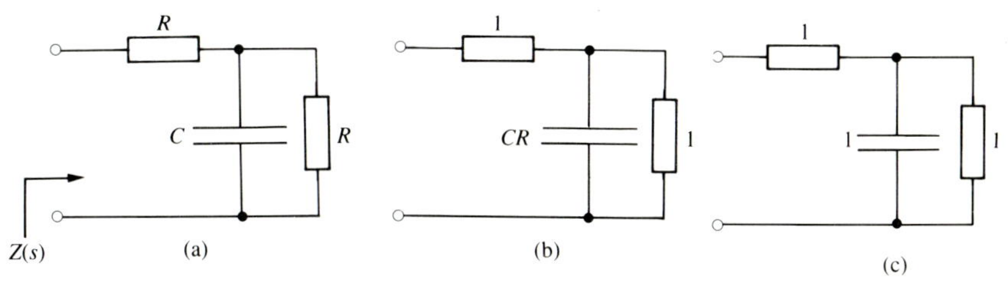

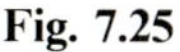
Fig. 7.25 □

From the preceding theory, and Example 7.6, the rules for 'denormalizing' the component values of a normalized network have emerged:

To increase the impedance level of a network by a factor η and to increase the critical frequency by a factor γ, then:

each resistor value should be multiplied by η
each inductance value should be multiplied by η/γ
each capacitance value should be multiplied by $1/\eta\gamma$.

Example 7.7 A normalized low-pass filter network is shown in Fig. 7.26.

It is, in fact, a second-order 'Butterworth' filter with a Q-value of $1/\sqrt{2}$, for which the normalized voltage transfer function is

$$\mathscr{H}(s) \equiv \frac{V(s)}{E(s)} = \left(s^2 + \frac{s}{Q} + 1\right)^{-1}.$$

Determine the component values for a filter with a critical frequency (i.e. a cut-off frequency in this case) of $10^5/2\pi$ Hz and a terminating impedance of 10 kΩ.

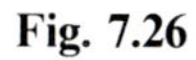
Fig. 7.26

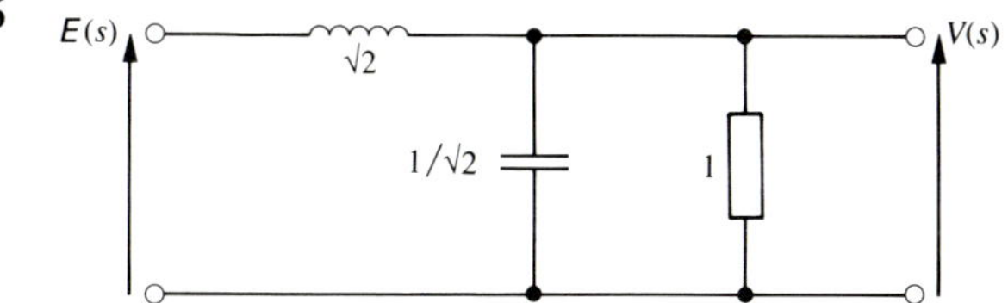

Note that in this example ω ($=2\pi \times$ frequency) $= 10^5$ rad s^{-1}.

resistor: $1 \rightarrow 10^4 = 10\ \text{k}\Omega$

inductor: $\sqrt{2} \rightarrow \sqrt{2} \times 10^4/10^5 = 0.141\ \text{H}$

capacitor: $1/\sqrt{2} \rightarrow 1/\sqrt{2} \times 10^4 \times 10^5 = 0.707\ \text{nF}$.

7.4.3 The magnitude and phase of system functions: Bode diagrams

It was emphasized in Section 6.1 that pure sinusoidal waveforms are not encountered in practice. Nevertheless, for many practical purposes it is useful to know the response of a system as a function of the frequency of a sinusoidal excitation. For instance, such knowledge is essential in assessing whether the bandwidth of a system is great enough for it to transmit a required number of the higher-frequency components in the spectrum of a non-sinusoidal excitation (see Section 6.2).

The response of a system to a sinusoidal excitation can be obtained by setting the complex frequency s equal to $\mathrm{j}\omega$ (see Table 6.1) in the expression for the system function $\mathscr{H}(s)$; this gives directly the ratio $\mathscr{R}(\mathrm{j}\omega)/\mathscr{E}(\mathrm{j}\omega)$, i.e. voltage gain, current gain, transimpedance, etc., depending on how the system function is defined in a particular situation.

Of course, the response of a circuit to a sinusoidal excitation could be obtained using steady-state a.c. circuit theory, but the practical point is that the system functions of a wide variety of networks are tabulated as s-domain functions, which in addition allows the calculation of responses to general, non-sinusoidal, excitations.

For compactness of presentation, the tabulated networks and their system functions are given in 'normalized' form, i.e. normalized to unit (angular) frequency and to an impedance level of 1 ohm.

Consider the circuit shown in Fig. 7.18(a): you will be able to show easily that the transfer impedance, $\mathscr{H}(s) \equiv V(s)/I(s)$, is given by

$$\mathscr{H}(s) = \frac{R}{2} \cdot \frac{(s + z_1)}{(s + p_1)},$$

where $z_1 \equiv (CR)^{-1}$, and $p_1 \equiv (2CR)^{-1}$.

Now in the polar form of complex numbers, the quotient

$$\frac{(a+\mathrm{j}b)}{(c+\mathrm{j}d)} = \frac{|(a+\mathrm{j}b)|\,\mathrm{e}^{\mathrm{j}\phi_{\mathrm{N}}}}{|(c+\mathrm{j}d)|\,\mathrm{e}^{\mathrm{j}\phi_{\mathrm{D}}}}$$

$$= \frac{|(a+\mathrm{j}b)|\,\mathrm{e}^{\mathrm{j}(\phi_{\mathrm{N}}-\phi_{\mathrm{D}})}}{|(c+\mathrm{j}d)|}, \tag{7.81}$$

where ϕ_{N}, ϕ_{D} are the 'arguments' (phases) of the complex numbers $(a+\mathrm{j}b)$, $(c+\mathrm{j}d)$ respectively. Hence, on setting $s = \mathrm{j}\omega$,

$$|\mathscr{H}(\mathrm{j}\omega)| = \frac{R}{2}\cdot\sqrt{\left(\frac{\omega^2+z_1^2}{\omega^2+p_1^2}\right)}$$

and

$$\underline{/\mathscr{H}(\mathrm{j}\omega)} = \tan^{-1}(\omega/z_1) - \tan^{-1}(\omega/p_1).$$

Now in Section 7.4.1 attention was drawn to the fact that, in general, system functions can be expressed as quotients of polynomials in the complex frequency s and, further, that these polynomials can be factorized:

$$\mathscr{H}(s) = K\cdot\frac{(s+z_1)(s+z_2)\cdots(s+z_n)\cdots}{(s+p_1)(s+p_2)\cdots(s+p_n)\cdots}.$$

The reader will have noticed in passing that the factors can have the following general forms:

K (a constant) $\qquad s$

$(s+s_1) \qquad (s+s_1)(s+s_2)$*

(*NB this form has been met as $\left(s^2+\dfrac{sR}{L}+\dfrac{1}{LC}\right)$, i.e. $\left(s^2+\dfrac{\omega_0 s}{Q}+\omega_0^2\right)$; see eqn (7.62), for example.)

It follows, by extending the previous arguments (see eqn (7.81)), that

$$|(\mathscr{H}(\mathrm{j}\omega)| = \frac{K\cdot[\text{product of the magnitudes of the terms } (s+z_n)]}{[\text{product of the magnitudes of the terms } (s+p_n)]}$$

and

$$\underline{/\mathscr{H}(\mathrm{j}\omega)} = [\text{sum of the arguments of the terms } (s+z_n)]$$
$$- [\text{sum of the arguments of the terms } (s+p_n)]. \tag{7.82}$$

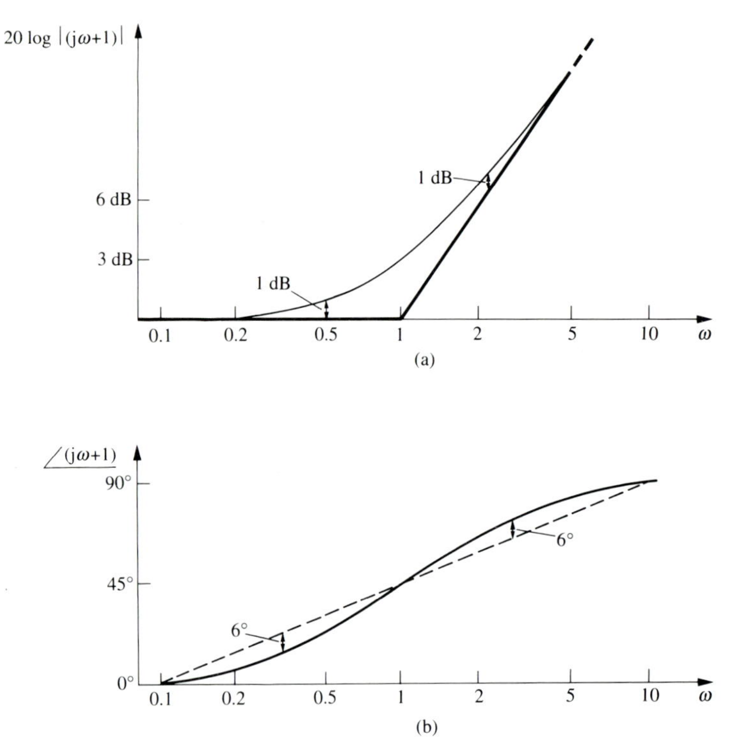

Fig. 7.27 The pair of Bode plots for a normalized system function $(s + 1)$: (a) the magnitude (amplitude) plot; (b) the argument (phase) plot.

It is common practice to express $|\mathscr{H}(j\omega)|$ in decibels, and so

$$20 \log |\mathscr{H}(j\omega)|$$
$$= 20\{\log K + [\text{sum of the logarithms of the terms } (s + z_n)]\}$$
$$- 20\{[\text{sum of the logarithms of the terms } (s + p_n)]\}. \quad (7.83)$$

For steady-state sinusoidal excitation, a first-order term of the form $(s + s_1)$ becomes $(j\omega + s_1)$ and has a magnitude $(\omega^2 + s_1^2)^{1/2}$. Apart from a scale factor (s_1), this magnitude can be written as $(\omega^2 + 1)^{1/2}$, where ω has now been normalised through $\omega \equiv \omega/s_1$, and the argument of the term is $\tan^{-1}(\omega)$. The magnitude (in decibels) and the argument are plotted in Figs. 7.27(a) and (b).

Now for $\omega \ll 1$, $|(j\omega + 1)| \approx 1$, and for $\omega \gg 1$, $|(j\omega + 1)| \approx \omega$; these approximations are indicated by the bold straight lines in Fig. 7.27(a). It is left as an exercise to show that the straight lines approximating to the magnitude plot in Fig. 7.27(a) differ from the 'true' curve by only about

1 dB at the frequencies an octave below and an octave above the 'corner' frequency, i.e. at normalized frequencies of 0.5 and 2.0 respectively. Notice also that the high-frequency slope (or 'roll-off') is 20 log ω, i.e. 20 dB per decade, or 6 dB per octave. In addition, the straight-line approximation to the phase-angle variation (Fig. 7.27(b)) has a slope of 45° per decade and its maximum deviation from the true curve is approximately 6°.

The magnitude (expressed in logarithmic form) and phase plots in the frequency domain for a system function are commonly referred to as **Bode plots**; the utility of a pair of such plots is that the individual terms in eqns (7.82) and (7.82), whose forms are well known (see Fig. 7.28), can be simply combined *additively*. The addition is particularly simple to perform if the straight-line approximations just referred to are sufficiently accurate for the purpose in mind.

In the case of a second-order term having the normalized form $(s^2 + s/Q + 1)^{-1}$ (see Fig. 7.28), the magnitude is

$$\{(1 - \omega^2)^2 + \omega^2/Q^2\}^{-1/2}$$

and the argument is

$$\tan^{-1}\left\{\frac{-\omega}{Q(1 - \omega^2)}\right\}.$$

For $\omega = 1$ (the resonance frequency) the magnitude is equal to Q, and for $\omega \to 0$ and $\omega \to \infty$ the magnitude tends to 1 and ω^{-2}, respectively. Hence the high-frequency roll-off is 40 dB per decade (12 dB per octave), i.e. twice as steep as for a first-order term. The higher the Q-value, the higher is the peak of the response and the sharper is the swing in phase from 0° to −180°.

If $\mathscr{H}(s)$ happens to be a transfer function, then $(s + 1)^{-1}$ and $(s^2 + s/Q + 1)^{-1}$ are both low-pass filter functions (see Fig. 7.28) with a critical frequency (normalized angular frequency, remember) at $\omega = 1$; note that the high-frequency roll-off of the second-order low-pass filter is twice as steep as that of the first-order filter.

Example 7.8 Obtain the frequency response of the transfer function

$$\mathscr{T}(s) \equiv \frac{(s + 100)}{s(s + 100)}$$

by constructing the Bode diagram.

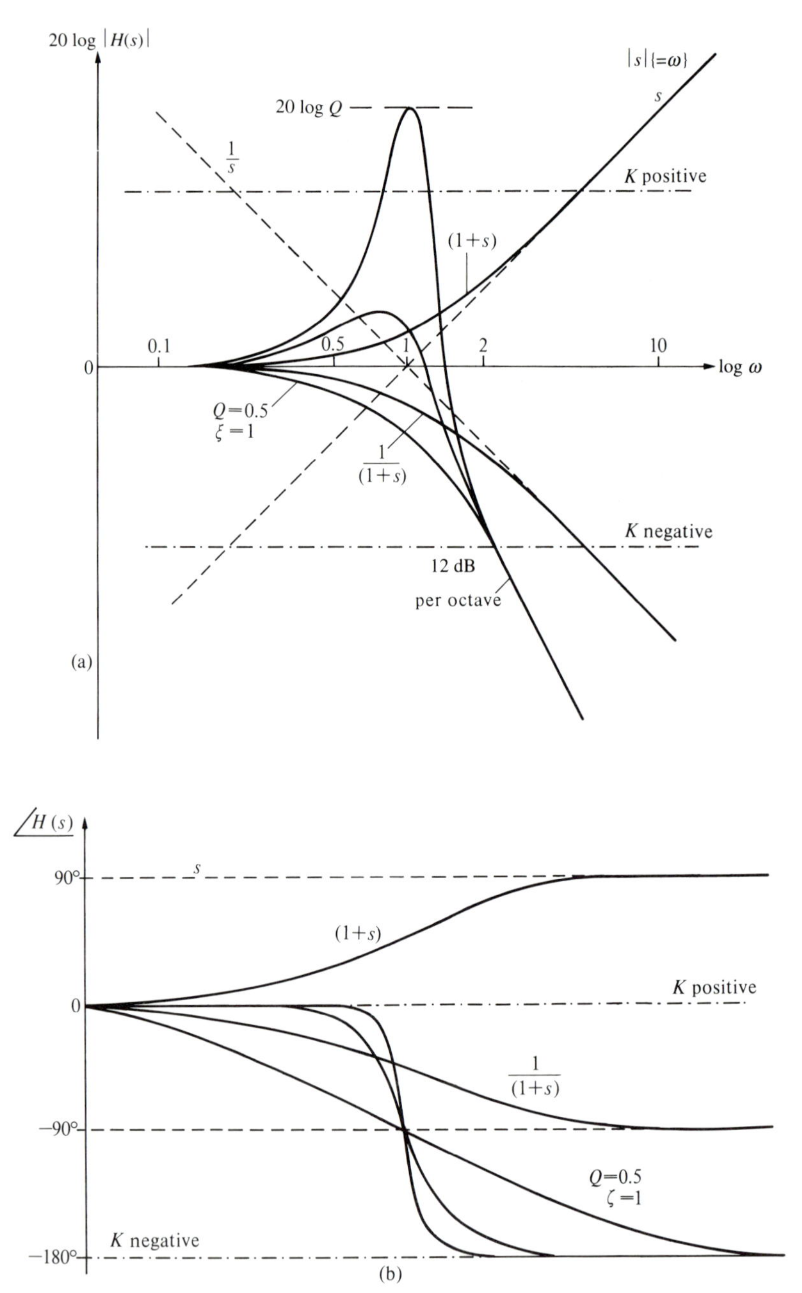

Fig. 7.28 Bode plots: (a) amplitude plots; (b) phase plots.

The first thing to notice is that $\mathscr{T}(s)$ can be rewritten as

$$\mathscr{T}(s) = \frac{\left(1 + \dfrac{s}{10}\right)}{10s\left(1 + \dfrac{s}{100}\right)},$$

where the two terms in brackets have corner frequencies of 10 and 100 respectively. So the respective magnitude responses on the Bode diagram are as shown in Fig. 7.29(a). Notice that the factor 1/10 is represented by the 'horizontal' straight line at -20 dB. The overall response is indicated by the bold line. □

Consider the normalized transfer function

$$\mathscr{T}(s) = \frac{K \cdot s}{\left(s^2 + \dfrac{s}{Q} + 1\right)}, \tag{7.84}$$

where K is a scale factor. What is the form of the Bode plot for this system function? The reader will be able to show easily that

$$\mathscr{T}(j\omega) = \frac{\dfrac{\omega^2}{Q} + j\omega(1 - \omega^2)}{\left\{(1 - \omega^2)^2 + \dfrac{\omega^2}{Q^2}\right\}}.$$

Hence

$$|\mathscr{T}(j\omega)| = \frac{K \cdot \omega}{\sqrt{\left[(1 - \omega^2)^2 + \dfrac{\omega^2}{Q^2}\right]}} \tag{7.85}$$

and

$$\underline{/\mathscr{T}(j\omega)} = \tan^{-1}\left\{\frac{\omega(1 - \omega^2)}{\omega^2/Q}\right\}. \tag{7.86}$$

From eqns (7.85) and (7.86),

$$\omega \to 0: \quad |\mathscr{T}(j\omega)| \approx K \cdot \omega \qquad \underline{/\mathscr{T}(j\omega)} \approx \tan^{-1}\left\{\frac{1}{\omega/Q}\right\} \to +90^\circ$$

$$\omega = 1: \quad |\mathscr{T}(j\omega)| = K \cdot Q \qquad \underline{/\mathscr{T}(j\omega)} = 0^\circ$$

$$\omega \to \infty: \quad |\mathscr{T}(j\omega)| \approx K/\omega \qquad \underline{/\mathscr{T}(j\omega)} \approx \tan^{-1}\left\{\frac{-\omega}{1/Q}\right\} \to -90^\circ.$$

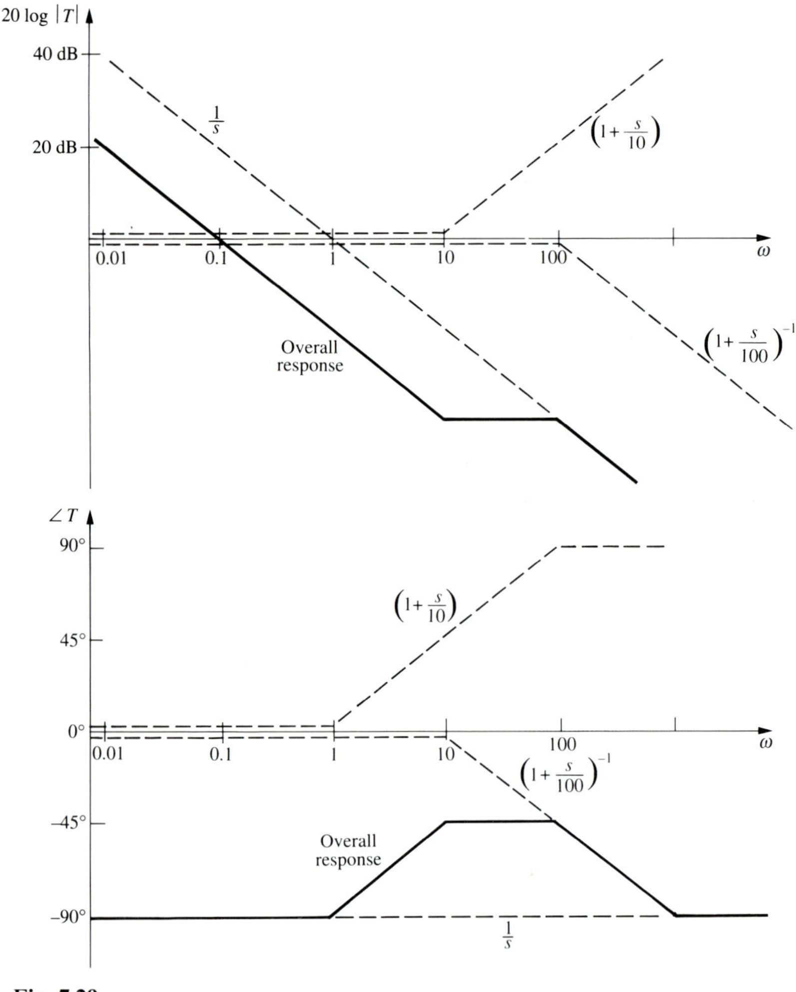

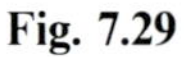
Fig. 7.29

The Bode plots (eqns (7.85) and (7.86)) are sketched in Fig. 7.30; this transfer function is a band-pass function.

In Fig. 7.31 are given the transfer functions and pole–zero diagrams for the four basic types of filter functions.

7.5 General aspects of filters

Filters are extensively used in all branches of electronics, ranging from low-pass instrumentation filters, having cut-off frequencies of a fraction

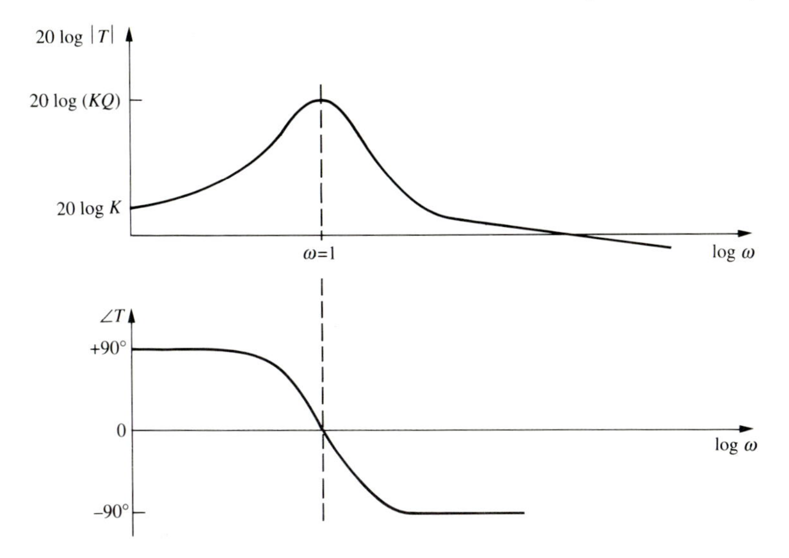

Fig. 7.30 The Bode plots for the second-order band-pass transfer function of eqn (7.84).

of a hertz, to narrow-band filters for use in the e.h.f. range, with centre frequencies ~ 10 GHz and pass band widths ~ 10 MHz. Filters may be low-pass, high-pass, band-pass, band-stop and even all-pass!. Additionally, filters may be passive or may contain active elements. The circuit elements used in the realization of practical filters may be lumped resistors, capacitors and inductors, and active devices, in discrete or integrated form, or they may be 'distributed elements' (see Chapter 8). Also, filter functions may be realized using digital as well as analogue circuits.

Hence filter design is an enormous field and a vast amount of information is available to the designer in the form of databases, algorithms and computer-aided design procedures (see the References given at the end of this chapter). Although only a glimpse of the field can possibly be given here, it is useful to be able to appreciate the ways in which filter characteristics are specified and to be aware of the (small) number of classes of filter that are mainly used. In addition, the consideration of some design problems will illustrate, incidentally, the use of the frequency and impedance normalization/denormalization procedures introduced in Section 7.4.2.

The ideal characteristics for the four types of filter are shown in Fig. 7.32. Readers will suspect, and rightly, that such ideal characteristics having perfectly sharp cut-offs, infinite attenuation in the stop band, and zero attenuation in the pass band (i.e. so-called 'brickwall' characteristics)

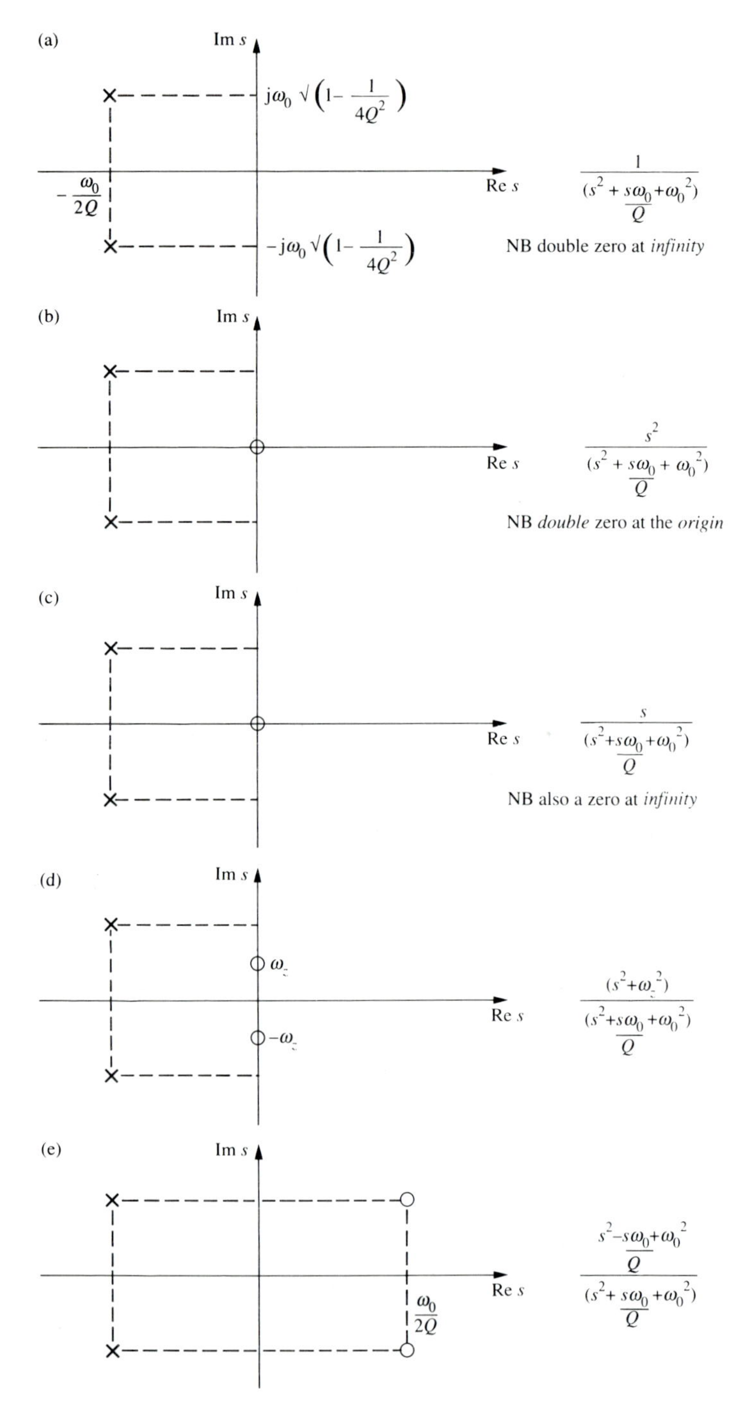
(a)
Im s
$j\omega_0 \sqrt{\left(1-\frac{1}{4Q^2}\right)}$
$-\frac{\omega_0}{2Q}$
Re s
$-j\omega_0 \sqrt{\left(1-\frac{1}{4Q^2}\right)}$
$\frac{1}{\left(s^2+\frac{s\omega_0}{Q}+\omega_0^2\right)}$
NB double zero at *infinity*
(b)
Im s
Re s
$\frac{s^2}{\left(s^2+\frac{s\omega_0}{Q}+\omega_0^2\right)}$
NB *double* zero at the *origin*
(c)
Im s
Re s
$\frac{s}{\left(s^2+\frac{s\omega_0}{Q}+\omega_0^2\right)}$
NB also a zero at *infinity*
(d)
Im s
ω_z
$-\omega_z$
Re s
$\frac{(s^2+\omega_z^2)}{\left(s^2+\frac{s\omega_0}{Q}+\omega_0^2\right)}$
(e)
Im s
$\frac{\omega_0}{2Q}$
Re s
$\frac{s^2-\frac{s\omega_0}{Q}+\omega_0^2}{\left(s^2+\frac{s\omega_0}{Q}+\omega_0^2\right)}$

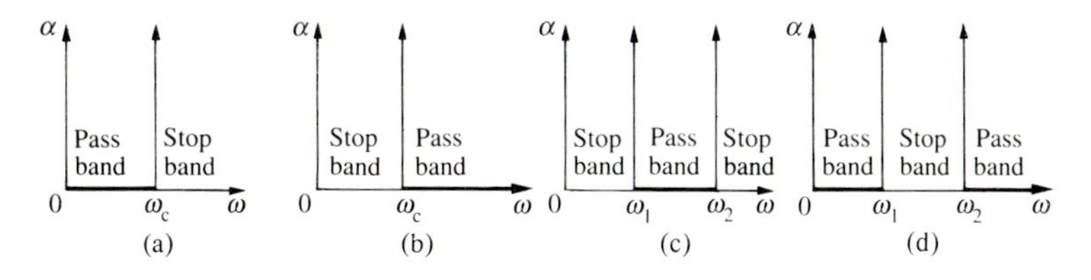

Fig. 7.32 Ideal, 'brickwall', filter characteristics (α is the attenuation): (a) low-pass; (b) high-pass; (c) band-pass; (d) band-stop.

cannot be realized in a practical circuit. Hence a major feature of the design process is the specification of characteristics that approximate closely enough for practical purposes to the ideal characteristic.

It will be indicated later that high-pass, band-pass, and band-stop filter networks can be derived from a low-pass network, so it is useful consider first the way in which low-pass characteristics are specified.

For a filter, the system function of interest is the transfer function $\mathscr{T}(s)$:

$$\mathscr{T}(s) \equiv \mathscr{R}(s)/\mathscr{E}(s),$$

where $\mathscr{R}(s)$ is the response of the network to an excitation $\mathscr{E}(s)$. The attenuation $\alpha(\omega)$ is specified through

$$\alpha(\omega) = -20 \log|\mathscr{T}(j\omega)| \text{ dB} \tag{7.87}$$

(NB $|\mathscr{T}(j\omega)| \leqslant 1$, so that α is a positive number.)

In the stop band of the ideal characteristic, $|\mathscr{T}(j\omega)| = 0$, which implies that $\alpha(\omega)$ is infinite. If the filter specification is relaxed from the ideal, then it is possible to realize a filter using a finite number of circuit elements. For instance, a minimum stop band attenuation of 60 dB is more than sufficient for most practical purposes. In addition, a small value of attenuation in the pass band can be allowed; the non-abrupt transition between the pass band and the stop band follows inevitably and must be accepted (see Fig. 7.33(a)).

Because of the almost limitless variety of filter specifications that arise in practice, compactness of presentation is obtained by displaying filter data in relation to a normalized, low-pass filter network (or '**prototype filter**') for which the **critical angular frequency** is taken as 1 radian per second and the impedance level of the network is taken as 1 Ω (see Fig.

Fig. 7.31 The transfer functions and pole–zero diagrams for second-order filter functions: (a) a low-pass filter; (b) a high-pass filter; (c) a band-pass filter; (d) a band-stop filter (sometimes called a 'notch', or 'band elimination' filter; (e) 'all-pass' (delay) filter.

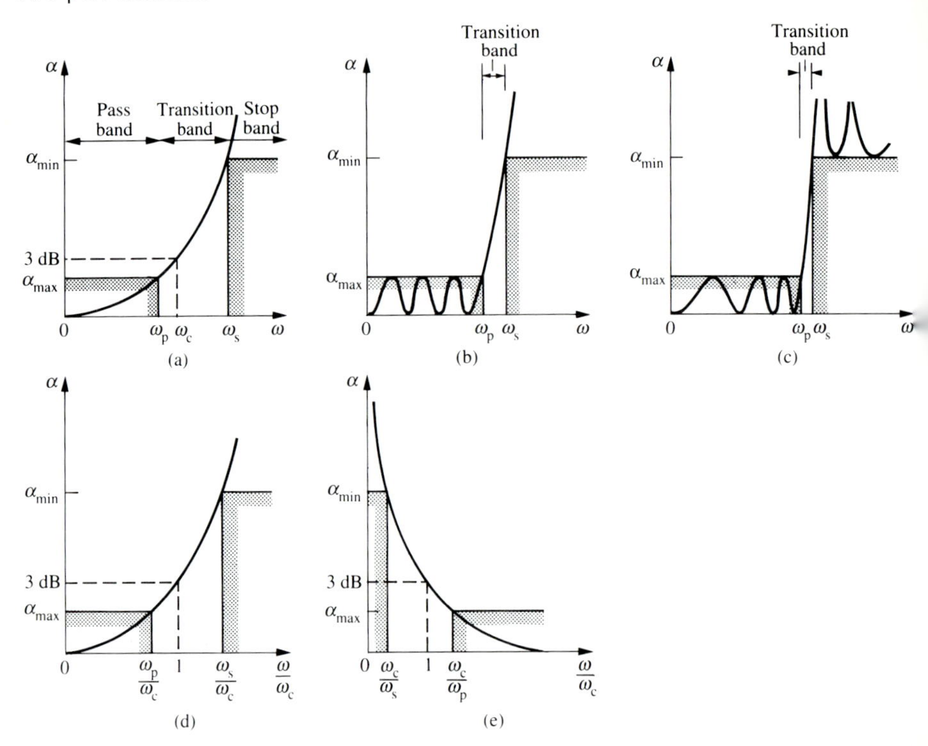

Fig. 7.33 Definitions of the pass band, transition band, and stop band for three types of low-pass filter characteristic: (a) 'maximally flat' (or Butterworth); (b) 'equi-ripple' (or Chebyshev); (c) 'elliptic' (or Cauer); (d) normalized low-pass Butterworth characteristic; (e) normalized high-pass Butterworth characteristic.

7.33(d)). It turns out that normalized high-pass, band-pass and band-stop networks can be obtained from a prototype network by simple transformations of the circuit elements (see below). In principle there is an infinity of ways of approximating to the ideal filter characteristic, but only a few are widely used in practice, of which the principal types are called 'maximally flat' (or Butterworth), 'equi-ripple' (or Chebyshev), and 'elliptic' (or Cauer) (see Fig. 7.33(a), (b), (c)).

The **filter specification** (i.e. the values of α_{min}, α_{max}, ω_p, ω_s) determines the order, n, of the filter, where n is equal to the number of reactive elements in the prototype filter network. A database will yield the network of the prototype filter network, which can then be denormalized to the required practical values of the critical frequency and impedance level using the rules given in Section 7.4.2.

If a high-pass, band-pass or band-stop filter specification has to be met, then there are simple procedures for obtaining the values of n, ω_p, ω_s for the equivalent prototype characteristic (the values of α_{min}, α_{max} are unchanged); see Example 7.9 below. The circuit elements of the prototype filter can be transformed according to simple rules to yield a normalized high-pass, band-pass or band-stop network as required. The design of a Butterworth low-pass filter illustrates the general procedure.

Example 7.9 Design a low-pass Butterworth filter to the following specification (refer to Fig. 7.33):

$$\alpha_{max} = 0.5 \text{ dB} \qquad \alpha_{min} = 20 \text{ dB}$$
$$\omega_p = 10^3 \text{ rad s}^{-1} \qquad \omega_s = 2 \times 10^3 \text{ rad s}^{-1} \qquad \text{impedance level } 1 \text{ k}\Omega$$

For a Butterworth low-pass filter of order n, the modulus of the transfer is given by

$$|\mathscr{T}(j\omega)| = (1 + \omega^{2n})^{-1/2}, \tag{7.88}$$

so that

$$\alpha_n(\omega) = 10 \log\left\{1 + \left(\frac{\omega}{\omega_c}\right)^{2n}\right\} \tag{7.89}$$

where ω_c is the frequency for which $\alpha = 3$ dB ('half-power point'). From this equation and using $\alpha(\omega) = \alpha_{max}$ at $\omega = \omega_p$ (see Fig. 7.33) it follows immediately that

$$\omega_c = 1.234 \times 10^3 \text{ rad s}^{-1}$$

and

$$n = \frac{\log\{[10^{\alpha_{min}/10} - 1]/[10^{\alpha_{max}/10} - 1]\}}{2 \log(\omega_s/\omega_p)}. \tag{7.90}$$

For the example under consideration, this yields $n = 4.83$, and since obviously the number of circuit elements in the actual filter must be an integer, n is taken as 5 (i.e. the next-higher integer). From tabulated data (see Table 7.2) for Butterworth filters, the required 5th-order prototype network for a passive filter is obtained; see Fig. 7.34(a).

Since the tabulated data that yielded the element values were based on the assumption that $\Omega_c = 1$ rad s^{-1}, the frequency denormalization ratio to be used is 1234:1. The impedance denormalization ratio is 10^3:1 and

Table 7.2. Element values for doubly-terminated prototype Butterworth filter networks (for $n = 2$ to $n = 6$) normalized to a critical frequency (half-power frequency) of 1 rad s^{-1}.
Note that each series element is an inductance and each shunt element is a capacitance. The choice between a minimum inductance or a minimum capacitance realization (odd-order networks) may be dictated by practical considerations.

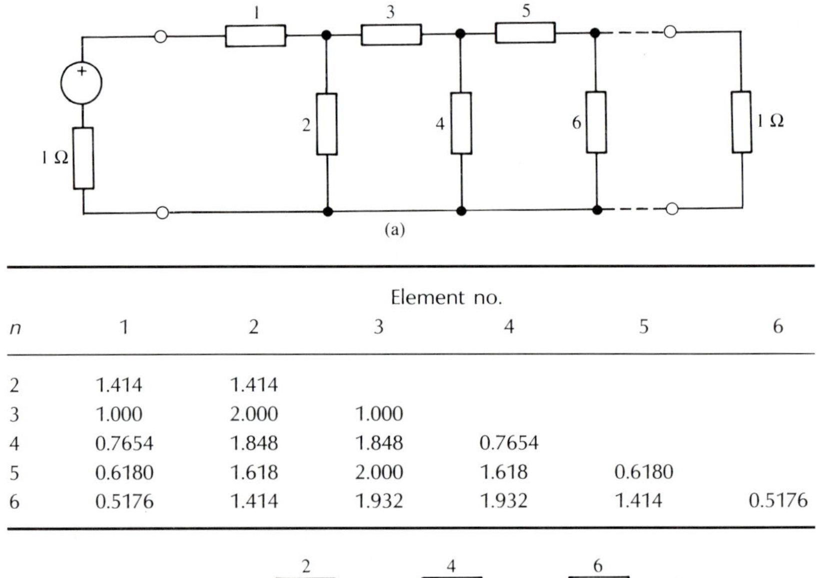

(a)

n	Element no. 1	2	3	4	5	6
2	1.414	1.414				
3	1.000	2.000	1.000			
4	0.7654	1.848	1.848	0.7654		
5	0.6180	1.618	2.000	1.618	0.6180	
6	0.5176	1.414	1.932	1.932	1.414	0.5176

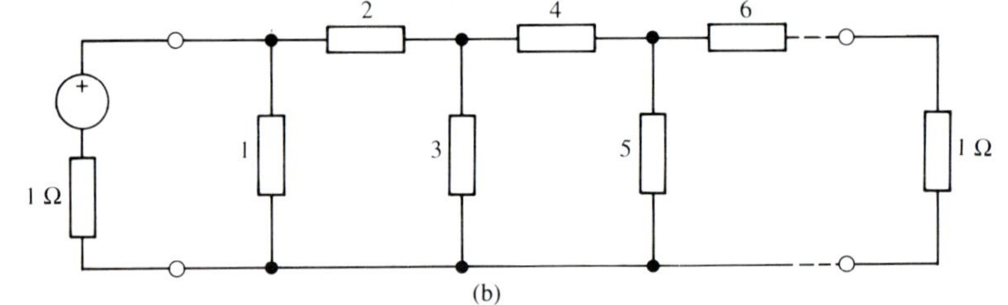

(b)

so it follows that in the required low-pass filter,

$$C_1 = C_5 = \frac{0.618}{1234 \times 10^3} = 0.5\ \mu\text{F}$$

$$C_3 = \frac{2.000}{1234 \times 10^3} = 1.62\ \mu\text{F}$$

$$L_2 = L_4 = \frac{1.618 \times 10^3}{1234} = 1.31\ \text{H}. \qquad \square$$

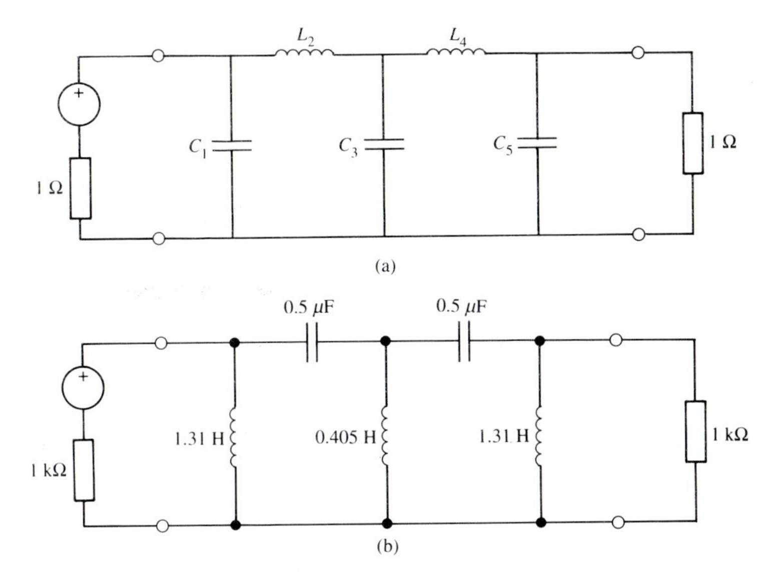

Fig. 7.34 (a) 5th-order Butterworth prototype filter network; $C_1 = C_5 = 0.618$ F; $C_3 = 2.000$ F; $L_2 = L_4 = 1.618$ H. (b) Denormalized high-pass Butterworth filter derived from the prototype filter network of (a).

Example 7.10 Design a Chebyshev filter to meet the following prototype filter specification:

$$\Omega_p = 1 \text{ rad s}^{-1} \qquad \Omega_s = 2.33 \text{ rad s}^{-1} \qquad \alpha_{max} = 0.5 \text{ dB} \qquad \alpha_{min} \geqslant 2 \text{ dB}.$$

For a Chebyshev low-pass filter, the modulus of the transfer function is given by

$$|\mathcal{T}(j\omega)| = (1 + \varepsilon^2 C_n^2(\omega))^{-1/2}, \tag{7.91}$$

so that

$$\alpha_n(\omega) = 10 \log\{1 + \varepsilon^2 C_n^2(\omega)\} \text{ dB}, \tag{7.92}$$

where ε is the pass-band 'ripple' ($= \alpha_{max}$) (see Fig. 7.33(b)). $C_n(\omega)$ are so-called Chebyshev polynomials, but their mathematical form does not need to be given here.

It can be shown that

$$\varepsilon^2 = 10^{\alpha_{max}/10} - 1 \tag{7.93}$$

and

$$n = \frac{\cosh^{-1}[\{(10^{\alpha_{min}/10} - 1)/(10^{\alpha_{max}/10} - 1)\}^{1/2}]}{\cosh^{-1}(\omega_s/\omega_p)}. \tag{7.94}$$

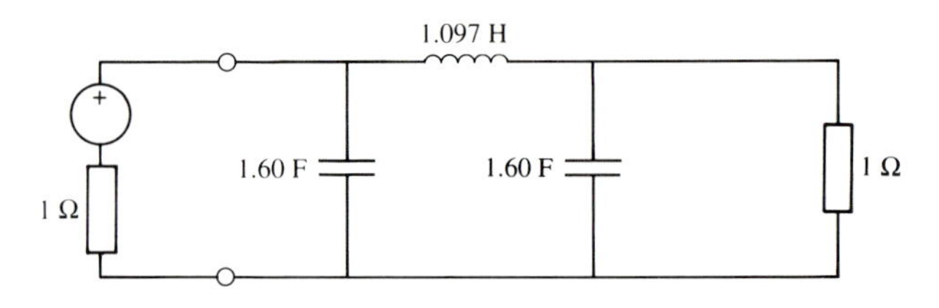

Fig. 7.35 A 3rd-order prototype Chebyshev filter network.

For the given specification these formulae yield

$$\varepsilon = 0.349 \qquad \text{and} \qquad n = 2.87 \text{ (round up to } n = 3).$$

From tabulated data for Chebyshev filters, the prototype 3rd-order network is shown in Fig. 7.35; this can be denormalized to meet a practical specification.

A normalized low-pass filter filter transfer function $\mathscr{T}(s)$ can be transformed to a normalized high-pass function having the same critical frequency by replacing s by $1/s$ wherever it occurs in the function. This is equivalent to saying that in the prototype network an inductance of value L is replaced by a capacitance of value $1/L$, and similarly a capacitance of value C should be replaced by an inductance of value $1/C$ (see Table 7.3). These new capacitance and inductance values can again be denormalized by following the rules already established. For example, in Problem 7.20 it is required to find the 5th-order high-pass Butterworth filter network with the same values of α_{min}, α_{max} and critical frequency as specified for the low-pass filter in Example 7.9. Using the rules outlined above, the denormalized network shown in Fig. 7.34(b) is obtained.

Table 7.3. Filter element transformations.

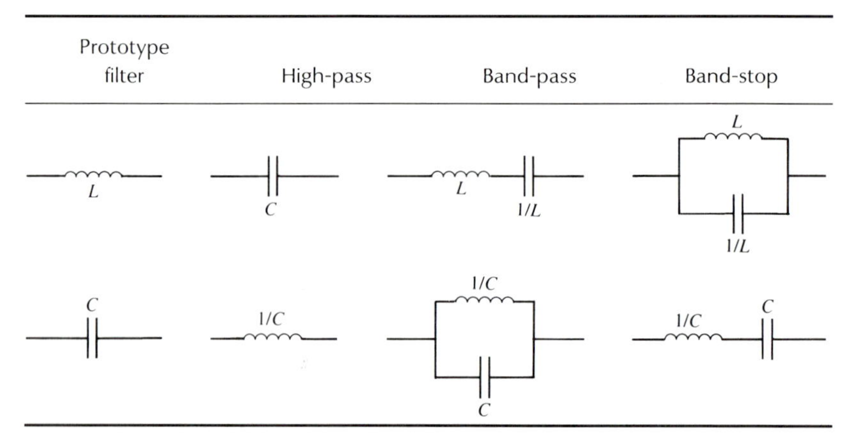

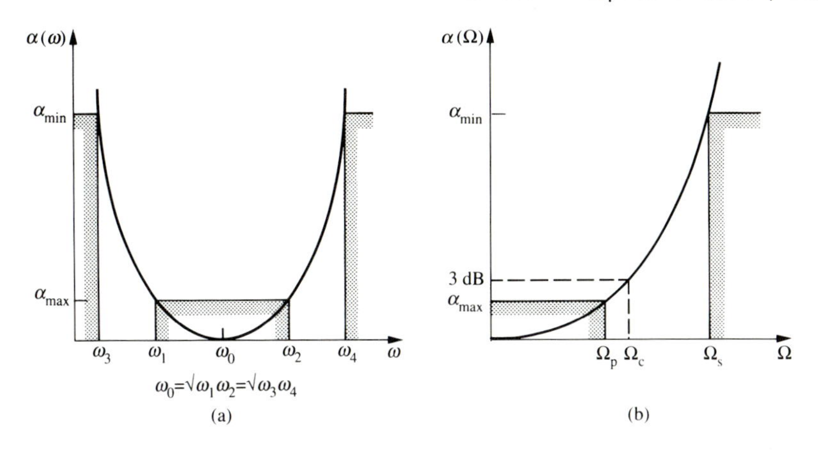

Fig. 7.36 (a) Band-pass filter characteristic. (b) Butterworth prototype filter characteristic derived from (a).

The procedure for designing a band-pass filter is somewhat more complicated, as might be expected. Consider the band-pass filter characteristic shown in Fig. 7.36(a) and note particularly that the bandwidth is specified as $(\omega_2 - \omega_1)$ and that $\omega_0 = \sqrt{(\omega_1\omega_2)} = \sqrt{(\omega_3\omega_4)}$. For the case of a Butterworth filter it can be shown that in the derived prototype characteristic (Fig. 7.36(b)),

$$\Omega_s = (\omega_4 - \omega_3)/(\omega_2 - \omega_1). \tag{7.95}$$ □

Example 7.11 As an example, consider the following band-pass filter specification:

$$\omega_3 = 250 \text{ rad s}^{-1} \qquad \omega_1 = 500 \text{ rad s}^{-1} \qquad \omega_2 = 10^3 \text{ rad s}^{-1}$$

$$\omega_4 = 2 \times 10^3 \text{ rad s}^{-1} \qquad \alpha_{max} = 0.5 \text{ dB} \qquad \alpha_{min} \geqslant 20 \text{ dB}.$$

Using eqn (7.95) yields $\Omega_s = 3.5 \text{ rad s}^{-1}$, and taking $\Omega_p = 1 \text{ rad s}^{-1}$ it follows that $n = 2.67$ (round up to $n = 3$). Note that $\omega_0 = 707 \text{ rad s}^{-1}$. Finally, using eqn (7.89) together with $\alpha(\Omega) = 0.5$ dB at $\Omega = 1 \text{ rad s}^{-1}$, then $\Omega_c = 1.42 \text{ rad s}^{-1}$. Tabulated data for Butterworth prototype filters yield the circuit shown in Fig. 7.37(a).

A prototype filter transfer function $\mathscr{T}(s)$ can be transformed into a normalized band-pass function by replacing s by $(s + 1/s)$ wherever it occurs in the function. In terms of the elements of the elements of the prototype network this transformation is given in Table 7.3, and the resulting normalized band-pass network is shown in Fig. 7.37(b). Denormalization factor is equal to 707/1.42, since $\omega_0 = 707 \text{ rad s}^{-1}$ and $\Omega_c = 1.42 \text{ rad s}^{-1}$. The required filter circuit is shown in Fig. 7.37(c). □

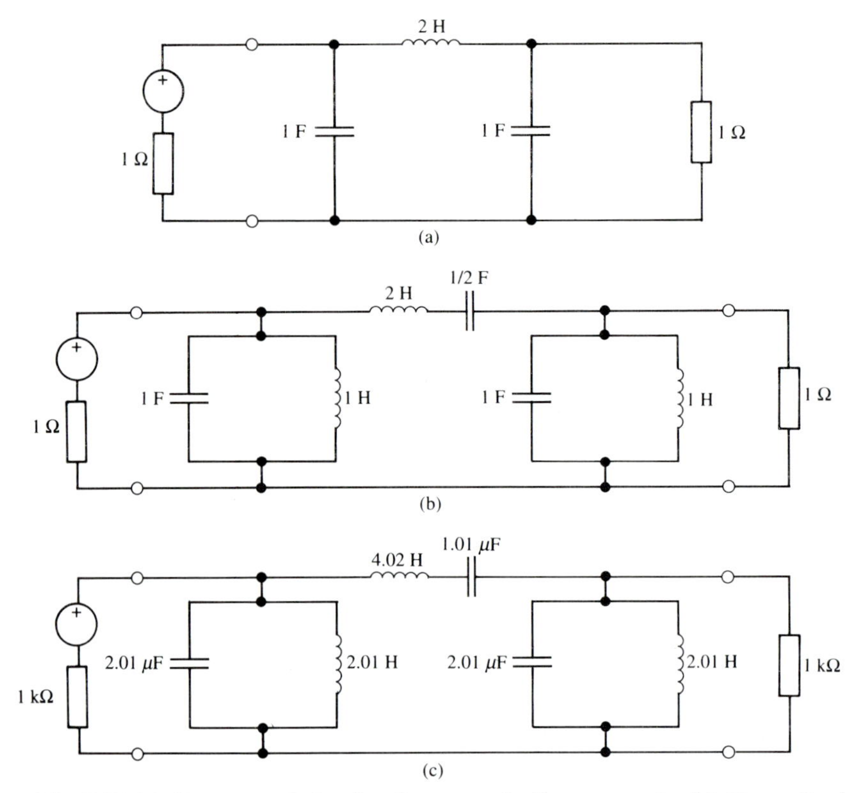

Fig. 7.37 (a) Prototype 3rd-order Butterworth filter network. (b) Normalized band-pass filter derived from (a). (c) Denormalized band-pass filter network for Example 7.11.

For a band-stop filter the same procedure is followed, in essence, except that the circuit elements of the equivalent prototype network are transformed as indicated in Table 7.3.

So far in this discussion of filters, attention has been focused on the magnitude of the transfer function $\mathscr{T}(s)$. In practice the frequency dependence of $\underline{/\mathscr{T}(s)}$ is often of great importance also, since for instance it is the source of differential propagation delays between the Fourier components of a non-sinusoidal signal.

For the faithful transmission of non-sinusoidal signals there are two requirements, ideally:

(i) a 'flat' magnitude response for ω ranging from 0 to ∞;
(ii) a linear phase response for ω from 0 to ∞.

Condition (i) follows from earlier discussions (particularly in Chapter 6),

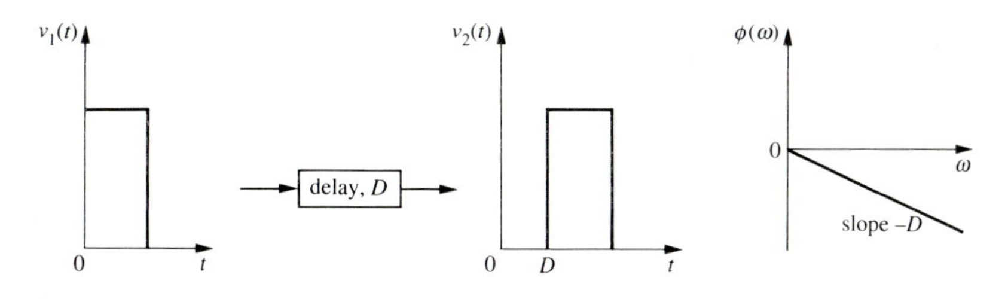

Fig. 7.38 A delayed pulse and a linear phase characteristic.

but (ii) may not appear obvious. So consider a square pulse propagating through a two-port network that imposes a delay D but does not distort the shape of the pulse (see Fig. 7.38).

It is required that

$$v_1(t) = v_2(t - D_p),$$

i.e. $v_2(t)$ is a delayed, but undistorted, version of $v_1(t)$ (note that D has been replaced by D_p to distinguish it from D_g, which will be introduced later).

Let one of the components in the Fourier spectrum of the input pulse be

$$v_1(t) = A \cdot \sin(\omega t + \theta).$$

For this component the output from the network will be

$$v_2(t) = A \cdot \sin\{\omega(t - D_p) + \theta\}$$

or

$$v_2(t) = A \cdot \sin\{\omega t + \theta - \omega D_p\}$$

So the input and output signals differ only in the phase angle $\phi \equiv -\omega D_p$.

If all of the Fourier components are delayed by the same amount D_p, so that the output is a delayed replica of the input, then D_p must be independent of ω. Hence $\phi \propto \omega$, i.e. a *linear* phase characteristic as specified in condition (ii) above.

$$\phi = -\omega D_p.$$

In phasor representation,

$$v_1 = A\angle\theta \qquad \text{and} \qquad v_2 = A\angle\theta - \omega D_p,$$

so that

$$\frac{v_2}{v_1} = \angle -\omega D_p \qquad \text{or} \qquad \frac{v_2}{v_1} = \mathrm{e}^{-\mathrm{j}\omega D_p}.$$

Strictly speaking, the delay D_p relates only to pure sinusoidal components of a signal, i.e. it is the so-called 'phase delay'. A transient signal can be constructed, in mathematical terms, from a group, or 'packet', of Fourier components. The 'dispersion' (or distortion) of a transient signal, due to differential propagation delays between the components of the signal, is related to the 'group delay', D_g, which is defined by

$$D_g = \frac{-\partial\phi(\omega)}{\partial\omega}.$$

For D_g to be independent of ω, then again it is required, ideally, that $\phi(\omega)$ is linearly dependent on ω so that $\partial\phi(\omega)/\partial\omega$ is a constant. So, if $\phi(\omega) = $ (a constant) $\times\ \omega$, then both D_p and D_g are independent of ω.

Unfortunately, networks that meet the requirements on the magnitude response of a filter do not have linear phase responses. Generally, Butterworth filters have more acceptable phase responses than Chebyshev and Cauer filters. 'All-pass' filters are an attempt to synthesize filter networks having a flat magnitude response (hence the name) and a linear phase response, but inevitably they fall short of the ideal.

It is common practice to design filters for their magnitude response and then to use an additional, compensating, circuit to produce an overall phase response that is an acceptable compromise from a practical point of view, e.g. the filter characteristic is flat enough, and the phase characteristic is linear enough, over a limited range of frequency.

Chebyshev and Cauer filters are superior to Butterworth filters in that they provide a sharper cut-off (i.e. a narrower transition band) for a given order of filter. At the critical frequency, the slope of a Chebshev filter is approximately n times as steep as that of the Butterworth filter, and the 'roll-off' at the stop band edge is $6(n - 1)$ dB/octave better.

Cauer filters have an even sharper cut-off than Chebyshev filters and, additionally, the zeros of transmission in the stop band at $2\omega_c$, $3.14\omega_c, \ldots$ can be exploited. For instance, if ω_c is the carrier frequency of a radio transmitter, then a Cauer filter would be useful for filtering out harmonics.

There is an advantage in using a network of lowest-possible order to realize a specified filter characteristic, in that the smallest number of circuit components is required (see Table 7.4). However, Butterworth filters are much less sensitive to the tolerances on practical circuit components, and on the Q-factors of inductors, than are Chebyshev and Cauer filters. This may make the Butterworth filter the preferred practical solution despite its higher order.

So far only passive filter networks have been considered. The reader will have noticed that the values of the inductances in the filter realizations of Examples 7.9 and 7.10 were rather large. Only at very high frequencies ($\sim$GHz) is it a practical proposition to fabricate inductors in integrated circuits. At lower frequencies the circuit properties of inductances can be

Table 7.4. Order of network required to realise the prototype filter specification: $\omega_s/\omega_p = 1.5$; $\alpha_{min} \geqslant 50$ dB; $\alpha_{max} = 0.5$ dB.

Filter type	Order, n
Butterworth	17
Chebyshev	8
Cauer	5

realized by using operational amplifiers in appropriate negative-feedback configurations. Sallen and Key circuits, for example, are very well known in this respect. Even-order filters are realized by cascading second-order active filter sections and, if an odd-order response is required, an additional first-order passive section can be connected in cascade to provide an overall odd-order response. Also, inductances can be simulated in ladder networks by using **general impedance converters** (GIC); one advantage of this approach is that the resulting filter has the low sensitivity to component variations, and to variations in other quantities, that is a characteristic of passive networks.

In summary it can be said that in the design of filters a large number of practical features have to be considered, together with their cost implications, and engineering compromises have then to be made. The problems arising from component variations increase out of proportion to the number of components (so keep n as small as possible). Furthermore the circuit properties of components vary significantly with frequency. Butterworth filters are the most tolerant of component variations and can be realized with Q-factors of ~ 3. Also, for a given bandwidth, they have a better phase characteristic and hence transient response than Chebyshev and Cauer filters of the same complexity. Cauer filters have the steepest cut-off for a given order of filter network.

Further reading

Biey, M. and Premoli, A. (1985). *Tables for active filter design.* Artech House, New York.

Chen, Wai-Kai (1986). *Passive and active filters.* Wiley, New York.

Matthaei, G., Young, L., Jones, M. E. T. (1980). *Microwave filters, impedance matching networks and coupling structures.* Artech House, New York.

Schaumann, R., Ghausi, M. S. and Laker, K. R. (1990). *Design of analog filters.* Prentice-Hall International, New York.

Van Valkenburg, M. E. (1982). *Analog filter design.* Holt, Rinehart and Winston, New York.

Problems

7.1 Show that the z- and y-parameters of a two-port network are related by:

$$z_{11} = \frac{y_{22}}{D_y} \qquad z_{12} = \frac{y_{12}}{D_y}$$

$$z_{21} = -\frac{y_{21}}{D_y} \qquad z_{22} = \frac{y_{11}}{D_y}$$

7.2 Derive eqns (7.19) relating the elements of equivalent T- and Π-networks as shown in Fig. 7.3.

7.3 Obtain the h-parameters of the Π-network shown in Fig. 7.39.

Fig. 7.39

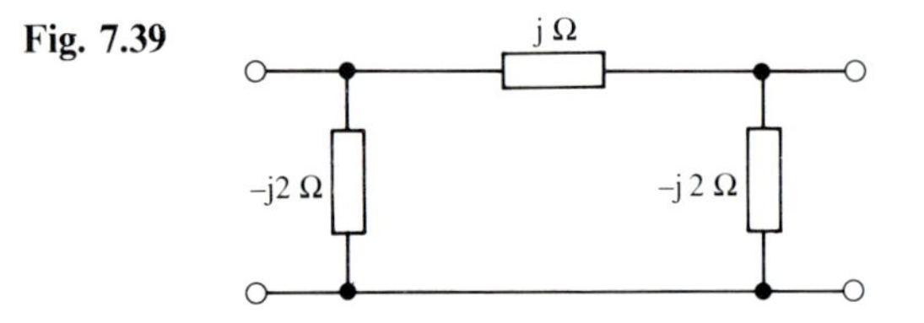

7.4 Obtain the expressions, eqns (7.20) and (7.24), for the characteristic impedance and admittance of symmetrical T- and Π-networks.

7.5 Obtain the expressions given in eqns (7.36) and (7.37) for the transmission matrices of the networks shown in Figs. 7.10(d) and (e).

7.6 Derive the relationships of eqns (7.42) for the elements of a symmetrical Π-network designed to match a source of internal resistance R_0 to a load resistance R_0 with a voltage attenuation ration n.

7.7 Design a matched symmetrical T-network to give an insertion loss of 14 dB when connected between a source of 50 Ω internal resistance and a 50 Ω load.

7.8 Obtain expressions for the y-parameters of the bridged-T network shown in Fig. 7.40.

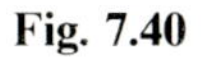

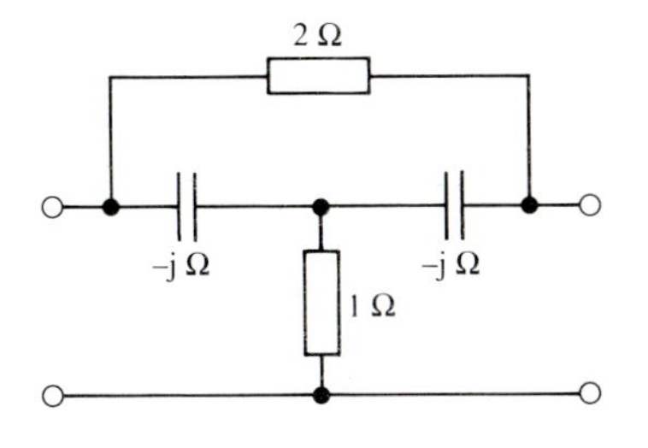

7.9 Write the mesh equations in matrix form for the network shown in Fig. 7.41 and then give the equation for determining the current through Z_2.

Fig. 7.41

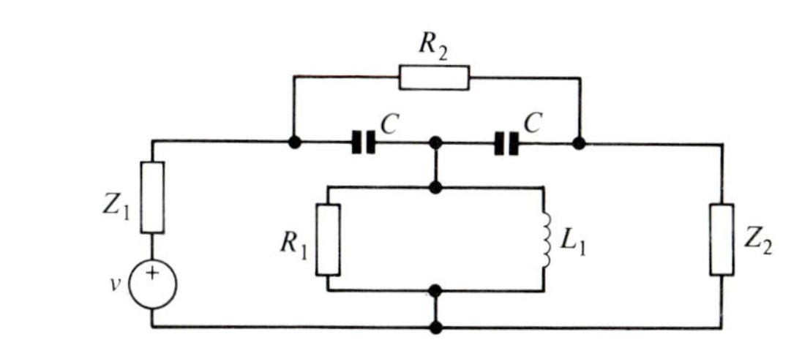

7.10 Write the nodal equations in matrix form for the network shown in Fig. 7.42.

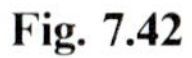

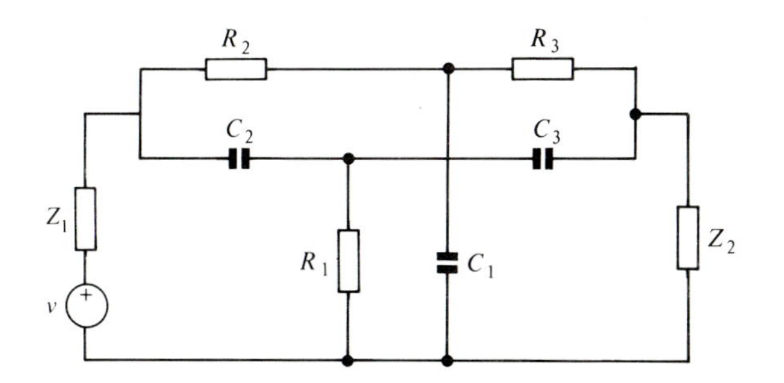

7.11 Obtain the transmission parameters in matrix form for the Π-network shown in Fig. 7.43.

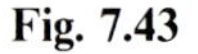
Fig. 7.43

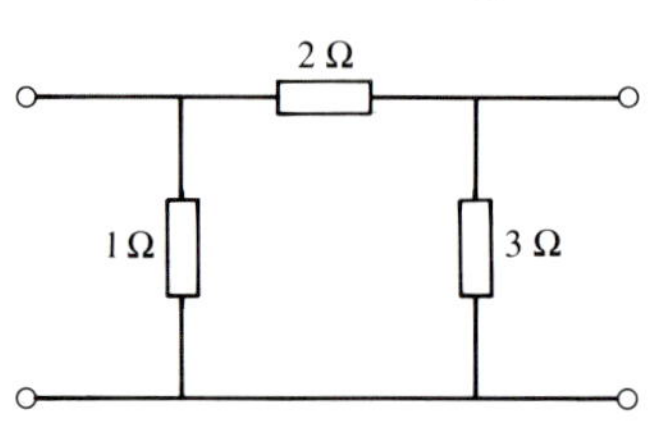

7.12 Derive expressions for the voltage and internal resistance of the Thévenin equivalent source at the output port of the network shown in Fig. 7.13(a).

7.13 Show that the design equations for the ⌈ matching network of Fig. 14(b) are

$$C = \frac{Q}{\omega_0 R_g} \qquad L = \frac{Q R_L}{\omega_0},$$

where

$$Q = \sqrt{\left(\frac{R_g}{R_L} - 1\right)} \text{ (NB } R_g > R_L\text{).}$$

7.14 Obtain the design equations, eqns (7.55), for the T matching network of Fig. 7.16(a).

7.15 Obtain the values of the inductances and capacitances of a Π-network to match a 10 Ω source to a 100 Ω resistive load at a frequency of 100 MHz; the effective Q-value of the network is to be 10.

7.16 Obtain the expressions in eqns (7.57(b)), (7.57(c)) and (7.57(d)) for the indicated system functions in Figs. 7.17(b), (c) and (d).

7.17 Obtain the voltage transfer function $V_0/(s)/V_{in}(s)$ for the network shown in Fig. 7.44.

Fig. 7.44

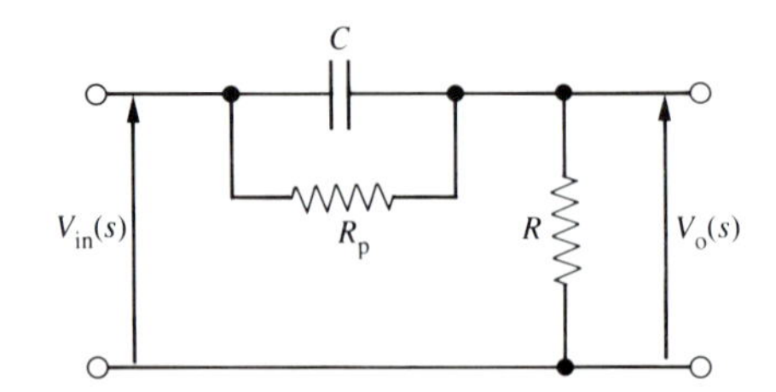

7.18 Show that in the Bode plot of Fig. 7.27 the straight-line approximation to the amplitude response differs from the true curve by ± 0.97 dB at the frequencies an octave above and an octave below the 'corner' frequency.

7.19 Find the zeros of the input admittance of the network shown in Fig. 7.45.

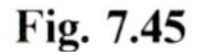
Fig. 7.45

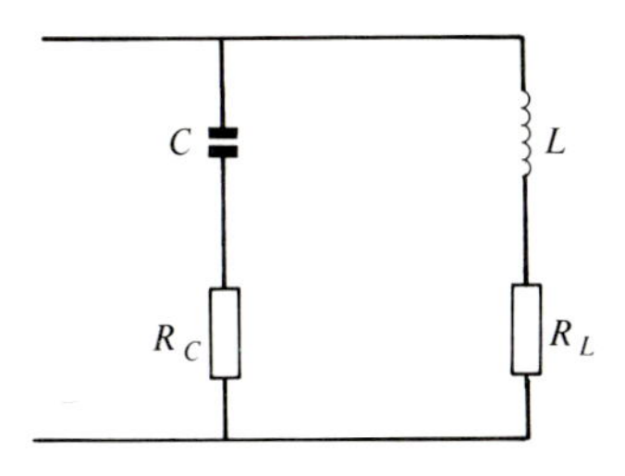

Hence find the condition that the circuit be non-oscillatory. What is the limiting value of R_C to meet the non-oscillatory requirement if $L = 100$ mH, $R_L = 10\ \Omega$, $C = 0.1\ \mu$F?

7.20 Confirm by calculation the values for the elements of the denormalized 5th-order Butterworth high-pass filter shown in Fig. 7.34(b).

Distributed circuits

8

8.1 Introduction

Practical experience supports the prediction of electromagnetic theory that disturbances (sine waves, 'steps', pulses of all shapes) in an electromagnetic field propagate at the speed of light in the medium or system concerned (*this statement is strictly true only under certain ideal conditions, but it is useful to adopt it for present purposes; see Section* 5.4.3).

Such disturbances constitute signals, and the statement includes in particular the transmission of signals along lines such as pairs of wires, twisted wire pairs, coaxial cables, tracks on circuit boards (particularly high-speed digital circuits), power supply tracks, waveguides, optical fibres, etc. Although the description of circuits has so far been in terms of voltages and currents, it should not be forgotten that the electrons that constitute the current in a circuit are 'driven' by the electric field that they experience (see Sections 1.2 and 5.2). This is a reminder that the voltages and currents derive from the pattern of electric and magnetic fields in and around the material parts of a circuit. All transmission lines consist of two conductors separated by a low-loss 'dielectric' medium (the insulating material between the conductors).

If a signal is propagating along a line, then a time-dependent potential difference exists between the two conductors, and equal and opposite time-dependent currents flow along the conductors. In the simplest conceptual situation (e.g. two parallel conductors) the time-dependent electric and magnetic fields (e.m. fields) that exist in the dielectric medium are transverse to the direction of propagation of the voltage/current signal along the conductors; it is useful to think of the e.m. field as being 'guided' by the two conductors. It is important to remember that the voltage–current pair and the electric field–magnetic field pair coexist; in time-dependent situations, one of these quantities is never found without the other three (see Section 5.4.3).

The e.m. field patterns are sketched in Fig. 8.1 for a coaxial line, twin conductors, microstrip, and rectangular waveguide. Waveguides are hollow metal pipes, of which the commonest form has a rectangular cross-section.

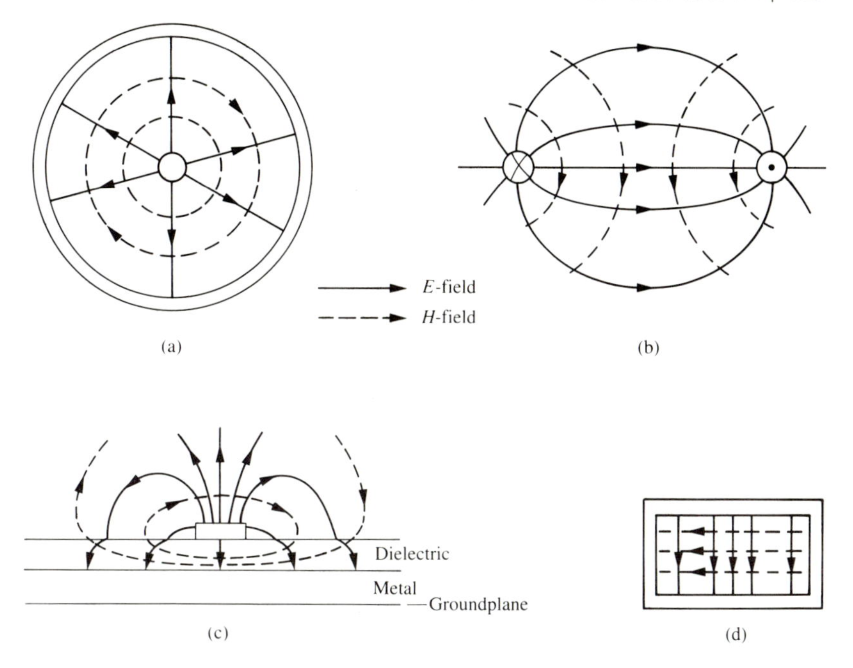

Fig. 8.1 The pattern of electromagnetic fields in (a) coaxial line, (b) twin conductors, (c) microstrip, (d) rectangular waveguide.

In the case of coaxial line and twin conductors, the electric field pattern between and around the conductors was derived in Section 1.5; notice that the magnetic field is everywhere perpendicular to the electric field (actually this is strictly true only for lossless lines, but it is an acceptable assumption for a wide range of practical lines). Such patterns, or 'modes of propagation', are called t.e.m. modes (transverse electric and magnetic), and the field pattern in such modes propagates along the line at the speed of light in the particular insulating medium. The derivation of the e.m. field pattern in microstrip is complicated by the fact that the field lines traverse two very different media: air and the substrate (commonly alumina). In waveguides there can be a component of the electric or the magnetic field in the direction of propagation, and the wavelength (λ_g) of waves in the guide is related to the 'free-space' wavelength (λ_0) by

$$1/\lambda_g^2 = 1/\lambda_0^2 - 1/\lambda_c^2 \qquad \text{(non-t.e.m. modes)},$$

where λ_c is directly related to the dimensions of the waveguide.

Before proceeding to more detailed and rigorous discussions in Sections 8.2 and 8.3, consider for the moment a system consisting of a pair of

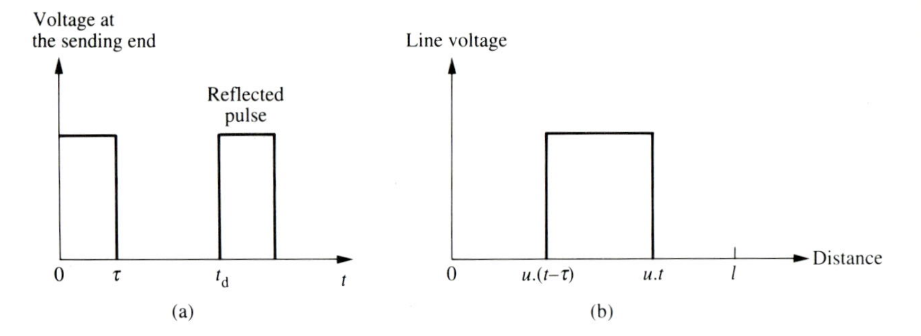

Fig. 8.2 Pulses on a line. (a) Outgoing and reflected pulses at the sending end ($t_d > \tau$). (b) Variation of the voltage along the line at time t, where $\tau < t < (t_d/2)$.

parallel wires, and assume that a rectangular voltage pulse of duration τ is applied between the two wires at the 'sending end' of the system. The pulse will travel along the wires and in general will be reflected at the termination (or 'receiving end') of the system so that it travels back towards the sending end, where the 'echo' can be detected. If the length of the system is l and the speed of propagation of the pulse is u, then the out-and-back time of flight (t_d) is equal to the total distance travelled divided by the speed, i.e. $2l/u$. Clearly, if the system is long enough so that $t_d > \tau$, then at the sending end the outgoing and reflected pulses will be resolved (see Fig. 8.2(a); it has been assumed that the pulse propagates without distortion). On the other hand, if $t_d < \tau$, then the leading edge of the reflected pulse will arrive back at the sending end of the line before the trailing edge of the outgoing pulse has left. In particular, if $t_d \ll \tau$, then the reflected pulse is only very slightly delayed compared with the outgoing pulse, which implies that, for the duration of the pulse, the voltage between the wires (and hence the associated electric field) at all positions along the system is established at the same, constant, value. This implies in turn that this system can be considered to behave as a single circuit element (or 'lumped' element). In contrast, if $t_d > \tau$, then at any instant of time the voltage is not the same at all positions along the system, the electric field between the wires varies with distance along the wires, and so the system has to be treated as a 'distributed' circuit (see Fig. 8.2(b)).

The distinction between lumped and distributed circuit elements can also be made from a consideration of sinusoidal voltages and currents. For instance, if the electric field E in a resistor is written as $E = \hat{E}\sin(\omega t - (2\pi x/\lambda))$, that is, a wave of amplitude $\hat{E}$, angular frequency ω and wavelength λ travelling in the $+x$-direction, then the variation (δE

say) of the electric field over an increment of distance δx is given by

$$\delta E = \frac{\partial E}{\partial x}\,\delta x = -\frac{2\pi \hat{E}}{\lambda}\cos\left(\omega t - \frac{2\pi x}{\lambda}\right)\delta x.$$

so

$$\frac{|\delta E|_{\max}}{\hat{E}} = 2\pi \cdot \frac{\delta x}{\lambda},$$

i.e. the variation in the electric field strength over the length of a circuit element depends on the ratio (length/wavelength).

If δx is taken as the length of a typical resistor, say 1 cm, and assuming a frequency of $3/2\pi$ MHz (i.e. about 500 kHz), then $|\delta E|_{\max}/\hat{E} = 10^{-4}$. For most practical purposes this would represent a negligible variation, so that the electric field could be regarded as constant over the length of the resistor, which could therefore be treated as a lumped circuit element. However, if the frequency is 10^4 times as high (5 GHz approximately), then $|\delta E|_{\max}/\hat{E} \approx 0.2$, which is a very significant variation indeed and the notion of a distributed resistance (and/or inductance and capacitance) has to be used. The same argument can be applied with even greater force to the connecting wires and cables in a system, since their linear dimensions are generally greater than those of individual circuit elements.

Another physical factor that has increasing significance as the operating frequency is increased is the radiation of energy as electromagnetic (e.m.) waves. Broadly speaking, the power radiated depends on the ratio $(l/\lambda)^2$, where l is the length of the circuit element in question. There are two major consequences of such radiation of power. First, it constitutes a significant additional source of power dissipation in the element and, as such, sets a limit to the quality factor of *LCR* networks. Second, it provides a mechanism that couples signals between components of a circuit, or a system, which is usually very undesirable; e.g. this is a cause of 'crosstalk' and also could provide positive feedback leading to oscillation. One of the greatest advantages of using coaxial cables to make interconnections (even at low frequencies where they can be treated as lumped elements) is that, since the inner conductor is completely surrounded by the outer, there is insignificant radiation of e.m. waves (no 'leakage'). In high-frequency circuits the possibility of unwanted coupling means that it is essential to pay great attention to the layout of the components and to provide screening between subsections of the system (see Chapter 9).

In summary, the concept of lumped circuit elements can be used when the functions of energy storage and dissipation can be associated uniquely, for practical purposes, with identifiable, localized, circuit elements, i.e. reactive and resistive components respectively.

If the ratio l/λ is not small, circuits are described in terms of distributed elements; the following list will serve as a reminder of the wavelengths of e.m. radiation in free space at a variety of frequencies.

Frequency	*Wavelength*	
50 Hz/60 Hz	6000/5000 km	a.c. 'mains'
1 kHz	300 km	audio
1 MHz	300 m	medium-wavelength radio
100 MHz	3 m	VHF
1 GHz	30 cm	SHF
10 GHz	3 cm	

So even at the low frequency of national power distribution systems, the wavelength is not much greater than the dimensions of the system. At the other end of this list, in the realm of microwave integrated circuits (MMICs), a lumped element description may be appropriate, since, even though the wavelength is only 3 cm, say, the dimensions of the circuit components are much smaller.

The term **transmission line** is usually reserved for structures that are at least a significant fraction of a wavelength in length and have uniform electromagnetic properties along their length. Examples are as follows.

coaxial lines: flexible, semiflexible, or rigid with solid insulation, perforated insulation or air-spaced.

parallel pair of conductors: parallel wires (air-spaced, insulated); parallel tracks on an insulating substrate.

strip-line: microstrip (in microwave integrated circuits): conducting strips insulated from a conducting ground plane.

waveguides: hollow metal pipes (commonly of rectangular cross-section).

8.2 Pulses on transmission lines

In transmission lines, the energy losses, and hence the attenuation of a pulse propagating along a line, arise from three physical mechanisms, namely the electrical resistance of the conductors, losses in the dielectric medium (see Section 5.1), and the leakage of e.m. radiation to the 'outside world'. For many types of line in practical use, the losses are relatively small and, for a wide range of practical purposes, such lines can be treated as 'lossless'; a fortunate consequence of this assumption is that the analysis of the propagation of signals is much simpler than it would otherwise be.

In order to discuss the propagation of pulses on a uniform transmission line, consider the model shown in Fig. 8.3. A battery of e.m.f. E_s, having internal resistance R_s, is connected via a switch to the 'sending end' of a transmission line represented by the two parallel wires of length l which

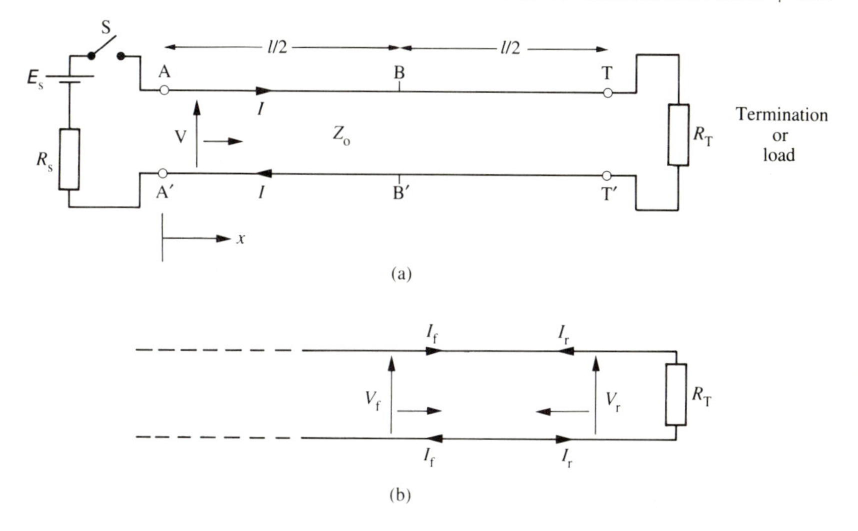

Fig. 8.3 (a) A representation of a source E_s of internal resistance R_s connected via a switch S to the 'sending end' of a uniform transmission line of length l, terminated by a load resistance R_T. (b) The model for the incident and reflected voltages and currents on the line.

are assumed to have zero resistance. Note also that in the case of a uniform line, signals propagate at a constant speed u.

On closing the switch S, a 'step' voltage is launched on the line and an electric field is set up between the two conductors. The step disturbance in the electric field is guided along the pair of conductors without attenuation, since the line is lossless, and is accompanied, as it must be, by an associated magnetic field. Now it should be remembered that there is stored energy associated with an electric field and with a magnetic field (see Sections 1.5 and 1.7), so, as the disturbance propagates along the line at a constant speed, the amount of energy stored on the line increases at a constant rate. This energy is supplied by the battery at a rate $E_s I$, where I is the current flowing in the conductors of the transmission line, and can be imagined to be stored in the distributed capacitance (C per unit length) and the distributed inductance (L per unit length) of the line.

If the speed of propagation of the step voltage along the line is denoted by u, then at a time t after the launching the disturbance, the source 'sees' a line of length ut only (NB the termination becomes 'visible' to the source only after a time $t = l/u$.

So, for $ut < l$, the source 'sees' a time-dependent capacitance C', where

$$C' = C \cdot u \cdot t. \tag{8.1a}$$

Similarly there is a time-dependent inductance L' given by

$$L' = L \cdot u \cdot t. \tag{8.1b}$$

If the line voltage is denoted by V, then the charge Q on the capacitance C' is given by $Q = C'V$, and the current through the capacitance is equal to $\mathrm{d}Q/\mathrm{d}t$. Hence it follows that the expression for the current I on the line is

$$I = \frac{\mathrm{d}}{\mathrm{d}t}(C'V) = V\frac{\mathrm{d}C'}{\mathrm{d}t}, \tag{8.2a}$$

since the voltage on the line has been assumed not to vary with distance along the line.

Similarly, the voltage associated with the rate of change of magnetic flux linked with the increasing effective length of line (see Section 1.7) is given by

$$V = \frac{\mathrm{d}}{\mathrm{d}t}(L'I),$$

and so it follows that the voltage between the lines is

$$V = I\frac{\mathrm{d}L'}{\mathrm{d}t}. \tag{8.2b}$$

From eqns (8.1) and (8.2) it follows that

$$u = \frac{1}{\sqrt{(LC)}} \tag{8.3}$$

and

$$Z_0 \equiv \frac{V}{I} = \sqrt{\left(\frac{L}{C}\right)}. \tag{8.4}$$

Z_0 is the so-called **characteristic impedance** of the transmission line. In the case of a lossless line, as is being considered at the moment, Z_0 is purely resistive, but in general, Z_0 is a complex impedance. The characteristic impedance is a very important property of a transmission line, as will be seen.

The values of L and C are determined by the geometry of the conductors and the relative permittivity (ϵ_r) and relative permeability (μ_r) of the insulating dielectric medium between and around the conductors. Expressions for C and L were obtained in Section 1.5 and 5.2.2, respectively, for a parallel pair of conductors and a coaxial line (apart from the multiplicative constants ϵ_r, μ_r).

From eqns (1.31) and (5.10) for a coaxial line, for instance, and using eqn (8.3), the speed of propagation u is given by

$$u = (\mu_r \mu_0 \epsilon_r \epsilon_0)^{-1/2}. \tag{8.5}$$

(This expression is true for t.e.m. modes on other types of line also.)

Although Z_0 has the dimensions of resistance, it must not be thought of as a circuit element in which energy is being dissipated. Rather, it represents the ratio (voltage/current) for an e.m. disturbance propagating along the line in question; the voltage and current are directly related to the electric field and the magnetic field, respectively, in the disturbance.

Expressions for the voltage V and current I on the line can be obtained from a consideration of the power flow from the source that is required to establish the electric and the magnetic fields on the line.

Using the expressions for the energy stored in a capacitance and an inductance, the rate of flow $\mathrm{d}W/\mathrm{d}t$ of energy into the line is given by

$$\frac{\mathrm{d}W}{\mathrm{d}t} = \frac{\mathrm{d}}{\mathrm{d}t}\left\{\frac{C'V^2}{2} + \frac{L'I^2}{2}\right\}.$$

The power ($E_s I$) delivered by the battery is equal to ($\mathrm{d}W/\mathrm{d}t + I^2 R_s$) and so, using eqns (8.1), it follows that

$$E_s I = \frac{u}{2}\{V^2 C + I^2 L\} + I^2 R_s$$

Using eqns (8.3) and (8.4), straightforward algebra leads to

$$V = \frac{E_s Z_0}{(Z_0 + R_s)}$$

(8.6) and

$$I = \frac{E_s}{(Z_0 + R_s)}.$$

What happens when the voltage step reaches the terminating resistance (or 'load') R_T? For the e.m. disturbance propagating along the line, the ratio (voltage/current) is equal to Z_0, as has just been seen, but for the load this ratio must be equal to R_T! How is this apparent incompatibility resolved?

The general mathematical solution of the problem includes, in fact, 'right-to-left' propagation of disturbances as well as the 'left-to-right' propagation that was assumed above without comment.

Assume that a right-to-left step, or 'reverse' step (V_r, I_r; see Fig. 8.3(b)), is launched at the load end of the line at the instant that the left-to-right step arrives; i.e. assume that there is a reflected voltage step. If the left-to-right step voltage, or 'forward' step, is now denoted by V_f, the following relationships hold:

$$\frac{V_f}{I_f} = Z_0 \qquad \frac{V_r}{-I_r} = Z_0, \tag{8.7}$$

where the sign convention that has been adopted for voltages and currents to help to keep track of successive reflections is as follows.

A positive line voltage is one for which the 'top' conductor is positive with respect to the 'bottom' conductor. A 'positive' current is one that flows left-to-right in the 'top' conductor and right-to-left in the 'bottom' conductor.

Now the voltage V_T across the load and the current I_T through it are related by

$$\frac{V_T}{I_T} = R_T, \tag{8.8}$$

where

$$V_{\mathrm{T}} = (V_{\mathrm{f}} + V_{\mathrm{r}}), \tag{8.9}$$

and, using the above convention,

$$I_{\mathrm{T}} = (I_{\mathrm{f}} + I_{\mathrm{r}}) \tag{8.9}$$

(see Fig. 8.3(b)).

A **voltage reflection coefficient** Γ_{T} for the termination is defined through

$$\Gamma_{\mathrm{T}} \equiv \frac{V_{\mathrm{r}}}{V_{\mathrm{f}}}. \tag{8.10}$$

It is left as a simple algebraic exercise to show that

$$\Gamma_{\mathrm{T}} = \frac{\left(\dfrac{R_{\mathrm{T}}}{Z_0} - 1\right)}{\left(\dfrac{R_{\mathrm{T}}}{Z_0} + 1\right)}. \tag{8.11}$$

So the answer to the question posed just above is that at the termination there is a reflected 'step' such that the resultant of the incident and reflected voltages divided by the resultant of the associated line currents is equal to the resistance of the load.

The power P_{T} absorbed in the load is equal to $V_{\mathrm{T}}I_{\mathrm{T}}$ and, using eqns (8.9), it therefore follows that

$$V_{\mathrm{T}}I_{\mathrm{T}} = (V_{\mathrm{r}} + V_{\mathrm{f}})(I_{\mathrm{f}} + I_{\mathrm{r}}),$$

and, using eqns (8.7),

$$P_{\mathrm{T}} = \frac{V_{\mathrm{f}}^2}{Z_0}\left(1 + \frac{V_{\mathrm{r}}}{V_{\mathrm{f}}}\right)\left(1 - \frac{V_{\mathrm{r}}}{V_{\mathrm{f}}}\right),$$

whence

$$P_{\mathrm{t}} = \frac{V_{\mathrm{f}}^2}{Z_0}(1 - \Gamma_{\mathrm{T}}^2)$$

or

$$P_{\mathrm{T}} = \frac{V_{\mathrm{f}}^2}{V_0} - \frac{V_{\mathrm{r}}^2}{Z_0}. \tag{8.12}$$

The latter form of eqn (8.12) expresses what could be deduced qualitatively simply by invoking the principle of conservation of energy, namely that the power absorbed in the load is equal to the difference between the power in the incident disturbance and that in the reflected disturbance. However, it is comforting that the adopted model together with the sign convention gives the same result.

Two extreme situations at the termination are of particular interest, namely $R_T = \infty$ (open-circuit) and $R_T = 0$ (short-circuit). For the open-circuit condition eqn (8.11) yields $\Gamma_T = +1$ and hence $V_r = V_f$ (and $V_T = 2V_f$), whilst for the short-circuit condition $\Gamma_T = -1$ and $V_r = -V_f$ (i.e. $V_T = 0$).

It is very important to note from eqn (8.11) that the reflection coefficient is zero if $R_T = Z_0$ and that in this circumstance all the power in the incident disturbance is absorbed in the load. In this case the load is said to be matched to the line.

Example 8.1 Determine the voltage and current on the line subsequent to the switch S being closed at time $t = 0$, given that $R_T = 2Z_0$ and $R_s = Z_0$ (i.e. the source is matched to the line in question).

Using eqn (8.11) the reflection coefficient $\Gamma_T = 1/3$, and from eqns (8.6), $V_f = E_s/2$. It therefore follows from eqns (8.9) that $V_T = 2E_s/3$. Hence, for time $t \leqslant l/u$, a voltage step of amplitude $E_s/2$ travels towards the termination until by $t = l/u$ a constant voltage $E_s/2$ is established along the length of the line. The reflected step of amplitude $(E_s/2)/3$ travels back towards the sending end of the line, combining additively with the already established constant voltage $E_s/2$, so that by time $t = 2l/u$ a constant voltage $2E_s/3$ is established on the line. Since the source is matched to the line, there are no further reflections at the sending end. The voltage on the line is sketched for successive times up to $t \geqslant 2l/u$ in Fig. 8.4(a).

The development of the current on the line is sketched in Fig. 8.4(b). For $t \leqslant l/u$ the line current I_f is $E_s/2Z_0$, and for $t \geqslant l/u$, the reflected (negative) step I_r has a magnitude of $I_f/3$. The reflection propagates back to the sending end and, for $t \geqslant 2l/u$, the steady current has a value of $E_s/3Z_0$. □

For capacitive and inductive terminations, the time-dependences of the voltages at the sending end are sketched in Fig. 8.5(a) and (b). In the case of the capacitive termination the capacitor acts initially as a short-circuit ($V_C = 0$) and ultimately as an open-circuit, i.e. when it has become fully charged to the line voltage ($I_C = 0$). The inductor behaves in the opposite sense: initially its acts as an open-circuit ($I_L = 0$) and ultimately as a

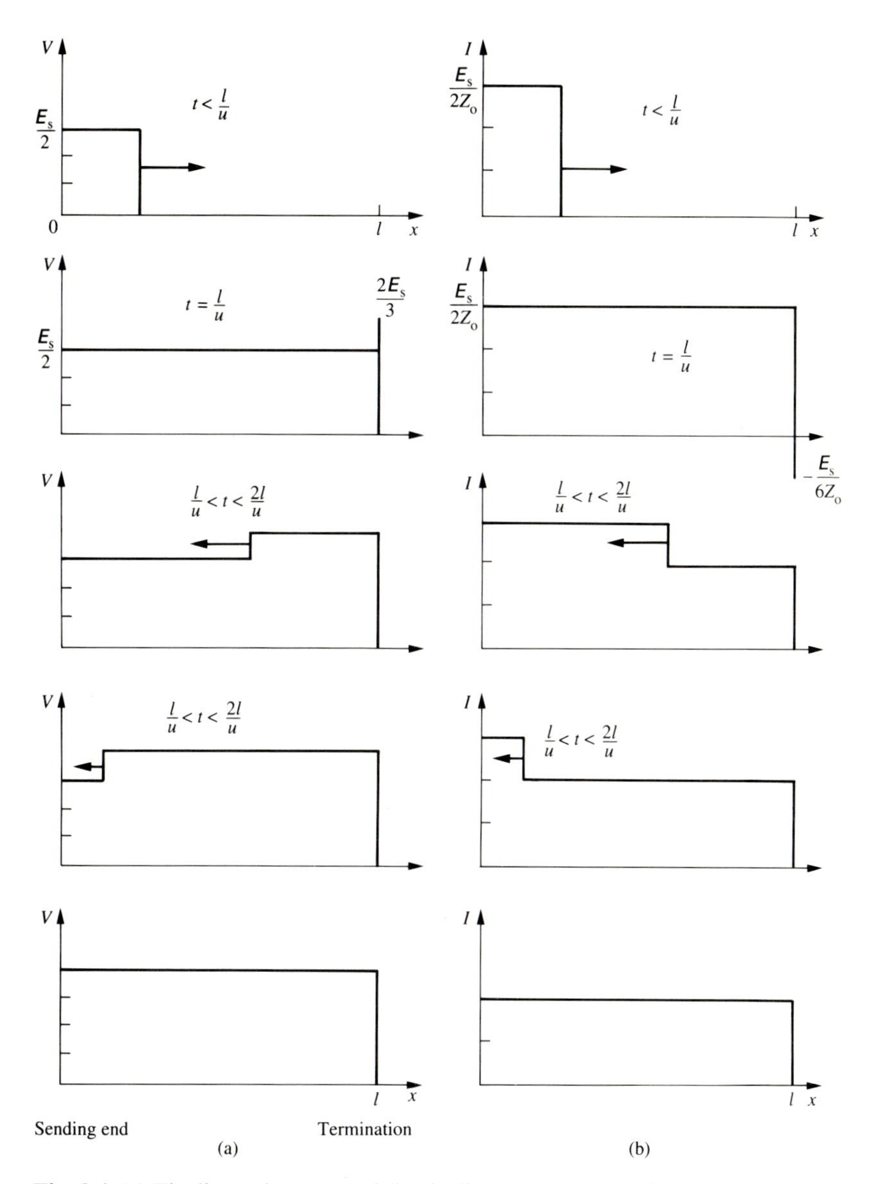

Fig. 8.4 (a) The line voltage V and (b) the line current I, as a function of distance along the lossless transmission line of Fig. 8.3(a) at successive times under the conditions $R_s = Z_0$ and $R_T = 2Z_0$.

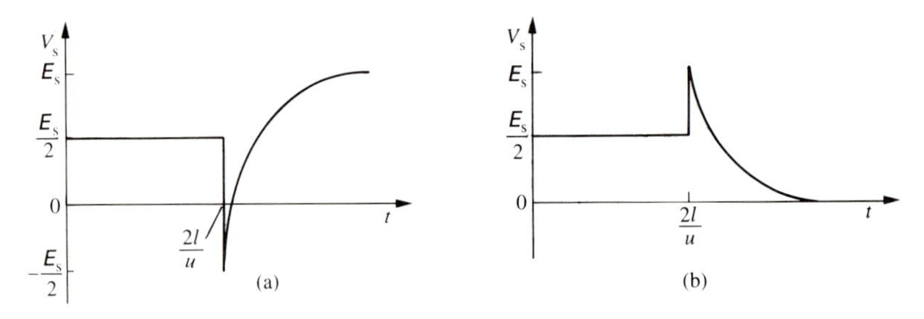

Fig. 8.5 The line voltage V_s at the sending end of the line for (a) a capacitive termination and (b) an inductive termination.

short-circuit when the current through it has finally attained the constant value (E_s/R_s) of the line current.

If the source is not matched to a line, then a useful means for assessing the successive reflections at the termination and at the sending end is a **lattice diagram**.

For a step voltage source of amplitude E_s, then the voltage V_1 launched on the line at the sending end is given by eqn (8.6):

$$V_1 = \frac{Z_0}{(Z_0 + R_s)} E_s.$$

Hence, if the reflection coefficient at the sending end is denoted by Γ_s, the voltage at the sending end develops (see Fig. 8.6) according to

$$V_{x=0} = V_1(1 + \Gamma_T + \Gamma_T\Gamma_s + \Gamma_T^2\Gamma_s + \Gamma_T^2\Gamma_s^2 + \cdots)$$

or
$$V_{x=0} = V_1 + V_1\Gamma_T(1 + \Gamma_s)(1 + \Gamma_T\Gamma_s + (\Gamma_T\Gamma_s)^2 + \cdots)$$

a geometrical progression

Now the series in parentheses is a geometrical progression and so, on using the expression for the sum to infinity of this series, namely $(1 - \Gamma_T\Gamma_s)^{-1}$, it follows that the ultimate (i.e. steady-state) value of the line voltage V_1 is given by

$$V_{x=0} = V_1 \cdot \frac{(1 + \Gamma_T)}{(1 - \Gamma_T\Gamma_s)}. \tag{8.13}$$

It can easily be shown, by substituting for Γ_T and Γ_s, that this expression for $V_{x=0}$ reduces to

$$V_{x=0} = E_s \cdot \frac{R_T}{(R_T + R_s)},$$

which corresponds to the steady-state situation (i.e. $t \to \infty$).

For a 'short' open-circuited line, say 10 cm in length, for which $u = 2c/3$, then $t_d = 1$ ns. Further, if $\Gamma_s = -1/2$, say, then after the 4th

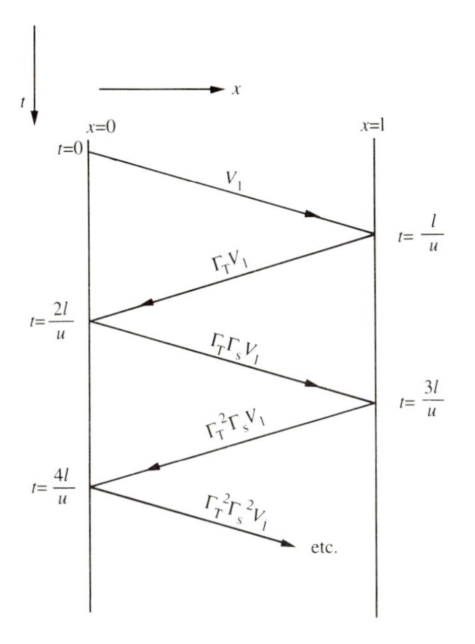

Fig. 8.6 A lattice diagram for a step voltage V_1 launched on a line of length l and on which the speed of propagation is u. The voltage reflection coefficients at the sending end and at the termination are denoted by Γ_s, Γ_T respectively.

reflection from the sending end $V_{x=0}$ has reached >98 per cent of its steady-state value, i.e. after only 4 ns! This emphasizes why these effects are not of practical significance, nor easily noticeable, in many circuit situations, especially if the rise times of the signals of interest are much greater than 1 ns. Also, of course, real lines are not lossless and successive reflections are attenuated in transit for this reason alone.

One reason for the interest in the successive reflections on a line is that a reflected pulse may cause the source to malfunction; for instance, in a digital system the specified tolerances on the working voltages of a logic gate device may be violated. Hence the terminations of lines in digital systems, both at the 'load' end and at the 'source' end must be designed carefully. It is worth noting that a closely packed sequence of memory elements connected to a line acts as an effective distributed capacitance of perhaps 10 pF cm^{-1}, and typical propagation delays in such 'loaded' lines are of the order of 1 ns for 10 cm of line.

It will be remembered that it is impossible to generate an ideal, 'mathematical' step voltage or current in a real circuit, and a better approximation to a real shape would be the 'truncated ramp' function, with rise time τ, which is sketched in Fig. 8.7(a). The analysis of the propagation and reflection is slightly more complicated than for an ideal step function, but the same principles are involved and the same general features emerge; see Fig. 8.7(b).

If a signal function has a characteristic time τ (say the duration, or the

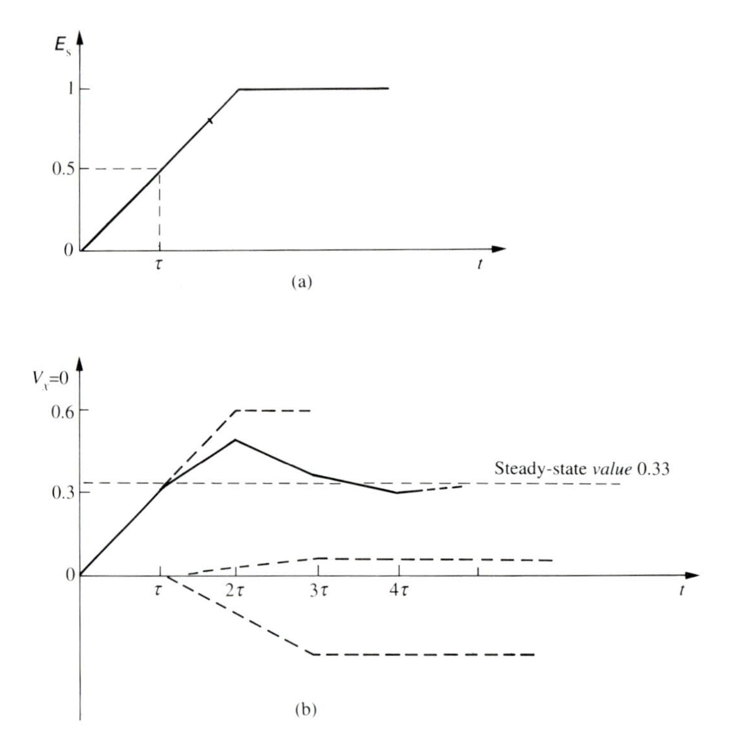

Fig. 8.7 (a) A truncated ramp function: rise time $\equiv \tau$. (b) The line voltage at the sending end of the (lossless) line, assuming $Z_0 = 75\ \Omega$, $R_T = 25\ \Omega$, and $R_s = 50\ \Omega$ (i.e. $\Gamma_T = -1/2$, $\Gamma_s = -1/5$), and also assuming that $t_d = \tau$..

rise time, of a pulse), then a convenient criterion for broadly discriminating between a lumped element description or a distributed element description of the circuit is the boundary condition $t_d = \tau$. If $t_d > \tau$, then a distributed element description must be used.

The speed of propagation of pulses usually lies in the range 2×10^8 m s^{-1} to 3×10^8 m s^{-1} (the speed of e.m. waves in free space), so that the corresponding propagation delays are in the range 5 ns m^{-1} and 3.3 ns m^{-1}.

In high-speed logic systems, the transitions between logic states occur in times of the order of nanoseconds, or less, and so the data buses have to be treated as distributed circuit elements. Logic gates are extremely non-linear devices with very different circuit properties in their two logic states, and this complicates the analysis of the propagation of the logic signals. Short-range interconnections between integrated-circuit packages may be realised by twisted-pair wire, ribbon cable, or printed-circuit board tracks. Also, the current pulses drawn by logic gates cause voltage spikes on the supply lines, which can interact with other gates and circuits; this

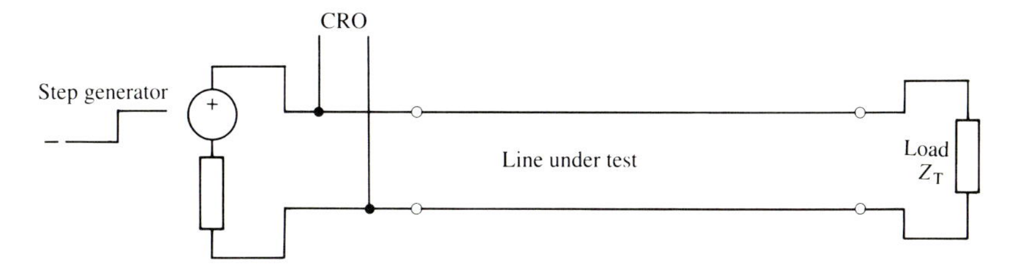

Fig. 8.8 The essential features of a time-domain reflectometer.

has to be guarded against by using reservoir capacitors and by terminating the supply lines correctly to suppress reflections. For longer-range interconnections, such as between networked computer terminals, coaxial cable is used. In this situation the required maximum information rate places a requirement on the operating bandwidth of the line, but other relevant factors are the attenuation per metre, mechanical flexibility (and cost!) of the available types of cable.

Measurements of the time-of-flight t_d and the amplitude and shape of reflected pulses can yield useful information about the characteristics of a line and of its termination; this is the technique of **time-domain reflectometry (TDR)**. The essential features of a system are shown in Fig. 8.8. A 'good' step voltage (rise time $\leqslant 0.1$ ns) is launched on the line under test and the incident and reflected waveforms are monitored using a CRO. The characteristic impedance of the line can be determined simply by measuring the value of the reflection coefficient for a value of terminating resistance not equal to Z_0: eqn (8.11) can be transposed to the form

$$Z_0 = R_T \frac{(1 - \Gamma_T)}{(1 + \Gamma_T)}. \tag{8.14}$$

For instance for $\Gamma_T = 1/3$, $R_T = 2Z_0$.

Any discontinuities in the line, such as connectors or faults, will generally cause reflections; the distance of a 'fault' from the sending end of the line can be determined if the speed of propagation of the step is known. The shape of a reflected signal gives information about any significant reactive component in the impedance of a reflecting element. The rise time of the voltage from the step generator determines the shortest distance between discontinuities on the line that can be resolved; shortest distance = (propagation speed × rise time)/2. For a high-resolution system the bandwidth B of the CRO amplifier must be great enough so that signal rise times as least as short as that of the original step voltage can be accommodated; the 'reciprocal spreading' criterion (see Section 6.6) requires that $B \approx 1/t_r$.

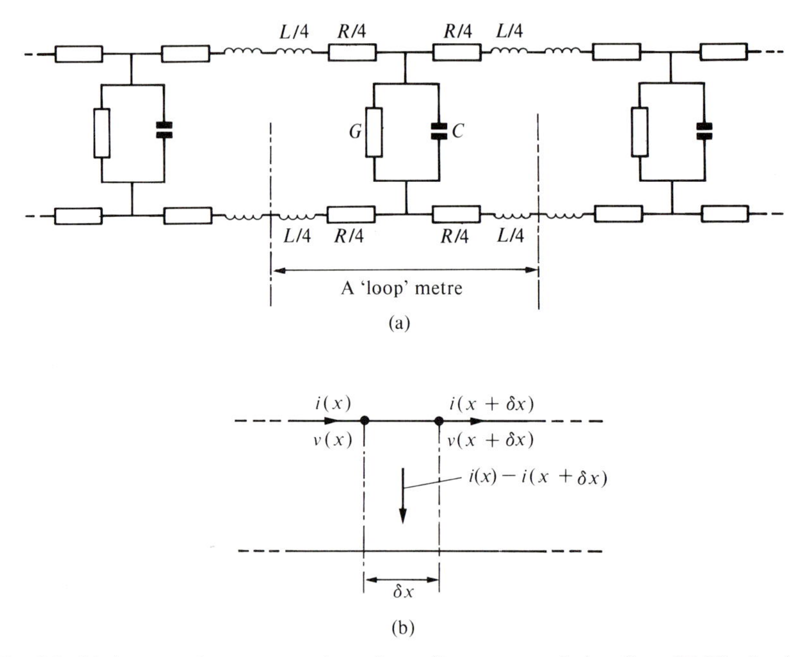

Fig. 8.9 (a) A general representation of a uniform transmission line, (b) The basis for a description of the variation of voltage and current along a line.

For $t_r \approx 1$ ns this requires $B \approx 1$ GHz, although for lines which are long enough, a bandwidth of 60 MHz to 100 MHz can be usefully exploited.

Real lines are lossy to some extent, of course, and also they are **dispersive**. The latter term means that the speed of propagation of a signal on the line is frequency-dependent. Hence the Fourier components of a signal travel at different speeds and the signal becomes distorted as it progresses along the line. The attenuation coefficient is frequency-dependent also, which is an additional source of distortion.

8.3 Continuous waves (c.w.) on transmission lines

8.3.1 Basic theory

The distribution of the energy storage and energy dissipation in a uniform, symmetrical line can be represented by a circuit of the form shown in Fig. 8.9. Here R is the distributed resistance per loop metre and L and C relate in a similar way to the distributed inductance and capacitance. The distributed shunt conductance G represents both the conductivity (usually negligible) and the dielectric losses of the insulating medium between and around the conductors.

If the line is imagined to extend in the x-direction, then the variation in the voltage between the conductors and the current along the conductors can be represented as in Fig. 8.9(b). Using this model, the expressions for the voltage $v(x, t)$ on the line, the characteristic impedance of the line, and the complex propagation coefficient are as given in eqns (8.24), (8.21) and (8.22) respectively. These equations can be obtained as outlined below.

Assuming that $\{\partial i(x)/\partial x\}\delta x$ can be neglected compared with $i(x)$, and that $\{\partial v(x)/\partial x\}\partial x$ can be neglected compared with $v(x)$, then the application of Kirchhoff's laws yields

$$v(x) - \left(v(x) + \frac{\partial v(x)}{\partial x}\,\delta x\right) = \left\{R\, i(x) + L\frac{\partial i(x)}{\partial t}\right\}\delta x$$

and

$$i(x) - \left(i(x) + \frac{\partial i(x)}{\partial x}\,\delta x\right) = \left\{G\, v(x) + C\frac{\partial v(x)}{\partial t}\right\}\delta x,$$

which give

$$-\frac{\partial v(x)}{\partial x} = R\, i(x) + L\frac{\partial i(t)}{\partial t}$$

(8.15) and

$$-\frac{\partial i(x)}{\partial x} = G\, v(x) + C\frac{\partial v(x)}{\partial t}.$$

For the c.w. situation considered in this section (i.e. for continuous sinusoidal waves), $v(x, t)$, $i(x, t)$ can be expressed as

$$v(x, t) = \hat{V}(x)\cdot \mathrm{e}^{\mathrm{j}\omega t} \qquad \text{and} \qquad i(x, t) = \hat{I}(x)\cdot \mathrm{e}^{\mathrm{j}\omega t},$$

whence, on substituting in eqns (8.15), it follows that

$$-\frac{\partial \hat{V}(x)}{\partial x} = (\hat{R} + \mathrm{j}\omega L)I(x) \qquad \text{and} \qquad -\frac{\partial \hat{I}(x)}{\partial x} = (\hat{G} + \mathrm{j}\omega C)V(x).$$

On differentiating the first of these two equations with respect to x and the second with respect to t, then

$$\frac{\mathrm{d}^2\hat{V}(x)}{\mathrm{d}x^2} = \gamma^2\hat{V}(x) \qquad \text{and similarly} \qquad \frac{\mathrm{d}^2\hat{I}(x)}{\mathrm{d}x^2} = \gamma^2\hat{I}(x), \tag{8.16}$$

where $\gamma \equiv \sqrt{[(R + \mathrm{j}\omega L)(G + \mathrm{j}\omega C)]}$. Historically, eqns (8.16) were known as the 'Telegrapher's equations'.

Table 8.1. Characteristics of some commonly used transmission lines

Type	C pF m^{-1}	Z_0 Ω	α dB m^{-1}
Twin axial data transmission cable	65	78	0.069 at 10 MHz
LAN cable (ethernet)	86	50	0.017 at 10 MHz
Radiofrequency coaxial cable	100	50	0.16 at 100 MHz
	51	75	0.12 at 100 MHz
Balanced twin feeder	13	300	0.16 at 1000 MHz

So in most of the following discussions the transmission lines in question will be assumed to be lossless.

If boundary conditions are defined in relation to the voltage and current at $x = 0$ (the 'sending end' of the line), then

$$V(x) = V_s; \qquad I(x) = I_s, \qquad \text{say, at } x = 0.$$

On substituting these conditions in eqns (8.19) and (8.20) it follows simply that

$$A = \frac{V_s + I_s Z_0}{2} \quad \text{and} \quad B = \frac{V_s - I_s Z_0}{2}. \tag{8.28}$$

Using these expressions for A, B in eqns (8.19) and (8.20), then

$$\begin{aligned} V(x) &= V_s \cosh \gamma x - I_s Z_0 \sinh \gamma x \\ I(x)_s &= I \cosh \gamma x - \frac{V_s}{Z_0} \sinh \gamma x, \end{aligned} \tag{8.29}$$

where the hyperbolic functions are defined through

$$\cosh \theta \equiv (e^{\theta} + e^{-\theta})/2, \qquad \sinh \theta \equiv (e^{\theta} - e^{-\theta})/2. \tag{8.30}$$

The impedance $Z(x)$ at any position (or 'plane') along the line is given by $Z(x) \equiv V(x)/I(x)$, the 'line impedance'.

8.3.2 A terminated line

Real lines are finite in length, and a vital question here is: what is the 'sending-end impedance' Z_s of a line of length l that is terminated by a load of impedance Z_T?

Consider the situation represented schematically in Fig. 8.10. If the line voltage and current at $x = l$ are denoted by V_T, I_T respectively, where

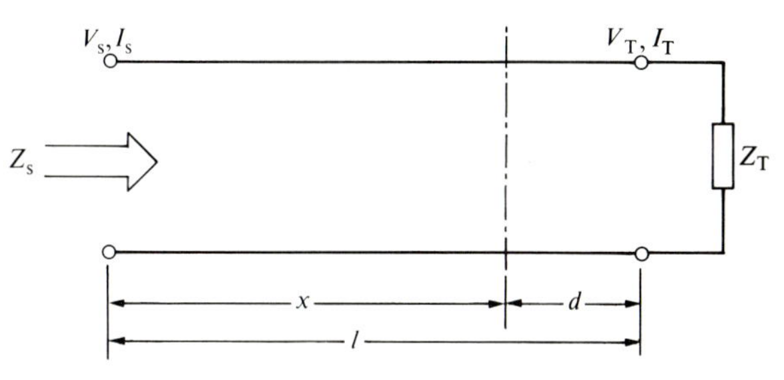

Fig. 8.10 The specification of the voltages, currents, and impedances on a general terminated line of length l.

$V_T/I_T = Z_T$ (note that the '$\hat{\ }$' has now been discarded), then from eqns (8.29)

$$V_T = V_s \cosh \gamma l - I_s Z_0 \sinh \gamma l$$

$$I_T = I_s \cosh \gamma l - \frac{V_s}{Z_0} \sinh \gamma l$$

or

$$Z_T = \frac{V_s \cosh \gamma l - I_s Z_0 \sinh \gamma l}{I_s \cosh \gamma l - \dfrac{V_s}{Z_0} \sinh \gamma l}.$$

Further, $Z_s \equiv V_s/I_s$ and so

$$Z_T = \frac{Z_s \cosh \gamma l - Z_0 \sinh \gamma l}{\cosh \gamma l - \dfrac{Z_s}{Z_0} \sinh \gamma l}.$$

This equation can easily be rearranged to give the very important expression

$$\frac{Z_s}{Z_0} = \frac{Z_T/Z_0 + \tanh \gamma l}{1 + \dfrac{Z_T}{Z_0} \tanh \gamma l} \tag{8.31}$$

where $\tanh \gamma l \equiv \sinh \gamma l / \cosh \gamma l = (e^{2\gamma l} - 1)/(e^{2\gamma l} + 1))$.

For the special case of a lossless line ($\alpha = 0$; $\gamma = j\beta$),

$$\frac{Z_s}{Z_0} = \frac{Z_T/Z_0 + j \tan \beta l}{1 + j \dfrac{Z_T}{Z_0} \tan \beta l} \tag{8.32}$$

(note that $\tanh j\theta = j \tan \theta$).

Now from eqns (8.28) the expression for the constant B can be rewritten as

$$B = \frac{I_s}{2}(V_s/I_s - Z_0) = \frac{I_s}{2}(Z_s - Z_0).$$

If $Z_T = Z_0$, then eqn (8.31) gives $Z_s = Z_0$ also, and in this circumstance $B = 0$, which means that there is no wave propagating in the $-x$ direction, i.e. all the power in the incident wave is absorbed in the load and there is no reflected wave. The load is said to be matched to the line in question. This parallels the situation for transient signals (pulses) on transmission lines, described in Section 8.2.

Two extreme cases are:

(8.32) open-circuit line: $Z_T = \infty$ and $Z_s\,(\equiv \text{'}Z_{oc}\text{'}) = Z_0 \coth \gamma l$

(8.33) short-circuited line: $Z_T = 0$ and $Z_s\,(\equiv \text{'}Z_{sc}\text{'}) = Z_0 \tanh \gamma l.$

From eqns (8.32) and (8.33) it follows that,

$$\tanh \gamma l = \sqrt{(Z_{sc}/Z_{oc})} \tag{8.35}$$

and

$$Z_0 = \sqrt{(Z_{sc} \cdot Z_{oc})}. \tag{8.36}$$

This means that the values of Z_0 and γ for a line (particularly its real part, the attenuation coefficient α) can be obtained from open- and short-circuit *measurements* made at the sending end of the line.

The concept of a characteristic impedance was discussed earlier in the context of symmetrical two-port networks (see Section 7.2, eqn (7.23); indeed, a uniform transmission can be modelled as a ladder network consisting of cascaded identical symmetrical T- (or Π-) sections.

Example 8.2 Short-circuit and open-circuit c.w. measurements at 100 MHz on a 25 m length of coaxial cable yielded the following results:

$$Z_{sc} = 19.5\underline{/+68^\circ}\ \Omega \qquad Z_{oc} = 290\underline{/-74^\circ}\ \Omega.$$

Determine the characteristic impedance Z_0 and the attenuation coefficient α for the cable.

Characteristic impedance:

$$Z_0 = \sqrt{(Z_{sc}Z_{oc})} = \sqrt{(19.5 \times 290\underline{/68^\circ + (-74^\circ)})}$$

or

$$Z_0 = 75.2\underline{/-3^\circ}\ \Omega.$$

Attenuation coefficient α:

NB the steps of the analysis, as spelled out below, may seem rather clumsy; an actual calculation (with the aid of a calculator/computer) is actually quite straightforward to perform!

$\left(\dfrac{Z_{sc}}{Z_{oc}}\right)$ is a complex number $\dfrac{|Z_{sc}|\angle Z_{sc}}{|Z_{oc}|\angle Z_{oc}}$, which can be written as $m\angle\phi$ (polar form), where $m \equiv |Z_{sc}|/|Z_{oc}|$ and $\phi \equiv \angle Z_{sc} - \angle Z_{oc}$. It turns out that $\sqrt{m}$ will be required, i.e. $m^{1/2}\angle\phi/2$. The cartesian form of m is $m = x + \mathrm{j}y$, where

$$x = m^{1/2}\cos\phi/2 \qquad \text{and} \qquad y = m^{1/2}\sin\phi/2.$$

Using $\tanh\gamma l = (\mathrm{e}^{2\gamma l} - 1)/(\mathrm{e}^{2\gamma l} + 1)$, it follows that

$$\mathrm{e}^{2\gamma l} = \frac{1 + x + \mathrm{j}y}{1 - x - \mathrm{j}y}$$

$$= \frac{r_1\angle\theta_1}{r_2\angle\theta_2},$$

where

$$\begin{aligned} r_1^2 &= (1 + x)^2 + y^2 \\ \tan\theta_1 &= y/(1 + x) \\ r_2^2 &= (1 - x)^2 + y^2 \\ \tan\theta_2 &= y/(1 - x). \end{aligned}$$

The above expression for $\mathrm{e}^{2\gamma l}$ can be written as $\mathrm{e}^{2\gamma l} = r\angle\theta$, say, where $r \equiv r_1/r_2$, $\theta \equiv \theta_1 - \theta_2$.

So, remembering that $\gamma = \alpha + \mathrm{j}\beta$, it follows that

$$\mathrm{e}^{2\alpha l}\cdot\mathrm{e}^{\mathrm{j}2\beta l} = r\angle\theta,$$

and so $2\alpha l = \ln r$ and $2\beta l = \theta \pm 2n\pi$. Hence

$$\alpha = \frac{\ln r}{2l}\ \mathrm{Np\ m^{-1}} \qquad \text{or} \qquad \alpha = \frac{8.68\ln r}{2l}\ \mathrm{dB\ m^{-1}}.$$

Turning to the particular problem, $m^{1/2} = 0.259$ and $\phi/2 = -3°$, which leads to $x = 0.259$, $y = -0.001$. Thence $r_1 = 1.259$, $\theta_1 = -0.005°$, $r_2 = 0.741$, $\theta_2 = -0.008°$, and so $r = 1.70$, $\theta = -0.013°$. Finally, $\alpha = 0.09\ \mathrm{dB\ m^{-1}}$.

□

A situation of considerable interest is when a line has a length equal to an odd integral number of quarter-wavelengths of the waves on the line: a **quarter-wavelength line** (or '$\lambda/4$' line). Assuming, for simplicity, that

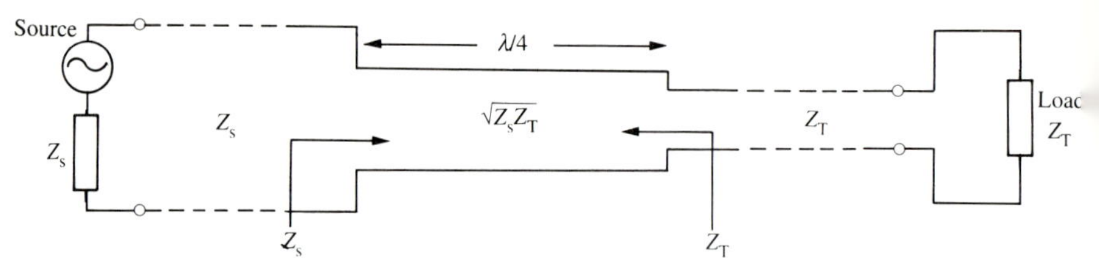

Fig. 8.11 Impedance matching using a quarter-wavelength line (or 'transformer').

losses in the line are negligible, then, for $l = \lambda/4$, $\beta l = \pi/2$ and $\tan \beta l = \infty$. In these circumstances eqn (8.32) becomes

$$\frac{Z_s}{Z_0} = \frac{Z_0}{Z_T} \quad \text{or} \quad Z_s = \frac{Z_0^2}{Z_T}. \tag{8.37}$$

Consider the situation illustration in Fig. 8.11, where a generator of impedance Z_s is connected to a load Z_T via a $\lambda/4$ line of characteristic impedance Z_0. If $Z_0 = \sqrt{(Z_sZ_T)}$, where Z_s, Z_T are the characteristic impedances of the lines connected to the generator and load respectively, then, as can be seen from eqn (8.36), the $\lambda/4$ line forms of power-matching transformer between the generator and the load, i.e. the impedance 'looking into' the $\lambda/4$ line is equal to Z_s. Also, and importantly, the impedance looking into the $\lambda/4$ line from the load end is equal to Z_T, i.e. the $\lambda/4$ is matched looking in at both ends.

Some aspects of the practical importance of power matching on transmission lines will be discussed in Section 8.4.

8.3.3 Standing waves on transmission lines

For the general case where the termination to a line is not matched to the characteristic impedance of the line, there is at least a partial reflection of the incident wave back towards the sending end; this situation is described by eqn (8.19).

A reflecting coefficient Γ_T for the termination (load) is defined in terms of the ratio of the 'right-to-left' and 'left-to-right' waves at $x = l$, the position of the load. Using eqn (8.19), and assuming a lossless line so that $\gamma = j\beta l$,

$$\Gamma_T \equiv \frac{B\,e^{j\beta l}}{A\,e^{-j\beta l}} \quad \text{or} \quad \Gamma_T = \frac{B}{A}\,e^{j2\beta l}. \tag{8.38}$$

From eqns (8.19) and (8.20), with $x = l$,

$$Z_T\left(=\frac{V_{x=1}}{I_{x=1}}\right) = Z_0 \cdot \frac{A\,e^{-j\beta l} + B\,e^{j\beta l}}{A\,e^{-j\beta l} - B\,e^{j\beta l}}$$

or

$$\frac{Z_T}{Z_0} = \frac{1 + \dfrac{B}{A}\,e^{j2\beta l}}{1 - \dfrac{B}{A}\,e^{j2\beta l}},$$

i.e.

$$\frac{Z_T}{Z_0} = \frac{1 + \Gamma_T}{1 - \Gamma_T}; \tag{8.39}$$

alternatively,

$$\Gamma_T = \frac{\dfrac{Z_T}{Z_0} - 1}{\dfrac{Z_T}{Z_0} + 1}. \tag{8.40}$$

Since Z_T is a complex impedance, in general, then Γ_T is a complex quantity; i.e. the reflected wave differs in amplitude and phase from the incident wave.

Substituting for Γ_T in eqn (8.19) from eqn (8.38) gives

$$\begin{aligned} V(x) &= A\,e^{-j\beta x} + A\Gamma_T\,e^{-j2\beta l}\,e^{j\beta x} \\ &= A(1 - \Gamma_T)\,e^{-j\beta x} + A\Gamma_T\,e^{-j\beta x} + A\Gamma_T\,e^{-j2\beta l}\,e^{j\beta x} \\ &= A(1 - \Gamma_T)\,e^{-j\beta x} + A\Gamma_T\,e^{-j\beta l}\underbrace{(e^{j\beta d} + e^{-j\beta d})}_{\text{`}2\cos\beta d\text{'}} \end{aligned} \tag{8.41}$$

where $d = (l - x)$; see Fig. 8.12. So, finally,

$$V(x) = A(1 - \Gamma_T)\,e^{-j\beta x} + 2A\cos\beta d \cdot \Gamma_T \cdot e^{-j\beta l}. \tag{8.42}$$

If Γ_T is written in polar form as $|\Gamma|\,e^{j\phi_T}$, and remembering that

$$v(x, t) = V(x)\,e^{j\omega t},$$

then

$$v(x, t) = A(1 - |\Gamma_T|\,e^{j\phi_T})\,e^{j(\omega t - \beta x)} + 2A\cos\beta d \cdot |\Gamma_T| \cdot e^{j(\omega t - \beta l + \phi_T)}. \tag{8.43}$$

a travelling wave — not a travelling wave, since there is no dependence on x.

The voltage has angular frequency ω and its amplitude varies periodically with position along the line through the factor $\cos\beta d$, since $d = 1 - x$.

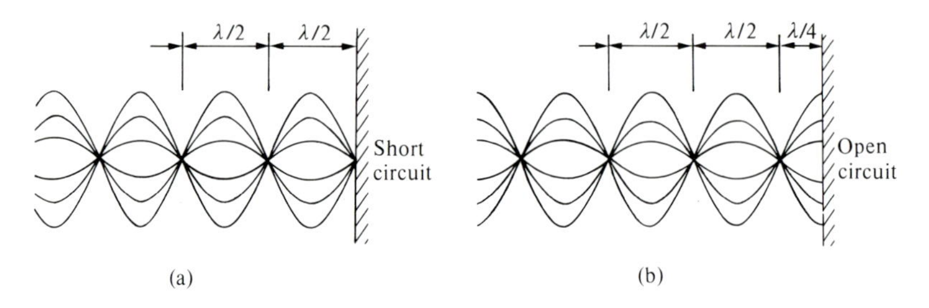

Fig. 8.12 The voltage standing-wave pattern for (a) an open-circuited line and (b) a short-circuited line.

A standing-wave pattern contains a lot of information about the properties of a terminated line but, before examining eqn (8.43) in detail, consider the case of an open-circuit termination (i.e. $|\Gamma_T| = 1$, $\phi_T = 0$). In this case,

$$v(x, t) = V(x)\,e^{j(\omega t - \beta l)},$$

where $V(x) = 2A\cos\beta d$. Hence $V(x) = 0$ (the 'nodes' in the pattern) for $\cos\beta d = 0$, i.e. for $\beta d = (2n + 1)\pi/2$ $(n = 0, 1, 2, 3, \ldots)$. Since $\beta = \omega/u = 2\pi/\lambda$, this condition becomes

$$d_n = \frac{n\lambda}{2} + \frac{\lambda}{4}$$

and the 'repeat' distance $d_{n+1} - d_n = \dfrac{\lambda}{2}$. The positions of the maxima ('*anti-nodes*') in the standing-wave pattern ($V(x) = 2A$) will obviously occur midway between the nodes (see Fig. 8.12(a)).

At a short-circuit termination, $\Gamma_T = e^{j\pi}$ $(= -1)$, since the incident and reflected voltages cancel out each other to give the required zero resultant voltage. So in this case the expression (eqn (8.43)) for $v(x, t)$ becomes

$$\begin{aligned} v(x, t) &= 2A\{e^{j(\omega t - \beta l + \beta d)} + \cos\beta d\, e^{j(\omega t - \beta l + \pi)}\} \qquad [\text{using } x = l - d] \\ &= 2A\, e^{j(\omega t - \beta l)}\{e^{j\beta d} - \cos\beta d\} \\ &= 2A\{\cos(\omega t - \beta l) + j\sin(\omega t - \beta l)\}j\sin\beta d. \end{aligned}$$

Now the voltage $v(x, t)$ must be real, and so on taking the real part of the last expression,

$$\begin{aligned} v(x, t) &= -2A\sin\beta d\cdot\sin(\omega t - \beta l) \\ &= 0 \qquad \text{for } \sin\beta d = n\pi \quad (n = 0, 1, 2, \ldots). \end{aligned}$$

So the nodes of the standing-wave pattern in the case of a short-circuit termination occur at $d = n\lambda/2$ (see Fig. 8.12).

If the impedance of the generator is not equal to Z_0, then the wave reflected from the termination will in turn be reflected at the sending end of the line to travel back in the $+x$ direction. Suppose that the reflection coefficient at the sending end is Γ_s; then the wave travelling in the $+x$ direction on the line has amplitude

$$A + \Gamma_s\Gamma_T A + \Gamma_s^2\Gamma_T^2 A + \Gamma_s^3\Gamma_T^3 A + \cdots = A', \qquad \text{say.}$$

The wave travelling in the $-x$ direction has amplitude

$$\Gamma_T A + \Gamma_s\Gamma_T^2 A + \Gamma_s^2\Gamma_T^3 A + \cdots = \Gamma_T A'.$$

Hence the effect of the reflection at the sending end is to change the amplitude of the wave travelling in the $+x$ direction, but otherwise the analysis is unchanged.

So far the extreme cases of an open-circuit and a short-circuit termination have been considered, together with the intermediate case of a matched termination.

To obtain an expression for the position of the nodes of the standing-wave pattern in the case of a general termination, consider eqn (8.41):

$$V(x) = A\,\mathrm{e}^{-\mathrm{j}\beta x}\{1 + |\Gamma_T|\,\mathrm{e}^{\mathrm{j}(\phi_T - 2\beta d)}\}, \tag{8.44}$$

since $l - x = d$. Bearing in mind that $v(x, t) = V(x)\,\mathrm{e}^{\mathrm{j}\omega t}$, then this expression represents the resultant of two phasors whose amplitudes are in the ratio $1{:}|\Gamma_T|$ and which have a phase difference of $(\phi_T - 2\beta d)$. The graphical representation of phasors was discussed in Section 3.2.1, and for the particular case being examined now, the diagram is shown in Fig. 8.13. The radius OT for $d = 0$ defines the position of the termination, and movement along the line 'towards the generator' (i.e. towards the sending end) is in the sense of increasing d, which is clockwise on the diagram. For a line that can be assumed to be lossless, T moves on a circular path.

As d varies from 0 to l, $V(x)$ varies periodically with the positions of the nodes, specified by

$$\phi_T - 2\beta d = -\pi \qquad \text{or} \qquad d = \frac{\phi_T\lambda}{4\pi} + \frac{\lambda}{4}. \tag{8.45}$$

The value of $V(x)_{\min}$ at a node is obtained by using eqn (8.45) in eqn (8.44), which yields

$$V(x)_{\min} = A\,\mathrm{e}^{-\mathrm{j}\beta x}\{1 - |\Gamma_T|\}.$$

Anti-nodes are specified by $(\phi_T - 2\beta d) = 0$, which gives

$$V(x)_{max} = A\,e^{-j\beta x}\{1 + |\Gamma_T|\}.$$

The ratio $V(x)_{max}/V(x)_{min}$ is called the **voltage standing-wave ratio** (v.s.w.r.), which will be denoted by r.

It follows from the expressions for $V(x)_{max}$ and $V(x)_{min}$ that

$$r = \frac{1 + |\Gamma_T|}{1 - |\Gamma_T|} \quad \text{and} \quad |\Gamma_T| = \frac{r - 1}{r + 1}. \tag{8.46}$$

$$1 \leqslant r \leqslant \infty$$

matched termination (r = 1) / open- or short-circuit termination (r = ∞)

Since the v.s.w.r. is a ratio, by definition, it is often quoted in decibels:

$$r\,(\text{dB}) = 20 \log\left(\frac{V(x)_{max}}{V(x)_{min}}\right) = 20 \log r.$$

Consider eqn (8.32) in the case of a lossless line. If the line is short-circuited ($Z_T = 0$) and Z_s is now taken to be the impedance 'looking in' to a line of variable length d, then the equation becomes

$$Z_s = j \tan \beta d, \tag{8.47}$$

i.e. the sending-end impedance is a pure reactance whose sign alternates between + (inductive reactance) and − (capacitive reactance); see Fig. 8.14.

Taking into account the losses in a real line, it can be shown quite easily that the line voltage and the line current, $V(x)$ and $I(x)$, are in-phase at the positions of both nodes and anti-nodes (i.e. the line impedance is

---

Fig. 8.13 (a) The resultant (MT) of a reference phasor of unit amplitude (MO) and a second phasor (OT) of amplitude $|\Gamma_T|$ and phase shift $(\phi_T - 2\beta d)$. As the point of interest is moved along the line (varying d) then, for a lossless line, the tip (T) of the resultant phasor moves on a circle of constant $|\Gamma_T|$. The minimum value of the resultant is MS and the maximum is MP. (b) Circles of constant normalized resistance. (c) Circles of constant normalized reactance. (d) This figure shows how the circles of constant resistance and constant reactance are superimposed to form a Smith chart.

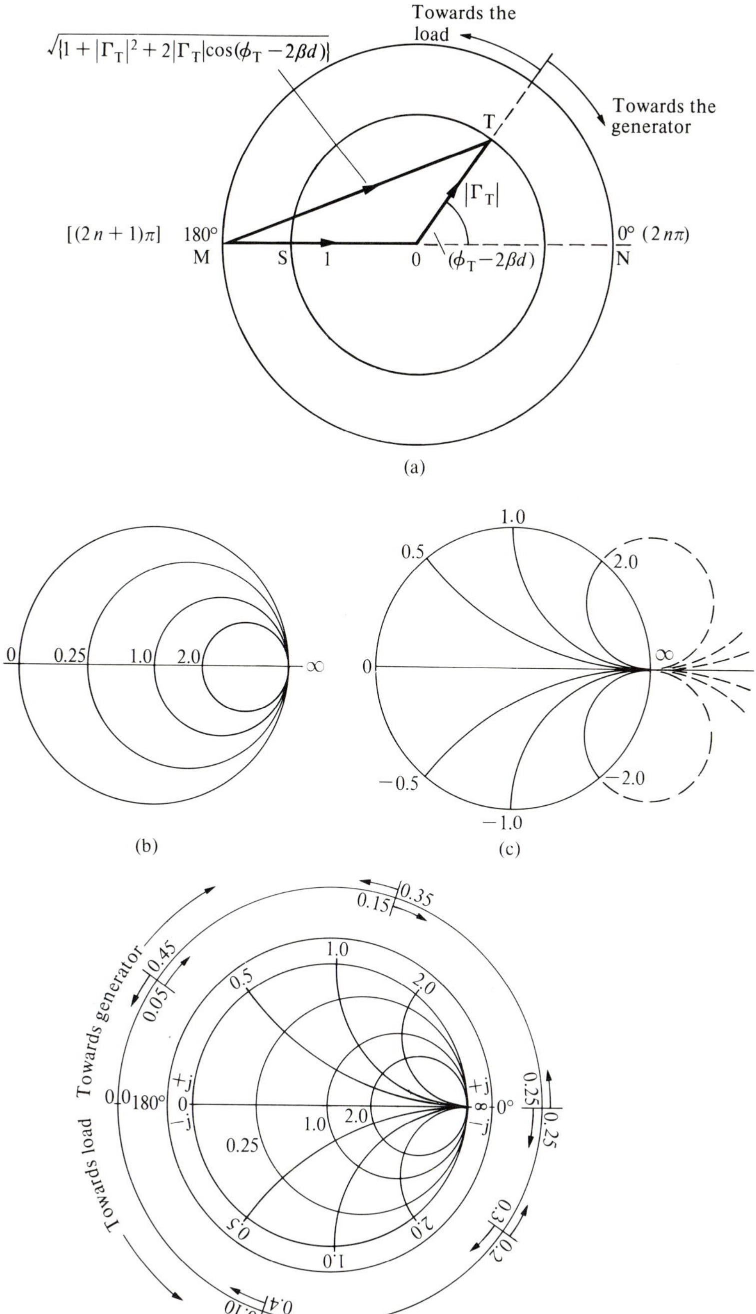

Towards the load

$\sqrt{\{1+|\Gamma_T|^2+2|\Gamma_T|\cos(\phi_T-2\beta d)\}}$

Towards the generator

T

$|\Gamma_T|$

$[(2n+1)\pi]$ 180°

M

S 1 0 $(\phi_T-2\beta d)$

0° $(2n\pi)$

N

(a)

0 0.25 1.0 2.0 ∞

(b)

1.0

0.5

2.0

0

∞

−0.5

−2.0

−1.0

(c)

0.35

0.15

Towards generator

0.45

0.05

1.0

0.5

2.0

+j

0.0 180° 0

−j

0.25

1.0

2.0

+j

∞ 0°

−j

0.25

0.25

Towards load

0.5

2.0

1.0

0.3

0.2

0.4

0.10

(d)

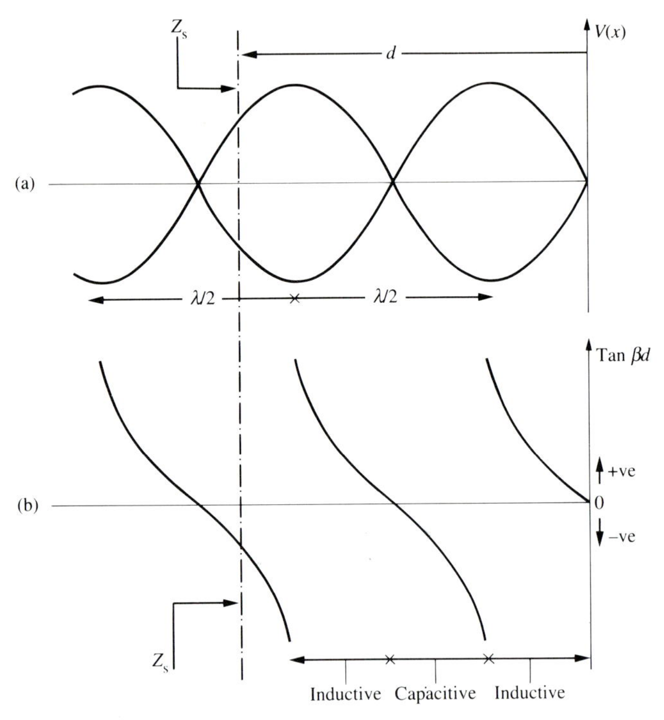

Fig. 8.14 (a) The voltage standing-wave pattern and (b) the line impedance for a short-circuited lossless line.

purely resistive at these positions), and hence that

$$\begin{aligned} &\text{at a node} && Z(x)_{\text{min}} = Z_0/r \\ &\text{at an anti-node} && Z(x)_{\text{max}} = rZ_0 \end{aligned} \tag{8.48}$$

Again the line impedance varies periodically between inductive and capacitive (see Fig. 8.15); this is what is predicted and described by eqn (8.31).

The value of the v.s.w.r. on a line can be of critical importance. If good power transfer to a terminating load is required, then, from what has gone before, it is clear that the the v.s.w.r. should be as close as possible to unity. In systems that handle high power transmissions, such as in a radar transmitter, a limiting physical factor may be the value of the amplitude of the electric field on the line at which the insulation breaks down: the 'breakdown field'. Since the power propagated along a line is proportional to the square of the amplitude of the electric field, the value of the breakdown field sets an upper limit to the power handling capacity of a line. Standing waves can produce peak voltages and currents that are

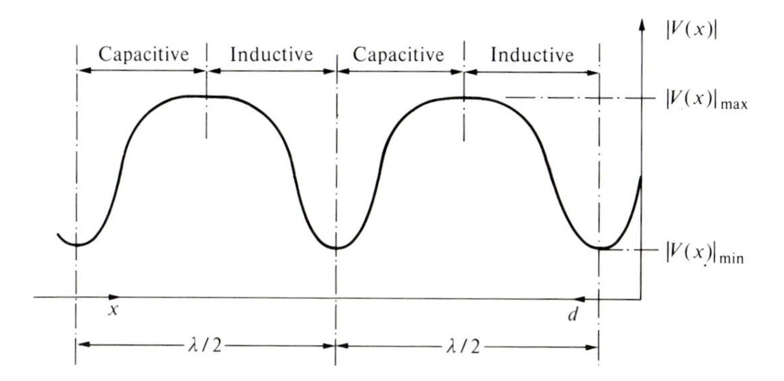

Fig. 8.15 A schematic illustration of the spatial variation along a line of the magnitude of the voltage in a standing wave pattern. The changes in sign of the reactive part of the line impedance are indicated also.

nearly twice as large as those that would exist if the line were terminated in a matched load.

It is interesting to note that a section of transmission line can be represented as a two-port network. Since sections of line and other elements of a system are for the most part connected 'in cascade', the so-called transmission parameters (or a-parameters) are commonly used. From eqns (8.29), with x replaced by by l, it follows that

$$V_s = V(l)\cosh \gamma l + I(l)Z_0 \sinh \gamma l$$

$$I_s = \frac{V(l)}{Z_0}\sinh \gamma l + I(l)\cosh \gamma l$$

and the 'transmission matrix' is

$$\begin{bmatrix} \cosh \gamma l & Z_0 \sinh \gamma l \\ \dfrac{\sinh \gamma l}{Z_0} & \cosh \gamma l \end{bmatrix}$$

or, for a lossless line,

$$\begin{bmatrix} \cos \beta l & \mathrm{j}Z_0 \sin \beta l \\ \dfrac{\mathrm{j}\sin \beta l}{Z_0} & \cos \beta l \end{bmatrix}$$

8.3.4 Waveguides

Electromagnetic waves, carrying energy and signals, can be guided conveniently along hollow metal pipes, although such structures become a practical proposition only at frequencies of about 3 GHz and above, since their lateral dimensions are of the same order of magnitude as the wavelength (e.g. 10 cm at 3 GHz; 3 cm at 10 GHz). Historically this was referred to as the 'microwave' region of the e.m. spectrum, a term still widely used, but the International Telecommunication Convention has designated frequencies in the region 3–30 GHz as SHF (super high frequency), and in the region 30–300 GHz as EHF (extremely high frequency). In these frequency regions the values of the attenuation constants of waveguides are of the order of 10^{-3} dB m^{-1}, which is considerably better than air-spaced coaxial line of equal cross-sectional area.

Obviously in the case of waveguides there is not a definable pair of conductors, and no 'flow and return' circuit, and hence line voltages and line currents cannot be uniquely defined. However, nominal line voltages and currents can be defined by equating the power flow expressed in terms of the nominal line voltage and current to the power flow expressed in terms of the electric and magnetic fields in the mode of propagation of interest; transmission line theory can then be exploited, using the nominal line voltage and current. There still remains the problem of representing waveguide components as transmission line elements. Any form of discontinuity in an otherwise uniform waveguide, whether it be an obstruction, an aperture in a containing conductor, a change in physical dimension, a bend, a twist, or a junction, will cause a partial reflection, at least, of an incident wave. A wide variety of 'obstacles' act as circuit elements in waveguides; obstacles such as metal posts or diaphragms containing apertures can be capacitive, inductive or resonant in nature and can form the basis of filters, for instance. In principle, field calculations can yield the equivalent circuit elements for the transmission line representation, but where such calculations are intractable, measurements have provided design data.

8.3.5 Impedance charts

Calculations on lines can be greatly facilitated by the use of *impedance charts* (or *transmission line charts*), of which the Smith chart is the most commonly encountered example. The theoretical derivation of this type of chart is given in many specialist texts (see, for example, the reference texts listed at the end of this chapter), and only an operational description will be given here.

Referring to Fig. 8.13, it can be seen that in moving along a (lossless) line the tip T of the vector OT traces a circle at a constant value of $|\Gamma_T|$ and hence a 'circle at constant v.s.w.r.' also; $r = V(x)_{max}/V(x)_{min} = MP/MS$. Traversal of a complete circle corresponds to moving through

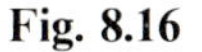

Fig. 8.16

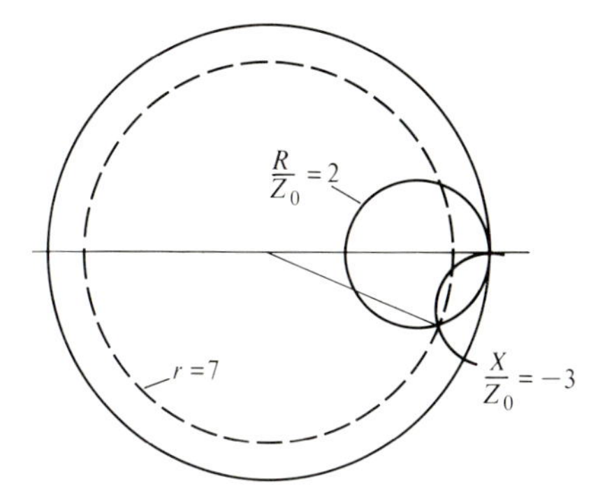

a distance $\lambda/2$ along the line; clockwise rotation corresponds to the sense 'towards the generator', and anti-clockwise to the sense 'towards the termination/load'.

It can be shown (see the cited texts) that two families of orthogonal circles having *normalized* resistance R/Z_0 and *normalized* reactance X/Z_0 as parameters, respectively, as depicted in Fig. 8.13(b) and (c), can be superimposed on the diagram of Fig. 8.13(a) to give the Smith chart (Fig. 8.13(d)). A transparent rotatable radial cursor, which is calibrated along its length in $|\Gamma_T|$ and v.s.w.r., is placed on the diagram with its central point ($|\Gamma_T| = 0$; $r = 1$) coincident with O.

The best way of continuing the description of the chart is by means of some illustrative examples.

Example 8.3 A line is terminated by a (normalized) impedance $(2 - j3)$. What is the v.s.w.r. on the line?

The intersection of the circles $R/Z_0 = 2$ and $X/Z_0 = -3$ lies on the v.s.w.r. circle $r = 7$ (see Fig. 8.16). □

Example 8.4 The v.s.w.r. on a line has the value 1.5. Find the normalized impedance at the position of a voltage maximum and at a voltage minimum.

The positions at which the voltage maxima and minima occur are specified by the points at which the v.s.w.r. = 1.5 circle intersects the 'horizontal' diameter of the Smith chart. In this case the points corresponding to the voltage maxima and minima have coordinates $(1.5 + j0.0)$ and $(0.66 + j0.0)$ respectively, which give the normalized impedances. Notice that $0.66 = 1/1.5$, in accordance with eqns (8.48); see Fig. 8.17.

Example 8.4 serves as a reminder that the positions of adjacent maxima and minima in a voltage standing-wave pattern are separated by a distance $\lambda/4$ along the line and also demonstrates that the line impedances

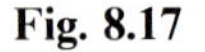

Fig. 8.17

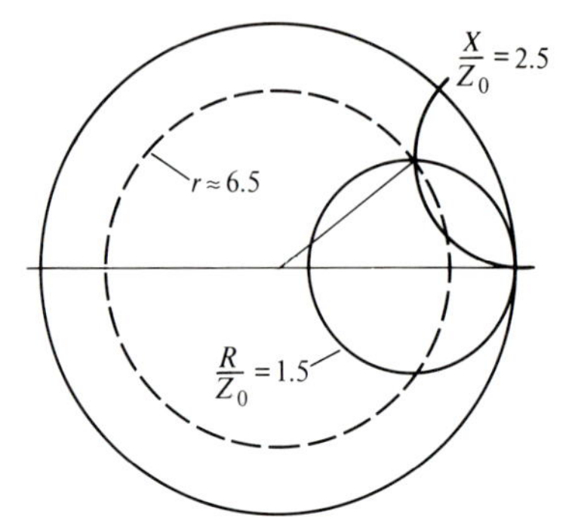

at the two points are the reciprocals of each other. It is left as an exercise to show from eqn (8.31) that, in general,

$$Z(d + \lambda/4) = Y(d)$$
$$Y(d + \lambda/4) = Z(d). \tag{8.49}$$

Hence a complex impedance can be converted to the corresponding admittance by rotating around the circle of constant v.s.w.r. on the Smith chart through an angle of 180° (i.e. through a distance of $\lambda/2$ along the line).

An impedance chart can be converted to an admittance chart by rotating the whole chart through 180°; but note that the left-hand end of the horizontal diameter still corresponds to the position of a voltage minimum in the voltage standing-wave pattern. □

Example 8.5 A line is terminated by a normalized impedance (0.8 + j0.5). Find

(a) the angle of the reflection coefficient;

(b) the v.s.w.r. on the line;

(c) the distance from the termination of the first minimum in the voltage standing-wave pattern.

(a) Find the point with coordinates (0.8 + j0.5) and then use the cursor to read the angle (96°) on the circular scale (see Fig. 8.18); this is the angle of the reflection coefficient (the modulus of the reflection coefficient can be read from the cursor and is about 0.3).

(b) At the point (0.8 + j0.5) the cursor scale gives $r \approx 1.8$.

(c) The reading on the wavelength scale at the angle of 96° is 0.116λ. To reach the position of the first minimum ($-180°$ or 0.5λ), it is necessary to move a distance along the line of ($0.5\lambda - 0.116\lambda = 0.384\lambda$) in the sense 'towards the generator'. □

Fig. 8.18

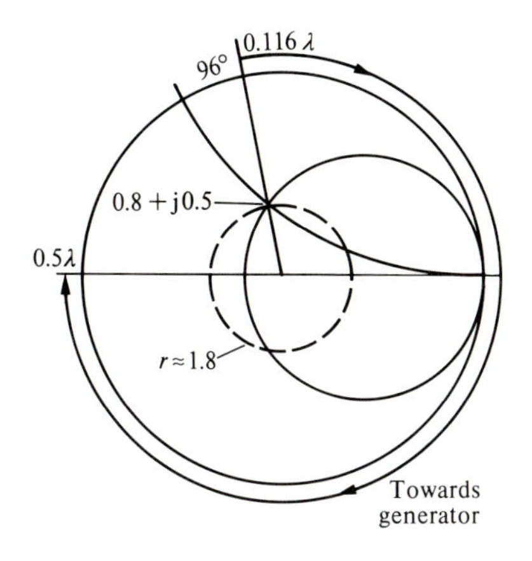

In telecommunications and radar systems there is generally a requirement that components and subsystems operate satisfactorily over a wide frequency bandwidth. Measurements on circuit elements, transmission lines and distributed systems generally are greatly facilitated by the use of a **network analyser**, which in essence enables swept frequency measurements of the reflection (and transmission) coefficients of passive and active circuit elements. The display can be of v.s.w.r., return loss ($\equiv -20 \log|\Gamma_T|$), or impedance/admittance on a Smith chart, for example. A relatively modest network analyser might cover the range 300 kHz – 3 GHz, whilst a computer-controlled automated network analyser can cover a much greater range, e.g. 500 MHz – 40 GHz. Instruments in the latter category are very sophisticated indeed and require a controlled environment to allow the highest-accuracy measurements to be made; for instance, thermal expansion or contraction in cables due to changes in the ambient temperature can be a significant source of measurement uncertainty. Standard terminations (short-circuit, matched, open-circuit) can be used to check the calibration of the instrument. Frequency-domain data can be transformed to the time domain, using an inverse Fourier transform, and hence reflections due to connectors, defects in interconnecting lines and leakages of e.m. fields can be corrected for; this allows the device under test (DUT) to be 'de-embedded' from the lines that connect it to the analyser.

8.3.6 Resonant lines

From eqn (8.32) the sending-end impedance of a short-circuited lossless line is given by

$$Z_s = jZ_0 \tan \beta l.$$

For $\beta l = n\pi$, then Z_s goes to zero, and for $\beta l = (2n + 1)\pi/2$, Z_s goes to infinity. These two situations mirror the characteristics of high-quality-factor (high-Q) series LCR and parallel LCR lumped element networks respectively (see Sections 4.2.1 and 4.2.2). The two conditions on βl correspond to $l = n\lambda/2$ ($n = 1, 2, \ldots$) and $l = (2n + 1)\lambda/4$ ($n = 0, 1, 2, \ldots$) respectively (see Fig. 8.12). The resonant frequencies in the two cases are

$$\omega_0 = \frac{2n\pi u}{2l} \text{ ('series')} \qquad \omega_0 = \frac{(2n+1)\pi u}{2l} \text{ ('parallel')}. \tag{8.50}$$

To take account of the losses in a real line, consider eqn (8.31), with $Z_T = 0$. The sending-end impedance Z_s is given by

$$\begin{aligned} Z_s &= Z_0 \tanh(\alpha l + j\beta l) \\ &= \frac{Z_0(\tanh \alpha l + \tanh j\beta l)}{1 + \tanh \alpha l \cdot \tanh j\beta l}. \end{aligned} \tag{8.51}$$

This expression will now be compared with the driving-point impedance of a series lumped element LCR circuit (see eqn (4.9)), which is

$$\begin{aligned} Z &= R + j\left(\omega L - \frac{1}{\omega C}\right) \\ &= R + j\omega L \frac{(\omega^2 - \omega_0^2)}{\omega^2}, \end{aligned}$$

where the resonant frequency ω_0 is defined through $\omega_0^2 = (LC)^{-1}$.

The quality factor Q of the resonant circuit was defined through $Q = \omega_0 L/R$, and so

$$Z = R\left\{1 + \frac{j\omega Q}{\omega_0}\left(\frac{\omega_0^2 - \omega^2}{\omega^2}\right)\right\}$$

or

$$Z = R\left\{1 + \frac{j\omega Q(\omega + \omega_0)(\omega - \omega_0)}{\omega_0 \omega^2}\right\}.$$

Now, for frequencies close to resonance, $(\omega + \omega_0) \approx 2\omega_0$ and $(\omega - \omega_0) \equiv \delta\omega$, so that

$$Z \approx R\left\{1 + j2Q\frac{\delta\omega}{\omega_0}\right\}. \tag{8.52}$$

Returning to eqn (8.51) for the short-circuited line, and assuming a low-loss line (i.e. αl small), then $\tanh \alpha l \approx \alpha l$ and $\tanh j\beta l = j \tan \beta l$, so that

$$Z_s \approx \frac{Z_0(\alpha l + j \tan \beta l)}{1 + j\alpha l \tan \beta l}.$$

Consider the series resonance case. For a frequency $(\omega_0 + \delta\omega)$ close to resonance, then $\{\beta l + \delta(\beta l)\} = n\pi + n\pi \frac{\delta\omega}{\omega_0}$ (since $\beta l = \omega l/u$, $\delta(\beta l) = \frac{1}{u}\delta(\omega)$, and by making use of eqns (8.50)), and so

$$\tan\left(n\pi + n\pi\frac{\delta\omega}{\omega_0}\right) = \tan n\pi\frac{\delta\omega}{\omega_0} \approx \frac{n\pi\delta\omega}{\omega_0},$$

since $\tan n\pi = 0$. For $\frac{\delta\omega}{\omega_0} \ll 1$,

$$\tan(n\pi\delta\omega/\omega_0) \approx n\pi\delta\omega/\omega_0.$$

Hence, for the case of $n = 1$, say, the expression for Z_s becomes

$$Z_s \approx Z_0\alpha l\left\{1 + j\frac{2\pi\delta\omega}{\alpha l\omega_0}\right\}. \tag{8.53}$$

So, on comparing eqns (8.52) and (8.53), it follows that the effective resistance and Q-factor of the short-circuited transmission are

$$R \approx \alpha l Z_0 \qquad \text{and} \qquad Q \approx \pi/\alpha l. \tag{8.54}$$

Analogous expressions can be obtained for the 'parallel resonance' case.

Example 8.6 Calculate the series resonance frequency and Q-factor of a resonant half-wavelength transmission line for which

$$u = 2.5 \times 10^8\ \mathrm{m\,s^{-1}}, \qquad l = 0.125\ \mathrm{m}, \qquad \alpha = 0.3\ \mathrm{dB\,m^{-1}}.$$

From eqns (8.50), and taking $n = 1$, $\omega_0 = \pi u/l$. Hence $f_0 = 1$ GHz.

$\alpha = 0.3/8.68\ \mathrm{Np\,m^{-1}}$ and so, using eqns (8.54), $Q = 727$.

Note that by careful design and by the use of very low-loss lines, even higher Q-factors can be obtained. However, it should be remembered that the above expression for the Q-factor relates to the unloaded situation (see Section 4.3). □

8.4 Some practical aspects of transmission lines

From an electrical point of view the important properties of a transmission line are its attenuation coefficient, characteristic impedance, and power handling capacity. Other important features that come into consideration are the lateral dimension (diameter of a coaxial line, separation between the conductors in the case of parallel conductors, width in the case of waveguide), mechanical flexibility or rigidity, and durability.

There are many different forms of transmission line, depending on the technical requirements (and cost). Some disparate examples are twisted wire pair and coaxial land lines for telephony, armoured submarine cable, balanced lines and waveguide as feeds to antennas, general-purpose '50 ohm' coaxial cable (and the high-power lines of national power distribution systems; these are not under discussion here).

The attenuation coefficient arises from the resistance of the conductors and the losses ('dielectric losses'—see Section 5.1) in the insulating medium in and around the conductors. At frequencies about 1 MHz, roughly speaking, the skin effect (see Section 5.2.1) becomes significant and the effective resistance becomes proportional to the square root of the frequency. The dielectric losses are proportional to frequency, and in the case of solid insulation such as polyethylene begin to exceed the resistive losses at frequencies of the order of a few gigahertz; the upper frequency limit to the economical use of flexible coaxial cables is approximately 5 GHz. At about 100 MHz the resistive losses are likely to be about fifty times the dielectric losses.

For a coaxial line it can be shown that the resistive contribution to the attenuation coefficient is a minimum when the ratio of the radii of the outer and inner conductors is given by $b/a = 3.592$. Hence, from eqn (8.26), with $\epsilon_r = \mu_r = 1$ (i.e. 'air-spaced'), the value of the characteristic impedance is

$$Z_0 = 77\ \Omega. \tag{8.55}$$

The function of the solid insulation, commonly polyethylene ($\epsilon_r \approx 2.3$), is to separate the inner and outer conductors (whilst maintaining flexibility), and in this circumstance,

$$Z_0 = 51\ \Omega. \tag{8.56}$$

'50 ohm' general-purpose coaxial cable is very widely used. A commonly used low-loss cable for use in radio systems has cellular polyethylene as insulator, giving 'air-spacing' effectively (hence the lower power losses), and has a characteristic impedance of 75 Ω.

If an antenna is balanced with respect to ground, as many are, then the downlead (in the case of receivers) or the feed (in the case of transmitters) should both be **balanced lines**; otherwise the ideal, symmetric, radiation pattern of the antenna will be distorted and, in the case of receivers, interference can get on to the line and cause serious problems. Twin conductor pair lines are balanced lines, but coaxial cable and microstrip are not. A popular balanced twin feeder cable has conductors separated by about 9 mm, with thin polyethylene as separator, and has a characteristic impedance of 300 Ω. Elements designed to connect an unbalanced line to a balanced load are known as **baluns** ('balanced to unbalanced').

The power-carrying capacity of a line is determined by a combination of two factors. First, the electrical breakdown strength of the insulation limits the peak value of the electric field (and hence the voltage between the conductors); second, there is the rise in temperature due to the losses of energy. In c.w. operation, temperature rise will usually be the limiting factor, unless there is a large voltage standing-wave ratio, whereas for high-amplitude short duty-cycle pulses, electrical breakdown can set a limit.

For a coaxial line the maximum value of the electric field occurs at the outer surface of the inner conductor (see eqn (1.30), with ϵ_0 now replaced by $\epsilon_r\epsilon_0$). If this field is set equal to the electric breakdown strength (E_b, say), then the breakdown voltage (V_b) on the line is given by

$$V_b = aE_b \ln(b/a). \tag{8.57}$$

V_b is ~2 kV, typically, for 50 Ω and 75 Ω coaxial cable with polyethylene as insulation. An approximate guide to the maximum power rating for a temperature rise of 45 K above ambient is

$$P_{max} \approx 10^{10}/f;$$

e.g. for $f = 1$ GHz, then $P_{max} \approx 10$ W.

It should have become clear that impedance matching figures importantly in the design and operation of systems incorporating transmission lines. In Section 3.5 'conjugate matching' between a generator and a load was defined, for which, assuming a lossless interconnecting line, Z_s is arranged to be equal to Z_g^* (see Fig. 8.19(a)), i.e.

$$Z_s = R_g - jX_g$$

or

$$R_s + jX_s = R_g - jX_g.$$

This condition demands that $R_s = R_g$ and $X_s = -X_g$. Since the reactance of a circuit element is frequency-dependent, the conjugate matching condition can be met at only one frequency unless the reactive elements are adjustable. So reflectionless matching is used for broadband operation; i.e. the load and the generator are separately matched to the characteristic impedance of the interconnecting line (see Fig. 8.19(b)). A further reason for preferring the reflectionless match is that the reflected wave can react with the generator to produce '*frequency pulling*'.

For typical microstrip (see Fig. 8.1(c)) the thickness of the dielectric and the width of the conducting strip are each a few millimetres. The characteristic impedance decreases in magnitude with increasing width of

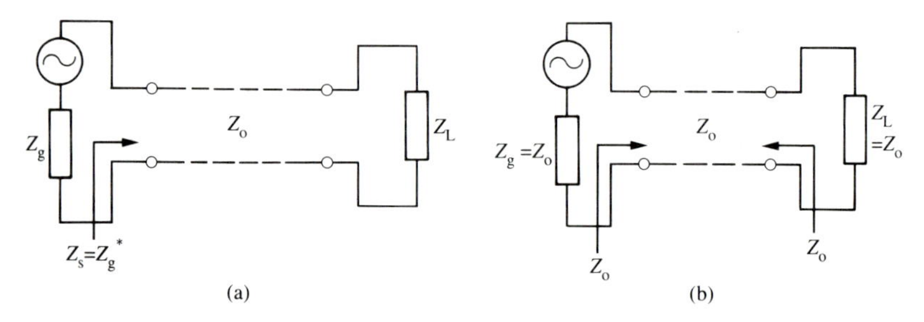

Fig. 8.19 (a) Conjugate matching (assuming a lossless line). (b) Reflectionless matching.

the strip, since the capacitance per unit length increases (see eqn (8.4)). For dimensions as just mentioned, Z_0 is $\sim 100\ \Omega$ and the attenuation coefficient is ~ 0.05 dB cm^{-1}.

For a length l of microstrip that is much less than $\lambda/12$, then the sending-end impedance Z_s is given by

$$Z_s \approx Z_0 \frac{\{Z_T/Z_0 + j\beta l\}}{\left\{1 + j\dfrac{Z_T}{Z_0}\cdot \beta l\right\}},$$

(see eqn (8.32)), since

$$\tan \beta l \approx \beta l \qquad \text{for } \beta l < \pi/6$$

(and $\beta = 2\pi/\lambda = \omega/u$, remember). Hence, if $Z_T \ll Z_0$, then $Z_s \approx j\omega l Z_0/u$.

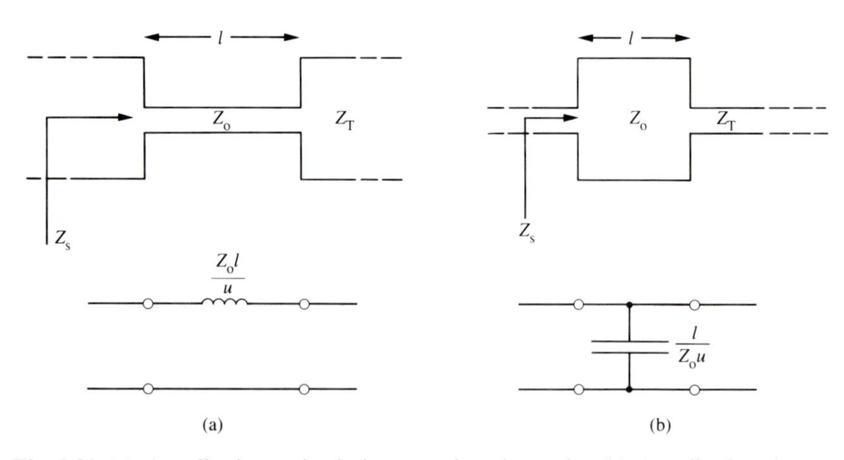

Fig. 8.20 (a) An effective series inductance in microstrip. (b) An effective shunt capacitance.

This indicates that an effective inductance of value lZ_0/u can be realized by a section of line terminated by a strip of much greater width (see Fig. 8.20(a)). Conversely, if $Z_T \gg Z_0$, then $Z_s \approx -jZ_0u/\omega l$ and a capacitance can be realized by a section of line terminated by a strip of much narrower width (see Fig. 8.20(b)).

Where the available active devices are smaller than such distributed passive elements, it is useful to be able to fabricate lumped resistors, inductors and capacitors in integrated circuit form: microwave integrated circuits (MIC). The state-of-the-art is constantly changing, but at present the upper frequency limit to this approach (determined by linear dimension $<\lambda/30$, roughly) is between 20 GHz and 60 GHz.

Further reading

Chipman, R. A. (1968). *Theory and problems of transmission lines.* Schaum's Outline Series, McGraw-Hill, New York. (This text gives the basic theory plus numerous worked examples and problems.)

Combes, P. F., Graffeuil, J. and Sautereau, J.-F. (1987). *Microwave components, devices and active circuits.* Wiley, New York.

Ramo, S., Whinnery, J. R., and Van Duzer, T. (1967). *Fields and waves in communication electronics.* Wiley, New York. (A good introduction to e.m. fields and waves precedes detailed treatments of transmission lines and microwave circuits.)

Problems

8.1 Obtain the expression (eqn (8.11)) for the reflection coefficient Γ_T of the termination of a transmission line:

$$\Gamma_T = \frac{\dfrac{R_T}{Z_0} - 1}{\dfrac{R_T}{Z_0} + 1}.$$

8.2 Obtain the expression (eqn (8.12)) for the power P_T absorbed in a resistive termination to a transmission line:

$$P_T = \frac{V_f^2}{Z_0}(1 - \Gamma_T^2).$$

8.3 The line voltage at the sending end of a transmission line of length 1.4 m, as displayed by a time-domain reflectometer, is shown in Fig. 8.21.

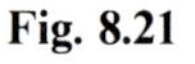

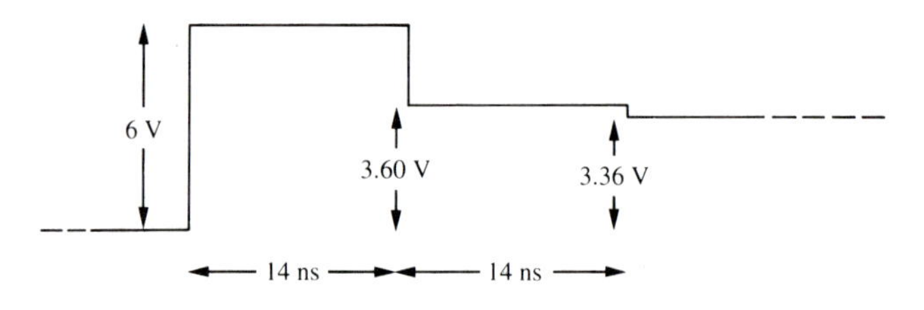

Assuming that the line is lossless, determine the characteristic impedance of the line, the resistance of the load and the speed of propagation of the 'step' on the line, given that the resistance of the source is 50 Ω and the source voltage is 10 V.

8.4 A wave frequency of $100/2\pi$ MHz propagates along a uniform transmission line for which $L = 0.5\ \mu\text{H m}^{-1}$ and $C = 200\ \text{pF m}^{-1}$. Calculate (a) the characteristic impedance of the line, (b) the phase change coefficient, (c) the wavelength of the wave on the line, and (d) the sending-end impedance (input impedance) of a quarter-wavelength section terminated by a load of impedance $-\text{j}50\ \Omega$.

8.5 A coaxial cable has the following characteristics: $L = 0.2\ \mu\text{H m}^{-1}$; $C = 100\ \text{pF m}^{-1}$; $R = 2 \times 10^{-2}\ \Omega\ \text{m}^{-1}$; $G = 10^{-8}\ \text{S m}^{-1}$. What are the values of Z_0, α, β, λ and u at a frequency of $100/2\pi$ MHz?

8.6 A transmission line with a characteristic impedance of 75 Ω is terminated by a load of impedance $(50 - \text{j}50)\Omega$. What is the voltage reflection coefficient of the load?

8.7 What is the normalized admittance corresponding to the normalized impedance $(1.5 + \text{j}2.5)$?

8.8 A lossless transmission has a distributed capacitance of $30\ \text{pF m}^{-1}$ and a distributed inductance of $0.2\ \mu\text{H m}^{-1}$. If the line is operated at $1000/2\pi$ MHz, what is the shortest length of short-circuited line that will have a capacitive input susceptance of magnitude 0.213 S?

8.9 The measured value of the v.s.w.r. on a lossless transmission line is 3.5 and the distance between successive minima in the v.s.w.r. pattern is 50 cm. If the first v.s.w.r. minimum from the termination is distant 35 cm, what is the normalized impedance of the terminating load?

Electromagnetic compatibility

9.1 Introduction

Electrical and electronic circuits and systems are becoming increasingly sophisticated and there is a tendency for them to be more densely packed; computing systems and aircraft control and navigation systems are two good examples. A consequence is that there is an increased risk that electromagnetic (e.m.) signals or disturbances in one part of a system will be picked up in another section. This is termed **electromagnetic interference** (e.m.i.) and is obviously undesirable, in that it can cause degradation of wanted signals, false indications, and actual system malfunctions. The problems are particularly acute if the system in question is designed to receive and process low-level signals.

If an interfering e.m. disturbance emanates from a system or source external to the system of particular interest, then it is useful to designate it as **intersystem interference**, in contrast to **intrasystem interference**, for which the source is inside the system itself. External sources may be man-made or natural; some examples of man-made sources are switches and relays (particularly those which handle high currents), electric motors, fluorescent lamps, corona discharges, and electric welding equipment. The most well-known natural sources of interference are electrical storms, solar activity, and cosmic radiation.

The definition of the boundary of a system is a matter of expediency; the reader may like to consider what could be defined as systems and subsystems in the following situations:

mains 'hum' on the sound output of a hi-fi audio system

interference and 'ghosts' on a TV picture

breakthrough of a radio transmission into a hi-fi audio system

uncontrolled switching on and off of a video cassette recorder due to 'spikes' on the 'mains' arising from the operation of relays controlling the motor of a domestic refrigerator

car ignition interference on a car radio

interference ('cross-talk') between VHF radio systems, e.g. police, ambulance service, taxi cabs the susceptibility of heart pacemakers to external e.m. fields

bit errors in digital information systems caused by spurious pulses.

If interconnected systems and sub-systems function as designed, with e.m.i. at a tolerably low level, then there is said to be **electromagnetic compatibility** (e.m.c.).

National and international standards for e.m.i. and e.m.c. have been established by advisory and regulatory bodies such as national measurement institutions, the professional engineering institutes and the International Electrotechnical Commission (IEC). It should not be assumed that the standards apply only to the most sophisticated equipment; for instance, 'electronic' weighing machines are very widely used in process industries and part of their overall specification is that they should operate satisfactorily in a radiofrequency field strength of 10 V m^{-1}.

In attempting to solve problems of e.m.i. it is necessary to identify the sources of interference and the pathways for interaction, and then to take remedial measures. Of course, with knowledge and practical experience, systems can be designed to have low susceptibility to known sources of interference, but there may be unanticipated sources and pathways! Sources are often designated as 'conductive' (e.g. switches, motors, power supplies) or 'radiative' (e.g. medical diathermy equipment, telecommunications equipment), but it should be realized that a particular item of equipment may be a significant source of both conducted and radiated interference. Pathways are also designated as conductive (e.g. wires, circuit components, cabinets, shields) or radiative. The technical problems associated with unavoidable, but reducible, random electronic noise are usually treated separately from e.m.i. which, broadly speaking, is non-random and impulsive. This does mean that a particular source of e.m.i. has a characteristic spectrum in the frequency domain, and filtering can significantly reduce the scale of the interference.

Clearly an understanding of the principles underlying e.m.i. and e.m.c., and the solving of practical and design problems, require a good working knowledge of a wide range of theoretical concepts and practical techniques that can be gained only through experience; many practical situations embrace a combination of conductive and radiative sources and pathways. However, it is also clear that a basic understanding of e.m. fields and conductive pathways is a prerequisite for a wider understanding, and it is these topics which will be the subject of this chapter.

9.2 Radiation of electromagnetic energy

All elements of a circuit that carry time-dependent currents radiate electromagnetic energy, although nothing like as efficiently as purpose-designed radiating antennas. The elements also act as receiving antennas, of course (the principle of reciprocity—see Section 2.6). An important distinction is made between two regions of the space around an antenna, namely the **far field** (or radiation field) and the **near field**.

For an element having dimensions much less than the wavelength of the e.m. radiation in question, then at distances from the element much greater than the wavelength, the e.m. field pattern in the radiation is very close to that of an ideal plane, transverse e.m. wave. As was seen in Section 5.4.3, the ratio of the electric field strength E to the magnetic field strength H in a plane e.m. wave, which is known as the wave impedance of the transmission medium, is given for non-conductors by

$$\frac{E}{H} = \left(\frac{\mu_r \mu_0}{\varepsilon_r \varepsilon_0}\right)^{1/2}. \tag{9.1}$$

In free space, for which $\mu_r = \varepsilon_r = 1$, $E/H = 377\ \Omega$ very closely, and this is known as the intrinsic wave impedance of free space. Equation (9.1) has very important implications for the screening of e.m. radiation, since an e.m. wave will be at least partly reflected at an interface between two media that have different values of wave impedance. The description of the relation between E and H for an e.m. disturbance propagating in a medium that has a significant electrical conductivity, and especially in metals, is more complicated, but again the feature of practical significance is that the wave impedance differs greatly from that of free space.

Assuming that the circuit elements of interest have dimensions much less than the wavelength of the e.m. waves, then far-field conditions obtain for distances $r \gg \lambda/2\pi$. Hence, except for UHF circuits, it will be near-field effects that are of interest largely. It should be noted that for $r < \lambda/2\pi$ the ratio E/H is different from $377\ \Omega$. For example, for an idealized source consisting of a straight element of current-carrying conductor (length $\ll \lambda$), then $E/H > 377\ \Omega$, whereas for an idealized small loop source, $E/H < 377\ \Omega$ (see Fig. 9.1). It should also be noted that in near-field situations there is a longitudinal component of E and of H.

In the realm of e.m.i. it is usual to include near-field effects under the general heading of radiative coupling. The mutual inductive coupling between the coils of a transformer is an extreme example of the near-field effect.

An electromagnetic pulse (e.m.p.) can be generated within a system (e.g. when an inductive load such as a solenoid or a motor is de-energized), or externally through lightning strikes or nuclear detonations. An electromagnetic pulse is characterized by very high electric and magnetic field strengths and by a fast rise time ($\sim 10^{12}\ \mathrm{V\,s^{-1}}$, say). The measures to

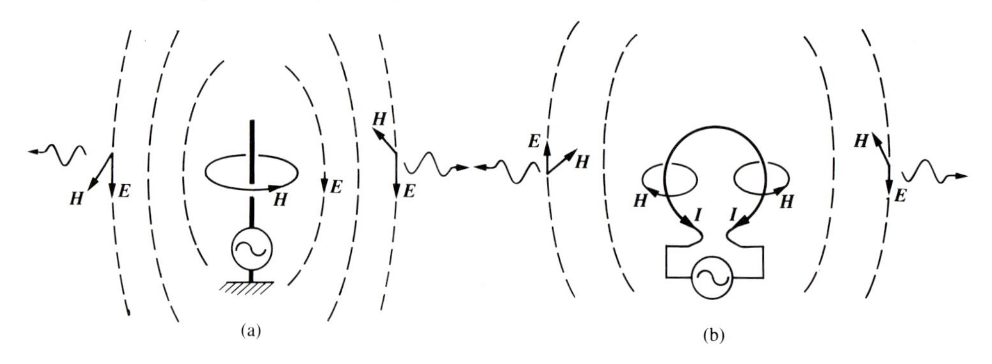

Fig. 9.1 A schematic illustration of the near-field and far-field regions for (a) an idealized short antenna, or 'monopole', and (b) a small loop antenna. At a distance of 0.02λ from these antennas the ratio E/H is $\sim 3000\,\Omega$ in case (a) (a high-impedance field) and $\sim 40\,\Omega$ in case (b) (a low-impedance field).

defend systems again e.m.p. rely on transient protective devices and will not be discussed further here.

9.3 Physical mechanisms for electromagnetic interference

Before considering the measures that may be taken to reduce e.m.i., it is necessary to recognize the possible pathways by which interference can enter (and leave) a system. It will be clear that in a real system there will be a complicated web of pathways; some of the principal ones are indicated schematically in Fig. 9.2. For illustrative purposes a system is depicted as consisting of a cabinet (usually of metal) that contains one or more subsystems (e.g. analogue amplifiers and/or digital signal processing units) together with one or more power supply units. It should be noted that every pathway is reversible.

The mains supply to the system will always carry impulsive, short-duration spikes (as high as a few hundred volts in some industrial situations) as well as other transient variations, and, depending on the properties of the power supply unit, these can propagate on to the power supply lines feeding the functional subsystems. Additionally there will be some radiation from the mains input line inside the enclosure formed by the cabinet. Unless good practice is followed regarding the connectors for the signal input and output lines, high-frequency interference signals may be coupled into, or out of, the system.

A metal cabinet constitutes a hollow conductor, and there cannot be a static electric field inside it due to an external static electric field. However, time-dependent e.m. fields (in a near-field regime) can be

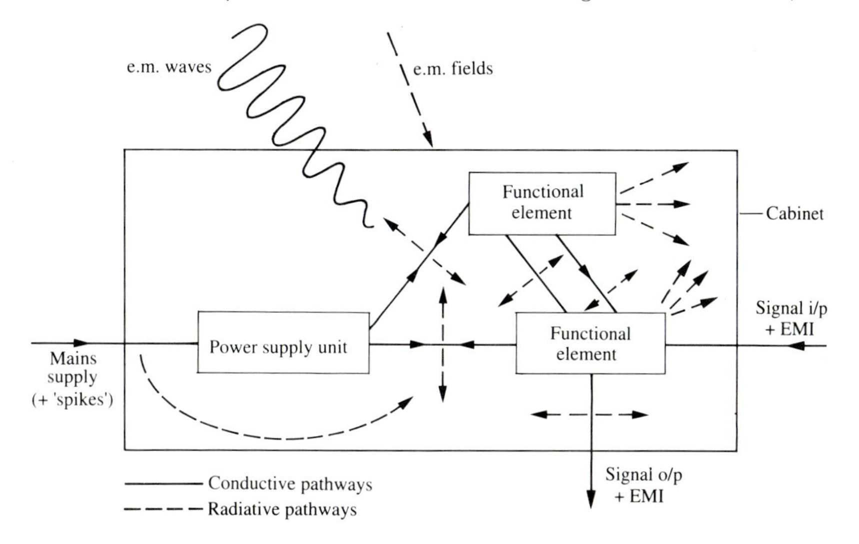

Fig. 9.2 A schematic illustration of the pathways for electromagnetic interference.

troublesome, particularly because of the magnetic component, which induces eddy currents in the conducting wall that, in turn, can cause radiation (near-field) into the enclosure. Although e.m. waves can penetrate a metal sheet to some extent, a more significant pathway may be apertures (holes and/or slots) in the cabinet wall. If practical considerations dictate that there must be apertures, then if possible their lateral dimensions should be made small compared with the wavelength of external e.m. radiations that are likely to cause problems. For the same physical reasons, subsystems within the cabinet that are sources of high-frequency interference should be contained within metal screens. Interconnecting wires and buses within the enclosure radiate e.m. energy, and this can cause practical problems.

So-called parasitic circuit elements can constitute undesirable conductive pathways. For instance, stray capacitance between conductors or between a conductor and a metal screen can provide an unwanted pathway for coupling between subsystems. Stray inductance or, more importantly, stray mutual inductance can also cause undesirable coupling. In high-impedance circuits, dirty or damp insulation, or low-quality insulation generally, can provide a leakage pathway of significance.

It should be remembered that equipment may have to function in a hostile environment. For instance, mechanical vibrations could cause wires in a system to vibrate, and if there is a magnetic field, an e.m.f. will be induced. Also the relative displacement of conductors due to vibrations will modulate the value of the stray capacitance, which may in turn become a source of interference. These last two examples could be classified as **microphonics**.

If a system contains light-sensitive devices, care must be taken to mitigate the effects of ambient light. Finally, a subtle effect which can be a nuisance is the demodulation of radiofrequency waves by non-linear elements such as 'dry' soldered joints and even rusty fixing bolts in a cabinet.

This brief sketch of the possible mechanisms should reinforce the earlier statement that a wide range of concepts and practical techniques are involved in identifying and counteracting e.m.i.

9.4 Conductive coupling

Consider the situation depicted in Fig. 9.3(a), where an interfering signal from the source V_I is coupled into the receptor circuit (in this case an amplifier) via the stray capacitance C_s.

As an order-of-magnitude estimate of a stray capacitance, consider two equal areas of metallization (5 mm × 5 mm, say) on opposite faces of a printed-circuit board of relative permittivity $\varepsilon_r \approx 5$ and thickness 1 mm. Using the equation for the capacitance of a parallel-plate capacitor (eqn (1.29)), the stray capacitance is roughly 1 pF.

For the purposes of analysis the situation of Fig. 9.3(a) can be represented as in Fig. 9.3(b), which emphasizes that it is a high-pass filter configuration. So the 'pick-up' at the input port of the amplifier is likely to be most troublesome in high-impedance circuits at high frequencies (see Example 9.1 below).

Example 9.1 From the equivalent circuit of Fig. 9.3(b): if R denotes the parallel combination of R_0 and R_{in}, then (after setting $V_0 = 0$)

$$\frac{V_{in}}{V_I} = \frac{R}{R - \dfrac{j}{\omega C_s}}$$

$$= \frac{R\left(R + \dfrac{j}{\omega C_s}\right)}{\left(R^2 + \dfrac{1}{\omega^2 C_s^2}\right)}.$$

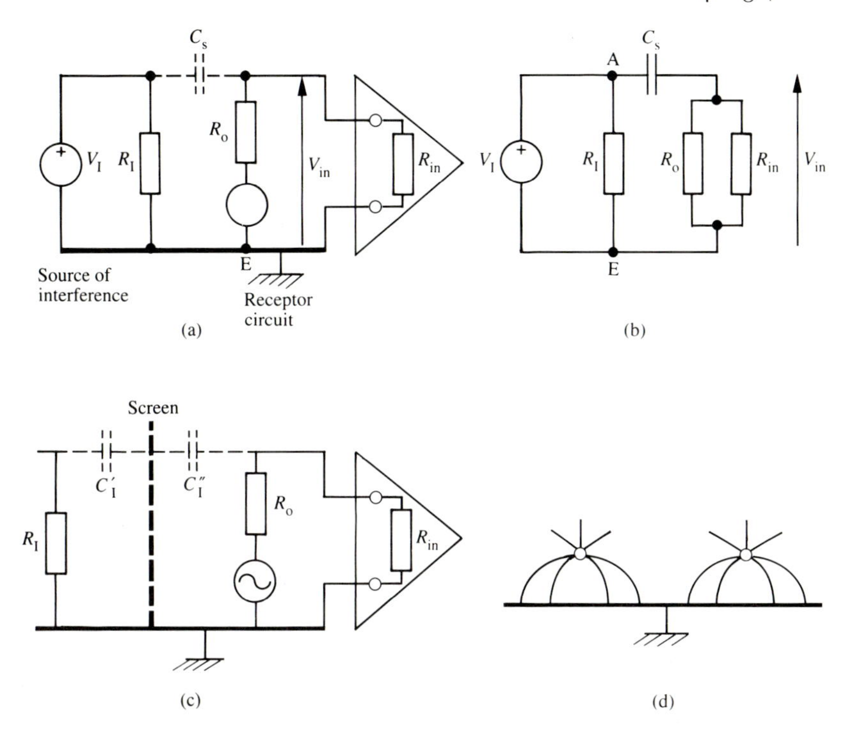

Fig. 9.3 E.m.i. via a stray capacitance. (a) The circuit and (b) the equivalent circuit. (c) Illustrates how the capacitive coupling between the two circuits can be reduced by the interposition of an earthed metal screen. The capacitance between two conductors (e.g. parallel wires or strips) can be reduced by the proximity of a metal ground plane, as indicated in (d).

So

$$\left|\frac{V_{in}}{V_I}\right| = \frac{\omega C_s R}{\sqrt{(1 + \omega^2 C_s^2 R^2)}}.$$

If $R_0 = R_{in} = 5\ \text{k}\Omega$, say, and $C_s = 0.1$ pF, then $|V_{in}/V_I| = 10^{-3}$ at a frequency of approximately 640 kHz. □

The interference can be reduced by decreasing the impedance level of the receptor circuit (R_0, R_{in} in Example 9.1) and/or by reducing the value of C_s. It is rarely a practical proposition to reduce C_s significantly by simply increasing the distance between the two circuits (unless they were extremely close initially), since $C_s \propto [\ln(\text{separation})]^{-1}$, roughly, and so does not decrease very rapidly (see eqns (1.31) and (1.38)) in respect of parallel wires and coaxial conductors).

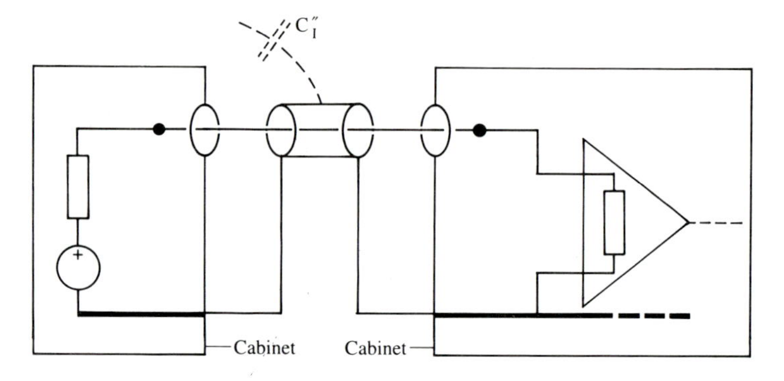

Fig. 9.4 Implementation of the screen of Fig. 9.3(c) using coaxial connecting line. C_I'' is the capacitance between the centre conductor and screen of the coaxial line.

A commonly used method to reduce interference coupling of this type is to interpose a grounded metal screen (i.e. with low resistance and low inductance) between the two circuits, as indicated schematically in Fig. 9.3(c); the interfering signal now goes to ground via C_I' and the low-impedance screen. Of course the high-frequency performance of the amplifier will be degraded owing to the presence of C_I'' in shunt with the input of the amplifier. The proximity of a conducting ground plane can also significantly reduce the capacitance between two wires or conducting tracks (see Fig. 9.3(d)). In high-impedance systems it is good practice to screen the subsystems from each other, using metal boxes, say, and to make the interconnections with coaxial cable, since ideally the current in a coaxial cable generates no e.m. field outside the cable (see Fig. 9.4).

Stray self-inductance is not a significant source of coupling between circuits, but mutual inductive coupling has to be guarded against; this will be considered in Section 9.5.

Resistive coupling can occur in two ways. First, imagine that in the circuit of Fig. 9.3(a) the stray capacitance C_I is replaced by a resistance R_I, say; in this circumstance the interfering signal would be coupled into the receptor circuit. However, the insulation between circuits is usually good (unless dirt, moisture or other contaminants degrades it) and so there is usually not a problem except in very high-impedance circuits such as electrometer circuits (for measuring electric charge), where the input resistance may be as high as $10^{12}\ \Omega$. Second, coupling exists if there is a resistance that is common to the two circuits in question; in practice this occurs most commonly through the small, but non-zero, resistance of supply rails and common grounds (see Section 9.6).

9.5 Radiative coupling

The near-field and far-field regions for e.m. radiation were described in Section 9.2, and clearly there is a gradation between them. However, for the descriptive purposes of this chapter, only the extreme situations will be assumed.

A vital point to remember is that e.m. disturbances are rapidly attenuated in conducting materials. For e.m. waves the so-called 'skin depth' δ (see Section 5.2.1), for which the amplitude of the e.m. field in a wave decays by a factor 1/e from its value at the surface of the material, is given by

$$\delta = (\pi \mu_0 \mu_r \sigma f)^{-1/2}. \tag{9.2}$$

Here σ is the electrical conductivity of the conducting material at the frequency (f) of interest. Some representative values of skin depth are displayed in Table 9.1.

At low frequencies, and particularly at the mains frequency of 50 Hz or 60 Hz, mutual inductive coupling via the magnetic field component has to be guarded against, especially where there are coils (inductors) or long, parallel runs of conductors. If two plane coils have their planes perpendicular to each other, then clearly no magnetic flux is linked with the receptor coil (see Fig. 9.5) and there will be no induced e.m.f. Hence careful attention should be paid to the siting (distance apart) and relative orientation of the sources of magnetic flux such as inductors and transformers, and of the receptor elements.

As was seen in Section 5.2.2, a parallel pair of wires has self-inductance. For balanced 'twin feeder' cable of characteristic impedance 300 Ω, with distributed capacitance of 13 pF m^{-1}, eqn (8.4) gives a value for the distributed inductance of about 1 μH m^{-1}. In the case of 'ribbon' cable,

Table 9.1. Skin depths

Material	Electrical conductivity σ 10^7 $(\Omega\,\mathrm{m})^{-1}$	δ (m) at a frequency of 50 Hz	1 kHz	1 MHz	3 GHz
Copper	5.8	9×10^{-3}	2×10^{-3}	7×10^{-5}	1×10^{-6}
Brass	1.6	2×10^{-3}	4×10^{-3}	10×10^{-5}	2×10^{-6}
Aluminium	3.5	10×10^{-3}	3×10^{-3}	9×10^{-5}	2×10^{-6}
Mumetal	0.16	0.4×10^{-3}	0.09×10^{-3}		
Supermalloy	0.13	0.2×10^{-3}	0.04×10^{-3}		

Fig. 9.5 Two plane coils arranged orthogonally so that (ideally) no magnetic flux is linked with the receptor coil.

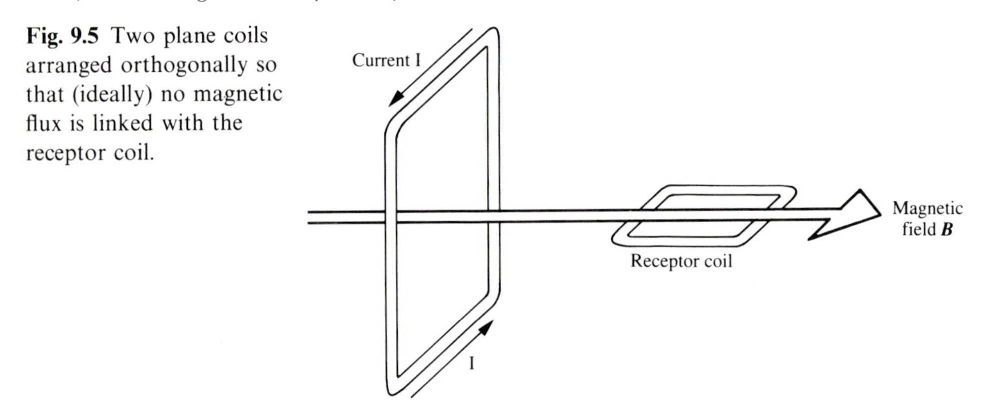

where the individual conductors have diameters of about 0.2 mm and the separation between centres of adjacent conductors is about 1 mm, eqn (5.11) gives an inductance value of about 1 μH m^{-1}. Since the coefficient of mutual inductance (M) between two circuits is given by $M = k\sqrt{(L_1 L_2)}$, where $0 < k < 1$, it follows that the coefficient of mutual inductance between two adjacent pairs of parallel wires can approach 1 μH m^{-1}. If one of the pairs of wires acts as a source, the mutual inductance due to coupling with the second pair can be reduced by keeping the second pair of wires as close together as possible, thus reducing the area for magnetic flux linkage. If it is practicable, the pair of conductors should be twisted together, which not only reduces the effective area but also means that the induced e.m.f. in the secondary circuit alternates in sign and so largely cancels out. However, for high-frequency circuits the latter procedure may not be viable, since unacceptable capacitance and series inductance is introduced. In these circumstances coaxial cable should be used; again the net induced e.m.f. is zero, since equal and opposite induced currents flow in paired diametrically opposite segments of the outer cylindrical conductor.

From Table 9.1 it can be seen that sheets, or boxes, of thin-gauge copper, aluminium or brass will provide effective screening against interfering time-dependent e.m. fields at frequencies greater than about 20 kHz. Note, however, that tin-plated steel is often used; although its electrical conductivity is lower, it does have a much higher value of relative permeability (refer to eqn (9.2)) and it also has a cost advantage.

At low frequencies a material having a high value of relative permeability, such as mumetal ($\mu_r \approx 8 \times 10^4$), can provide good screening against interfering magnetic fields. The disadvantage of such materials, quite apart from their relatively high cost, is that they have to be handled very carefully, since their magnetic properties are adversely affected by mechanical strain. In recent years, high-permeability materials with better mechanical properties have been developed and their cost is decreasing.

A rather subtle pathway for interference is the presence of non-linear elements in the receptor that may be intrinsic to some of the components of the circuit (particularly the active devices) or may arise from poor ('dry') soldered joints and even from corroded joints in metal screens and boxes. A high-frequency interfering signal is demodulated by the non-linear element and its 'envelope' is injected into the receptor circuit as interference.

High-frequency signals can propagate along pairs of conductors in transmission line modes, and even as surface waves, and so, if necessary, low-pass filters should be incorporated into signal lines and supply lines, particularly where they enter or leave a system. Shunt capacitors can be used to effect the filtering action and also ferrite 'beads' or tubes, which can be placed around a line, are a convenient way of realizing a high series inductance, and loss, at high frequencies.

9.6 Grounds, earths, and power supplies

In the context of circuits and systems the terms 'ground' and 'earth', and even 'common' are often used interchangeably. However, the concept of a reference potential ('zero' volts), which is what is really implied by these three terms, embraces some subtle features and requires careful consideration; poor design and practice with respect to grounding is a major source of e.m.i.

Grounding, shielding, and filtering have both distinctive and complementary roles to play in the reduction of e.m.i., but the hierarchy of precautionary procedures should be in the order just given. The correct implementation of well-designed grounds often significantly reduces the requirement for shielding and filtering, and effective grounding and shielding together should reduce the need for filtering. It is not good practice to rely on filtering to counteract the effects of poor grounding and shielding.

Life would be simple in some respects if all the subsystems of a system could be self-contained and electrically 'floating' with respect to earth (see Fig. 9.6(a)), with electrical isolation achieved by means of transformers or optical isolators. Of course the power supply of each subsystem would have to be isolated from earth (by a transformer); in some specialized equipment, batteries are used as power supplies. The subsystems do not float literally, and the necessary supporting structures should provide very good insulation from earth.

Static electrical charge can build up on floating systems, and this is a potential hazard, since spark discharges can eventually occur (see Section 1.4). Discharge can occur slowly and continuously, through corona

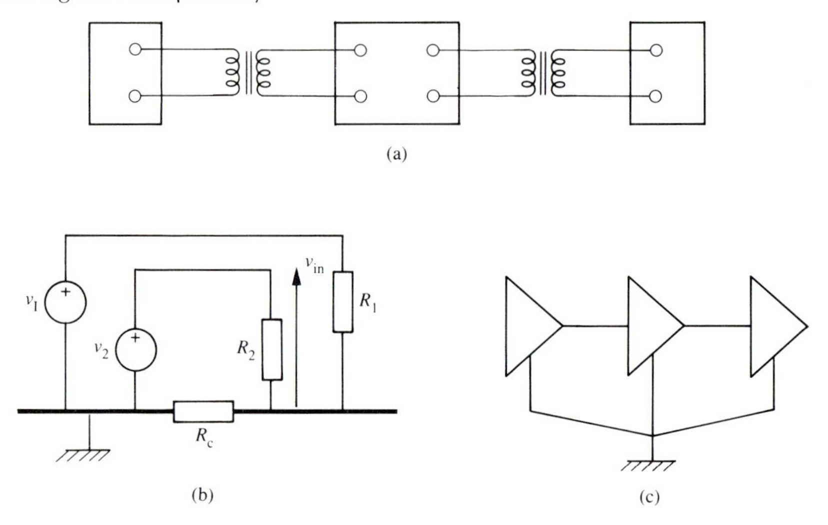

Fig. 9.6 (a) Isolated, 'floating' subsystems. (b) Interference arising from the resistance of the ground plane or line. (c) The use of a single ground point.

discharge and current flow over the surface of insulators, rather than through disruptive sparks, but this is a relatively noisy process (in the sense of electrical noise) and can be a source of significant interference. For this reason, and for other reasons of safety, the cabinets of subsystems are almost invariably connected to mains earth. If a circuit is floating within such a cabinet it is usually necessary to provide a high-resistance path to earth; this prevents the build-up of large static charges, by allowing the charge to leak to earth in a continuous and relatively noise-free manner. Some field-effect transistors (MOSFETs) have an extremely well insulated 'gate' contact ($\sim 10^{12}\ \Omega$). Before being connected into a circuit they are in a floating state, and the lack of an easy discharge path for static charges can lead to a build-up of charge and, eventually, to a damaging electrical breakdown in the material of the transistor. A voltage of 400 V is commonly taken as the threshold above which irreversible damage is likely to occur, and this can easily be generated through frictional charging of a person at a workplace. Hence such devices are supplied with their leads short-circuited.

Currents can flow in the ground line, or plane, owing to electromagnetic coupling to a 'ground loop' (see Fig. 9.7) and owing to power supply return currents, for example. Although the ground will be fabricated from a material of low electrical resistivity, there will be small potential differences due to the current flow; hence a ground plane, for example,

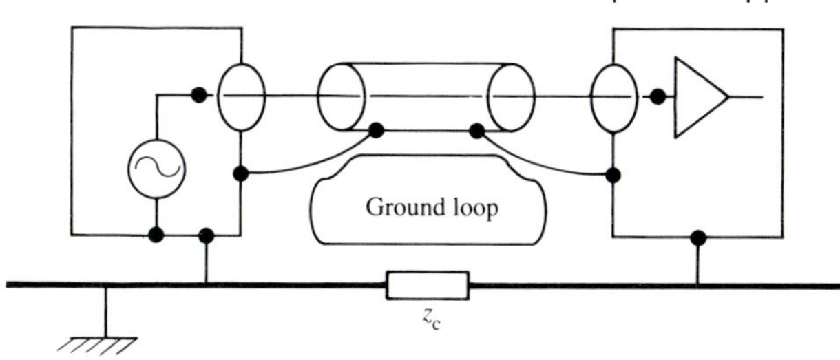

Fig. 9.7 A schematic illustration of a 'ground loop'.

will not be a true equipotential surface. Such potential differences will be coupled into the receptor circuit and will constitute interference.

Another undesirable consequence of the non-zero resistance (R_c) of the ground is illustrated in Fig. 9.6(b): v_I represents a source that is grounded as indicated with respect to a receptor circuit represented by the mesh containing the source v_2 and load resistance R_2. The latter resistance could be the input resistance of an amplifier, for instance, for which the input voltage is v_{in}. It is left as an exercise to show that

$$v_{in} = \frac{v_2 R_2 (R_1 + R_c)}{(R_1 + R_2)\left\{\dfrac{R_1 R_2}{(R_1 + R_2)} + R_c\right\}} - \frac{v_I R_c R_2}{(R_1 + R_2)\left\{\dfrac{R_1 R_2}{(R_1 + R_2)} + R_c\right\}}.$$

Since the parallel combination of R_1 and R_2 will be much greater than R_c in practical situations it follows that

$$v_{in} \approx v_2\left(1 + \frac{R_c}{R_1}\right) - v_I \frac{R_c}{R_1}, \tag{9.3}$$

where the second term represents the coupling of the interfering source into the receptor circuit. Clearly the interference would be removed if the grounding was rearranged so that the return current due to the source v_I did not flow through the mutual resistance R_c of the two meshes. This can be effected, both in this illustrative situation and in a wide range of practical situations, by using a **single-point ground** (see Fig. 9.6(c)). Caution must be exercised, however, in situations where high-frequency currents are flowing: if the connection to the single-point ground has a length that is a significant fraction of a wavelength, then it behaves as a distributed element and its impedance can become significantly large. For example, if the length of the connection is close to a quarter-wavelength, then it behaves like a short-circuited $\lambda/4$ line (see Section 8.3.6) and has a high input impedance. In such circumstances a multi-point ground must be used in order that the individual ground connections can be kept short, and also the ground line or plane should be of high quality with a very low impedance.

Not only ground lines but also power supply lines have a non-zero

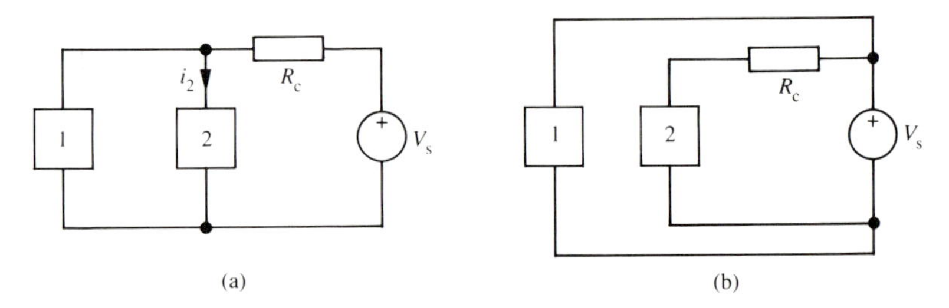

Fig. 9.8 (a) Power supply coupling. (b) Decoupling of the subsystems from the common power supply.

resistance and hence there is a possibility of interference coupling between subsystems which share a power supply (see Fig. 9.8(a)); this is often called power supply coupling. If i_2 denotes a variation in the current drawn by subsystem 2 (due to a signal that is being amplified, say), then the resultant fluctuation in the supply voltage level (V_1) to subsystem 1 is equal to $i_2 R_c$, and this in turn will generate an interfering signal on the signal line at the output of subsystem 1. The problem may be particularly acute if subsystem 1 is a sensitive pre-amplifier, say. Alleviation of this problem may be achieved by using separate supply lines, as indicated in Fig. 9.8(b), but note that the subsystems will have slightly different ground levels, owing to the small, but non-zero, resistances of the ground lines.

As was mentioned earlier, currents can flow in ground loops, owing to electromagnetic induction; e.m. fields at 50 Hz or 60 Hz due to mains transformers and mains supply lines are often the dominant source of interference by this mechanism. The interference can be reduced at source by the twisting together, or screening, of mains supply lines and by the careful routeing of these lines and by careful siting of transformers. If a ground loop exists, the magnitude of the induced e.m.f. is reduced if the area of the loop is made as small as possible. Ground loops can be eliminated if circuits are isolated through the use of transformers, or by optical isolators, as indicated in Fig. 9.6(a). However, a transformer has limited bandwidth, and stray capacitance both in transformers (between the primary and secondary windings) and in optical isolators can couple signals from one circuit to another. Ground loops in signal circuits should be broken if possible, e.g. by grounding only one end of a screened coaxial line (see Fig. 9.9); the screening against stray capacitance will be maintained. If the pair of conductors of a signal line are close together, the interfering signals picked up by each line, due to a particular source of interference, will be more or less of equal magnitude, i.e. the interference will be a 'common-mode' signal. Hence if the signal of interest is

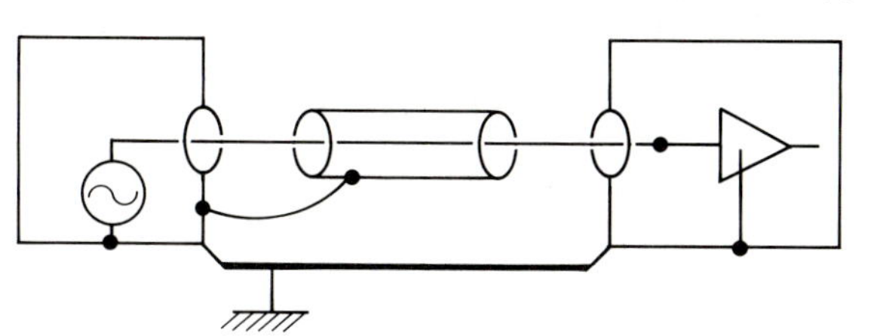

Fig. 9.9 Obviating a ground loop by grounding only one end of a coaxial line.

propagated via 'differential mode' devices (e.g. transformers, differential amplifiers)*, the interfering signal will be greatly attenuated.

* A differential voltage amplifier amplifies the difference between the two voltages applied to the input terminals and, ideally, the amplifier gain is zero for two equal voltages applied to the input terminals (a 'common-mode' signal). The 'common mode rejection ratio' (c.m.r.r.) is the ratio of the differential voltage gain to the common-mode voltage gain, and commonly can be as high as 10^5 (100 dB). A well-designed transformer should have a high value of c.m.r.r., since equal and opposite currents in its primary winding should produce cancelling magnetic fluxes in the transformer core.

Appendix: Partial fractions—general rules

If a function $P(x)$ is equal to the quotient of two polynomials in x:

$$P(x) = \frac{N(x)}{D(x)},$$

then $P(x)$ can be expressed in the form of an algebraic sum of terms called partial fractions. The individual terms in the sum can be obtained by following the rules given below.

Cases:

(i) Degree of $N(x) <$ degree of $D(x)$.

(ii) Degree of $N(x) \geqslant$ degree of $D(x)$.

Case (i).

(a) To every *linear* factor $(a_1x + b_1)$ of $D(x)$ there will be a corresponding partial fraction $A_1/(a_1x + b_1)$.

(b) To every *quadratic* factor $(a_2x^2 + b_2x + c_2)$ of $D(x)$ there will be a corresponding partial fraction $(A_2x + B_2)/(a_2x^2 + b_2x + c_2)$.

(c) To every repeated *linear* factor $(a_3x + b_3)^2$ of $D(x)$ there will be two corresponding partial fractions

$$\frac{A_3}{(a_3x + b_3)} + \frac{B_3}{(a_3x + b_3)^2}.$$

(d) To every repeated *quadratic* factor $(a_4x^2 + b_4x + c_4)^2$ of $D(x)$ there will be two corresponding partial fractions

$$\frac{A_4x + B_4}{(a_4x^2 + b_4x + c_4)} + \frac{C_4x + E_4}{(a_4x^2 + b_4x + c_4)^2}.$$

(e) To every thrice-repeated *linear* factor $(a_5x + b_5)^3$ of $D(x)$ there will be three corresponding partial fractions

$$\frac{A_5}{(a_5x + b_5)} + \frac{B_5}{(a_5x + b_5)^2} + \frac{C_5}{(a_5x + b_5)^3}$$

(f) To every *cubic* factor $(a_6x^3 + b_6x^2 + c_6x + d_6)$ of $D(x)$ there will be a corresponding partial fraction

$$\frac{A_6x^2 + B_6x + C_6}{(a_6x^3 + b_6x^2 + c_6x + d_6)},$$

and so on.

Case (ii).

(a) If $N(x)$ is of the same degree as $D(x)$ then a constant A will be added to the partial fractions as given in Case (i).

(b) If $N(x)$ is one degree higher than $D(x)$ then $(Ax + B)$ will be added to the partial fractions given in Case (i).

(c) If $N(x)$ is two degrees higher than $D(x)$ then $(Ax^2 + Bx + C)$ will be added to the partial fractions given in Case (i); and so on.

Answers to problems

1.1 6 W
1.2 6.2 μW; 1.2×10^9 W m^{-3}
1.3 3 N
1.4 -8×10^{-16} J (or -5000 eV)
1.5 -2.4 mV
1.6 Effective capacitance of the combination: $6C$
Charge: $2CV$, CV, $3CV$ on each of the two capacitors in series

2.1 1 kV; 10 Ω
2.2 123.3 V; 2.84 kW
2.3 $R_2 \geqslant 240$ kΩ
2.6 $I_0 = 8/19$ A; $R_0 = 19$ Ω
2.7 2.18 μA
2.8 (a) $E_0 = 12$ V; $R_0 = 4$ Ω
(b) $I_0 = 3$ A; $R_0 = 4$ Ω
2.9 0.21
2.10 Thévenin: 1.46 V; 3.1 kΩ. Norton: 0.48 mA; 3.1 kΩ
2.11 6 V battery: 3 W. 2 V battery: 1 W

3.2 1 mA
3.3 (a) CR_2; (b) $C(R_1 + R_2)$
3.4 (a) E; (b) E/R_2; (c) $L/(R_1 + R_2)$
3.5 (a) $(ac + bd) - j(ad - bc)$; (b) $13\angle\tan^{-1}(5/12)$; (c) $1\angle\tan^{-1}(4/3)$
3.6 (a) $-30°$ (or $+330°$); (b) $-120°$ (or $+240°$); (c) $165°$
3.7 $5\angle\tan^{-1}(0.75)$
3.8 $2.5\cos(\omega t + \phi)$, where $\phi = -103°$ or $+257°$
3.9 $(\phi_1 + \phi_2)$
3.10 $2\cos(\omega t - 60°)$
3.11 $i = \hat{V}\{(1/R^2) + (1/\omega^2 L^2)\}^{1/2} \sin(\omega t - \phi)$, where $\phi = \tan^{-1}(R/\omega L)$
3.12 334 V; $-72.6°$
3.13 15.7 mA; $-81°$
3.14 Power factor = 0.84; power = 41 W
3.15 (a) 100 Ω, (b) 55 μF
3.16 $0.1\sin(1000t - 2\pi/3)$ V; $0.112\sin(1000t - 63.4°)$ V
3.17 $G = (2R^2 + \omega^2 L^2)/\{R(R^2 + \omega^2 L^2)\}$; $B = (\omega L)/(R^2 + \omega^2 L^2)$
3.18 $\hat{V} = \hat{I}/\{(1/R^2) + \omega^2 C^2\}^{-1/2}$; $\tan\phi = -(\omega RC)$

3.19 $G = R(R^2 + \omega^2 L^2)^{-1}$; $B = \omega C\{R^2 - L/C + \omega^2 L^2\}(R^2 + \omega^2 L^2)^{-1}$
3.20 0.954; 262 W
3.21 3.2×10^{-3}; 29 mW
3.22 $L_{\text{eff}} = 1.11$ mH; $R_{\text{eff}} = 278\ \Omega$
3.24 $(1 + \omega^2 C^2 R^2)^{-1/2}$; $\tan^{-1}(-\omega CR)$
3.26 $Y = 1/R$
3.27 $2.70\angle -59.1°$ A; $8.1\angle -57.2°$ A
3.28 $0.16\angle -61.3°$
3.29 $0.72\angle -23.7°$ A; $(2.54 + j1.67)\ \Omega$
3.30 $4.47\angle 63.4°$ V; $(6 + j2)\ \Omega$
3.31 $0.66\angle -9.46°$ A; 26 W

4.2 $Y = R/(R^2 + \omega^2 L^2) + j\omega[C - \{L/(R^2 + \omega^2 L^2)\}]$
4.3 2.5 nF; 0.25 nF; 2π kΩ; 200π kΩ
4.4 $X = (1 - \omega^2 LC)/\{\omega C(\omega^2 LC - 2)\}$
$X = 0$: $\omega_0^2 = 1/(LC)$ $\quad B = 0$: $\omega_0^2 = 2/LC$
4.5 $C = 6.3$ nF; $R = 4.3\ \Omega$: 2.5 kΩ

5.2 9.4 mm, 0.94 mm, 94 μm, 9.4 μm, 0.94 μm
5.5 $E = -\hat{\boldsymbol{i}}\left(\frac{a}{y} - bz\right) + \hat{\boldsymbol{j}}\left(\frac{ax}{y^2}\right) + \hat{\boldsymbol{k}}bx$
5.6 $\nabla \cdot \boldsymbol{E} = -2(x^2y^2 + x^2z^2 + y^2z^2)$
$\nabla \times \boldsymbol{E} = -8xyz(\hat{\boldsymbol{i}}x + \hat{\boldsymbol{j}}y + \hat{\boldsymbol{k}}z)$
5.7 87 μV m^{-1}

6.4 $\hat{V}/2$; $\hat{V}/2$; $\hat{V}/2\sqrt{3}$
6.7 $v(t) = (AL/R)\cdot(1 - e^{-Rt/L})$

7.3 $h_{11} = j2\ \Omega$; $h_{12} = 2$; $h_{21} = -2$; $h_{22} = j\frac{3}{2}$ S
7.7 Series elements: $2R_0/3 = 33\ \Omega$
Shunt elements: $5R_0/12 = 21\ \Omega$
7.8 $y_{11} = y_{22} = (3 + j5)/6$ S; $y_{21} = y_{12} = -(1 + j3)/6$ S

7.9

$$\begin{bmatrix} \left(Z + R_1 - \dfrac{j}{\omega C}\right) & 0 & \dfrac{j}{\omega C} & -R_1 \\ 0 & \left(Z_2 + j\omega L_1 - \dfrac{j}{\omega C}\right) & \dfrac{j}{\omega C} & -j\omega L_1 \\ \dfrac{j}{\omega C} & \dfrac{j}{\omega C} & \left(R_2 - j\dfrac{2}{\omega C}\right) & 0 \\ -R_1 & -j\omega L_1 & 0 & (R_1 + j\omega L_1) \end{bmatrix}$$

"Δ"

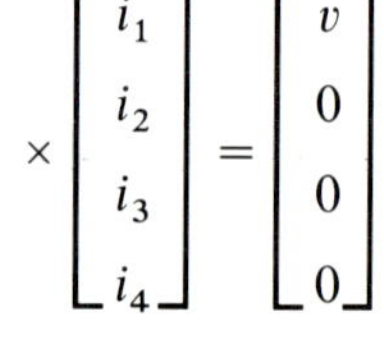

$$\times \begin{bmatrix} i_1 \\ i_2 \\ i_3 \\ i_4 \end{bmatrix} = \begin{bmatrix} v \\ 0 \\ 0 \\ 0 \end{bmatrix}$$

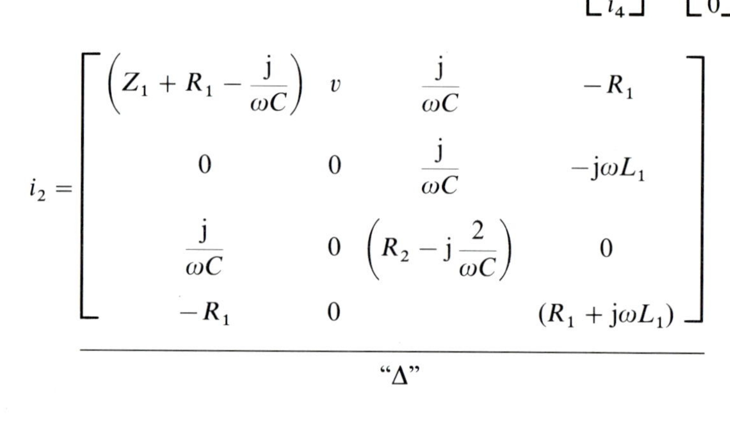

$$i_2 = \frac{\begin{bmatrix} \left(Z_1 + R_1 - \dfrac{j}{\omega C}\right) & v & \dfrac{j}{\omega C} & -R_1 \\ 0 & 0 & \dfrac{j}{\omega C} & -j\omega L_1 \\ \dfrac{j}{\omega C} & 0 & \left(R_2 - j\dfrac{2}{\omega C}\right) & 0 \\ -R_1 & 0 & & (R_1 + j\omega L_1) \end{bmatrix}}{\text{"}\Delta\text{"}}$$

7.10
$$\begin{bmatrix} \left(\dfrac{1}{Z_1}+\dfrac{1}{R_2}+\mathrm{j}\omega C_2\right) & -\mathrm{j}\omega C_2 & -\dfrac{1}{R_2} & 0 \\ -\mathrm{j}\omega C_2 & \left(\dfrac{1}{R_1}+\mathrm{j}\omega C_2+\mathrm{j}\omega C_3\right) & 0 & -\mathrm{j}\omega C_3 \\ -\dfrac{1}{R_2} & 0 & \left(\dfrac{1}{R_2}+\dfrac{1}{R_3}+\mathrm{j}\omega C_1\right) & -\dfrac{1}{R_3} \\ 0 & -\mathrm{j}\omega C_3 & 0 & \left(\dfrac{1}{Z_2}+\dfrac{1}{R_3}+\mathrm{j}\omega C_3\right) \end{bmatrix} \times \begin{bmatrix} v_1 \\ v_2 \\ v_3 \\ v_4 \end{bmatrix} = \begin{bmatrix} \dfrac{v_1}{Z_1} \\ 0 \\ 0 \\ 0 \end{bmatrix}$$

7.11 $\begin{bmatrix} \frac{5}{3} & 2 \\ 2 & 3 \end{bmatrix}$

7.12 $\dfrac{h_{21}v_s}{\{h_{22}(h_{11}+Z_s)-h_{12}h_{21}\}}$; $\dfrac{(h_{11}+Z_s)}{\{h_{12}h_{21}-h_{22}(h_{11}+Z_s)\}}$

7.15 Series inductances: 15.8 nH; 4.76 nH
Shunt capacitances: 0.481 nF; 0.159 nF

7.17 $\dfrac{V_0(s)}{V_{in}(s)} = \dfrac{R + sCR_pR}{R_p + R + sCR_pR}$

7.19 Zeroes: $s = -(CR_C)^{-1}, -R_L/L$
Poles: $s = -(R_C + R_L)/2L \pm \mathrm{j}[(1/LC) - \{(R_C + R_L)^2/4L^2\}]^{-1/2}$
Non-oscillatory for $(LC)^{-1} \leqslant (R_C + R_L)^2$
Limiting condition: $R_C \geqslant 1990\ \Omega$

8.3 $Z_0 = 75\ \Omega$; $R_T = 25\ \Omega$; $u = 2 \times 10^8\ \mathrm{m\ s^{-1}}$
8.4 (a) 50 Ω; (b) 1 rad m^{-1}; (c) 2π m; (d) j50 Ω
8.5 $Z_0 = 45\ \Omega$; $\alpha = 2.2 \times 10^{-4}$; $\beta = 0.45$ rad m^{-1};
$\lambda = 14.1$ m; $u = 2.2 \times 10^8\ \mathrm{m\ s^{-1}}$
8.6 $0.42\angle{-95^\circ}$
8.7 0.18 − j0.3
8.8 1.26 m
8.9 0.72 + j1.09

Index